Frontiers in Clinical Drug Research-
Central Nervous System
(Volume 2)

Editor

Atta-ur-Rahman, *FRS*

Kings College
University of Cambridge
Cambridge
UK

CONTENTS

PREFACE

The second volume of the eBook series entitled: "***Frontiers in Clinical Drug Research – Central Nervous System***" comprises five chapters in cutting edge fields. In the first chapter, Micheli *et al.* highlight the potential of nucleic acids as drugs for neurodegenerative diseases. Till date, various types of nucleic acid-based therapeutics have been proposed. This chapter demonstrates the rationale and current status of nucleic acid-based strategies for treating Huntington's disease and Parkinson's disease.

In the second chapter, Marcos Arturo Martínez Banaclocha discusses the current pharmacological interventions available for treating neurodegenerative diseases and brain aging. The author highlights the role of oxidised proteins at the cysteine residues in some neurodegenerative diseases and in aging.

Kazuo Abe focuses on non-motor symptoms in Parkinson's disease along with the pharmacological therapies to tackle the disease in the third chapter. In the fourth chapter, Anderson and Maes discuss melatonin interactions with the a7nAChR. In the last chapter of this eBook, Trevor R. Norman discusses another important disease of the Central Nervous System i.e. Major Depressive Disorder. The author highlights the pharmacological properties of some novel drugs and evaluates their efficacy critically.

I would like to thank all the contributors for their work and cooperation and would also like to appreciate all the technical staff of Bentham Science Publishers, especially, Mr. Mahmood Alam (Director Publications), Mr. Shehzad Naqvi (Senior Manager Publications) and Dr. Faryal Sami (Assistant Manager Publications) for their hard work and dedication.

Prof. Atta-ur-Rahman, FRS
Honorary Life Fellow
Kings College
University of Cambridge
Cambridge
UK

CONTRIBUTORS

Carla Emiliani	Department of Chemistry, Biology and Biotechnology, University of Perugia, Italy
George Anderson	CRC Scotland & London, Eccleston Square, London, UK
Kazuo Abe	Department of Community Health Medicine, Graduate School of Hyogo College of Medicine, Division of Neurology, Hospital of Hyogo College of Medicine, 1-1 Mukogawa-Cho, Nishinomiya-City, Hyogo 663 8131, Japan
Marcos Arturo Martínez Banaclocha	Pathology Service at the Lluis Alcanyis Hospital, Játiva, Valencia, Spain
Maria Rita Micheli	Department of Chemistry, Biology and Biotechnology, University of Perugia, Italy
Mario Polidoro	Department of Chemistry, Biology and Biotechnology, University of Perugia, Italy
Michael Maes	Deakin University, Department of Psychiatry, Geelong, Australia
Rodolfo Bova	Department of Chemistry, Biology and Biotechnology, University of Perugia, Italy
Trevor R. Norman	Department of Psychiatry, University of Melbourne, Austin Hospital, Heidelberg, Australia

Frontiers in Clinical Drug Research-
Central Nervous System
(Volume 2)

2

Frontiers in Clinical Drug Research – Central Nervous System

Volume # **2**

Prof. Atta-ur-Rahman

ISSN (Online): **2214-6318**

ISSN: Print: **2467-9623**

ISBN (eBook): 978-1-68108-189-2

ISBN (Print): 978-1-68108-190-8

BENTHAM SCIENCE Bentham *e* Books

CHAPTER 1

Nucleic Acids as Drugs for Neurodegenerative Diseases

Maria Rita Micheli*, Rodolfo Bova, Mario Polidoro and Carla Emiliani

Department of Chemistry, Biology and Biotechnology, University of Perugia, Italy

Abstract: The development of nucleic acid-based therapeutics has recently aroused an increasing interest due to the great promise they hold for the treatment of a wide range of both inherited and acquired disorders. A prominent part in this field is played by neurodegenerative diseases, for which therapeutic interventions are currently limited to palliative and symptomatic treatment. Advances in the elucidation of the molecular mechanisms involved in these disorders provide the basis for developing nucleic acid-based treatment strategies able to address the molecular cause of the disease.

Several types of nucleic acid-based therapeutics have been proposed. All those therapeutics may be grouped into two main classes, *i.e.*, protein coding nucleic acids (DNA molecules) and regulatory nucleic acids (DNA or RNA molecules). The use of nucleic acids belonging to the first class is aimed at providing the relevant cells with a protein that either permits to rescue a missing function or supplies a new function able to counteract or alleviate the disease. Regulatory nucleic acids are used in order to counteract the harmful effects of a specific gene in the relevant cells. They include several types of molecules that allow virtually any step of gene expression to be controlled. Regulation at transcriptional and post-transcriptional levels can be achieved by the means of oligonucleotides, catalytic nucleic acids, siRNAs and antisense RNAs, while protein synthesis and protein function can be inhibited by siRNAs and aptamers or decoys, respectively.

The delivery of genes coding for neurotrophic factors as neuroprotective/ neurorestorative agents is a favored strategy for the treatment of neurodegenerative diseases, but a number of alternative strategies have also been proposed, such as the use of aromatic-L-amino decarboxylase (AADC) encoding gene for Parkinson's disease. More recently, advances in gene silencing technology have led to the evaluation of strategies aimed at selectively interfering with the pathogenetic mechanisms underlying disease phenotype, as in the case of Huntington's disease where RNA interference technology could provide a tool to target the mutant *HTT* allele.

A prerequisite for the successful clinical application of nucleic acid-based therapeutics to the treatment of neurodegenerative diseases is the availability of safe and efficient

*Corresponding author Maria Rita Micheli:** Department of Chemistry, Biology and Biotechnology, University of Perugia, Italy; Tel: +39-75-5855758; E-mail: mariarita.micheli@unipg.it

Atta-ur-Rahman (Ed.)

systems for nucleic acid delivery to the central nervous system, or better to the relevant neuronal subpopulations depending on the specific disease. A few viral vectors, those based on the adeno-associated virus or lentiviruses in particular, have shown promise as neuron-targeted nucleic acid carriers and the clinical trials undertaken to date have employed viral vectors almost exclusively. However, viral vectors have several drawbacks, such as those resulting from a non-complete safety, that significantly limit their widespread clinical use. Consequently, a great deal of efforts to develop non-viral vectors for nucleic acid delivery to the central nervous system has been made in the recent years. Non-viral vectors offer several advantages including improved safety profiles, lower production costs and ability to target specific neuronal subpopulations, but their delivery efficiency has to be improved in order to thoroughly realize their potential in clinical settings.

This chapter illustrates the rationale and current status of nucleic acid-based strategies for the treatment of two neurodegenerative movement disorders, Huntington's disease and Parkinson's disease.

Keywords: CNS-targeted nucleic acid delivery, Huntington's disease, movement disorders, neurodegenerative diseases, nucleic acid therapeutics, Parkinson's disease.

INTRODUCTION

Neurodegenerative diseases are a broad spectrum of central nervous system (CNS) disorders that are characterized by a chronic and progressive course and share a common hallmark consisting of the selective death of specific neuronal populations. The clinical manifestation of each neurodegenerative disease depends on the specific type of neurons that is involved. These disorders include both inherited diseases caused by single gene mutations (*e.g.,* Huntington's disease and several spinocerebellar ataxias) and common diseases of more complex origin (*e.g.,* Alzheimer's disease and Parkinson's disease).

Common age-related neurodegenerative diseases, along with cerebrovascular disorders, are currently among the leading causes of death and morbidity in Western countries and their prevalence is destinated to further rise as a consequence of average lifespan increase. This type of diseases exerts a growing impact from the societal and economic point of view, which makes the development of strategies for early detection as well as effective and safe treatments more important than ever.

Notwithstanding the significant advance made in recent years towards the elucidation of the pathogenic mechanisms underlying these various neurological disorders, no cures currently exist and therapeutic interventions are still limited to palliative and symptomatic treatment. Therefore, much effort is being made to develop innovative therapeutic strategies and among these, nucleic acid-based strategies may have the potential for substantial advancements.

Nucleic acid-based therapeutic strategies, collectively known as gene therapy, refer to the delivery of nucleic acid molecules to target cell populations in order to achieve either the over-expression of a therapeutic gene, or inhibit the expression of an endogenous harmful gene, or even restore the function of a defective gene [1].

It was around the 1970s when the idea to deliver a therapeutic gene to treat human diseases came out [2] and the first foreseen application was the treatment of recessively inherited diseases by delivering a wild type copy of the defective gene responsible for the disease. Since then, the potential of gene therapy approach has been considerably increased by the development of novel kinds of therapeutic nucleic acids. Moreover, the range of candidate diseases for this treatment modality has expanded beyond that of inherited diseases to include more common diseases, such as cardiovascular disorders, cancer and degenerative disorders of CNS.

Two main nucleic acid-based approaches, "gene addition" and targeted inhibition of gene expression, are currently investigated for the treatment of neurodegenerative diseases. The earliest attempts at neurodegenerative disease gene therapy focused on the "gene addition" approach. Independent of the actual cause of neuron depletion, a therapeutic goal common to most neurodegenerative diseases is to preserve the viability and function of the residual neurons. Delivery of genes coding for neurotrophic factors as neuroprotective/neurorestorative agents is therefore an intensely investigated strategy. Further candidate therapeutic genes are chosen for each specific disease on the basis of available knowledge about the disease causing mutation or the pathogenetic mechanism underlying the disease.

More recently, advances in gene silencing technology have led to the evaluation of strategies aimed at selectively interfering with the pathogenetic mechanisms underlying disease phenotype. These strategies are applied to the targeting of gain-of-function mutant alleles in dominantly inherited diseases as well as genes known to contribute to the phenotype of more complex diseases.

This chapter illustrates the rationale and current status of nucleic acid-based strategies for the treatment of two neurodegenerative movement disorders, Huntington's disease and Parkinson's disease.

NUCLEIC ACID-BASED THERAPEUTICS

Several types of nucleic acid-based therapeutics have been proposed. All those therapeutics may be grouped into two main classes, *i.e.*, protein coding nucleic acids (DNA molecules) and non-coding nucleic acids (DNA or RNA molecules) with regulatory function (regulatory nucleic acids).

Protein Coding Nucleic Acids

The use of protein coding nucleic acids (usually cDNAs, *i.e.*, the double-stranded DNA copies derived from the gene mRNAs) is aimed at providing the relevant cells with a protein that either permits to rescue a missing function or supplies a new function able to counteract or alleviate the disease. Candidate therapeutic genes are obvious in the case of recessively inherited neurodegenerative diseases due to loss-of-function mutations, such as, for example, lysosomal storage diseases or recessively inherited forms of Parkinson's disease. For these disorders, the strategy relies on the delivery of the wild-type copy of the disease gene in order to supply the missing or defective protein. On the other hand, the choice of candidate therapeutic genes for non-inherited as well as dominantly inherited monogenic neurodegenerative diseases relies on the available knowledge about pathogenetic mechanisms.

A strategy that widely applies to the treatment of neurodegenerative diseases is the delivery of genes coding for neurotrophic factors (NTFs) as neuroprotective/neurorestorative agents. NTFs are secreted proteins expressed in both developing and adult nervous system. They regulate the development, maintenance, function and plasticity of the nervous system and exert their actions through binding and activating specific cell surface receptors. A single neuronal group can respond to several NTFs and a given NTF affects many neuronal types. Neurons can derive trophic support not only from innervated cells (retrograde transport), but also from afferent neurons (anterograde transport), or even themselves (autocrine mechanism). Therefore, the trophic requirement of a neuronal population is due to a complex interaction between different NTFs that contribute to the highly specific connectivity of the nervous system. Importantly, NTFs not only promote neuron survival (survival effect), but can also protect specific neuronal populations against different types of brain insults (neuroprotective effect) and repair already damaged neurons (neurorestorative effect). In consideration of these properties, NTFs have been considered ideal candidates as neuroprotective and neurorestorative agents.

A peculiar type of therapeutic genes are those coding for gene-engineered antibodies aimed at ablating the abnormal function of specific intracellular proteins [3, 4]. Natural antibodies can be genetically engineered to obtain the so-called "intrabodies" (iAbs), *i.e.,* smaller molecules which are more suitable to be intracellularly expressed. Among intrabody formats, the first choice is the single chain variable fragment (scFv) which consists of the variable domains of the immunoglobulin heavy (VH) and light (VL) chains kept together by a flexible polypeptide linker. The resulting molecule is a monovalent antibody fragment which still retains the binding specificity of full-length antibody but is encoded by a single gene. Intrabodies for a specific antigen can be isolated from a naïve human repertoire by a variety of *in vitro* selection platforms such as phage-, yeast-, ribosome-, or bacterial-display systems. Vector-mediated delivery of the selected intrabody encoding gene allows its expression within the relevant cells, with the potential for alteration of the folding, interactions, or subcellular localization of the target protein. Based on these properties, intrabodies are emerging therapeutic molecules for neurodegenerative diseases in which misfolded and aggregated proteins are involved, including Alzheimer's, Parkinson's, Huntington's and prion diseases.

Regulatory Nucleic Acids

Regulatory nucleic acids are used in order to counteract the harmful effects of a specific gene in the relevant cells. They include several types of molecules (Table 1) that allow virtually any step of gene expression to be controlled. Regulation at transcriptional and post-transcriptional levels can be achieved by the means of antisense oligonucleotides (ASOs), catalytic nucleic acids, short interfering RNAs (siRNAs) and antisense RNAs, while protein synthesis and protein function can be inhibited by siRNAs and aptamers or decoys, respectively. DNA molecules must be exogenously administered to the target cells while RNA molecules can also be generated within the target cells upon transfer of their coding DNA sequences.

For dominantly inherited neurodegenerative diseases, such as, for example, Huntington's disease, knowledge of the disease-causing mutation directly indicates the candidate target gene for therapeutic silencing. Candidate target genes for non-inherited diseases are less obvious, but they may be chosen on the basis of their involvement in cellular pathways known to contribute to the disease phenotype.

While regulatory nucleic acid therapeutics were originally developed as tools to suppress expression or function of specific disease-associated proteins, their field of application has recently been extended to novel targets. Strong evidence supporting the involvement of microRNAs in the pathogenesis of neurodegenerative diseases led to the development of strategies aimed at targeting specific disease-associated miRNAs [5-7].

Table 1: Therapeutic non-coding nucleic acids for suppressing protein expression or function.

		Target	Effect
Antisense oligonucleotides	Oligodeoxyribonucleotides	mRNA	degradation of mRNA - inhibition of translation
	Phosphorothioate oligonucleotides		
	2'-ribose modified oligonucleotides		
	Locked nucleic acids (LNAs)		
	Ethylene-bridged nucleic acids (ENAs)	Gene	inhibition of transcription
	Morpholinos (PMOs)		
	Peptide nucleic acids (PNAs)		
Catalytic nucleic acids	Ribozymes	mRNA	mRNA cleavage
	DNAzymes		
Small regulatory RNAs	siRNAs	mRNA	degradation of mRNA - inhibition of translation
	shRNAs		
	Artificial miRNAs		
Protein-binding oligonucleotides	Decoys	Transcription factor	Competition with natural binding site
	Aptamers	Protein	Block of protein function

Antisense Oligonucleotides

Antisense oligonucleotides (ASOs) are short, chemically synthesized, single-stranded DNA molecules that, upon cellular internalization, can selectively inhibit the expression of a target gene by interacting with its corresponding mRNA [8, 9]. The complementarity of ASO sequences to those of their target mRNAs allows the formation of DNA/RNA duplexes and, as a consequence, leads to degradation of the RNA strand of the duplexes by cellular RNase H enzyme, or inhibition of translation by steric hindrance. Synthetic oligodeoxynucleotides can also be applied to block transcription of a target gene through the formation of triple helix DNA structures in the promoter region. Further applications include modulation of pre-mRNA splicing and correction of point mutations.

In vivo application of oligodeoxynucleotides suffers from serious limitations such as poor tissue distribution, cytotoxicity and mostly, low stability due to degradation by DNases. A variety of chemical modifications have been, therefore, proposed in order to overcome these limitations [8, 9]. Modified oligonucleotides include phosphoro-thioate oligodeoxynucleotides, 2'-ribose modified oligonucleotides (2'-O-methyl- and 2'-O-methoxyethyl-RNA), bridged nucleic acids (locked nucleic acids and ethylene-bridged nucleic acids, LNAs and ENAs, respectively), morpholinos (PMOs, phosphorodiamidate morpholino oligomers) and peptide nucleic acids (PNAs). While chemical modifications lead to significant improvements, some drawbacks still remain. For example, PNAs strongly bind to their targets and are very resistant to degradation, but their bioavailability is modest when administered *in vivo*.

Catalytic Nucleic Acids

Ribozymes are antisense RNA molecules which include a catalytic core able to cleave the target mRNA once the RNA-RNA duplex has formed, thus preventing its translation [10]. A variety of RNAs endowed with enzymatic activity have been found in lower eukaryotes, viruses and some bacteria. Among these naturally occurring ribozymes, hammerhead and hairpin ribozymes provided the basis for the development of targeted gene silencing tools. Hammerhead ribozymes cleave RNA at the nucleotide sequence U-H (where H is A, C, or U) by hydrolysis of a 3'-5' phosphodiester bond, while hairpin ribozymes cleave at the nucleotide sequence C-U-G. Unlike ASOs that bind their targets in a stoichiometric manner, thanks to their enzymatic nature ribozymes act on multiple substrate molecules, which provides them with a higher efficiency. Ribozymes can be exogenously administered as chemically synthesized molecules or expressed intracellularly once delivered by a vector. In the first case, however, they suffer from the same limitations of ASOs, including low stability and limited tissue distribution and cell uptake.

Deoxyribozymes, or DNAzymes, are analogs of ribozymes in which the nucleic acid is DNA instead of RNA [10, 11]. Similar to ribozymes, they consist of a catalytic core flanked by sequences complementary to the target mRNA, but show an improved biological stability thanks to the replacement of RNA backbone by DNA.

Small Regulatory RNAs

RNA interference (RNAi) is a natural cellular process that regulates gene expression and provides an innate defense mechanism against invading viruses and transposable elements [12]. The idea to exploit this naturally occurring process to develop tools aimed at selectively silencing target genes of interest paved the way to a novel therapeutic strategy which potentially applies to a wide range of diseases.

RNAi is a sequence-specific gene silencing process which is triggered by the presence, in the cytoplasm, of small, double-stranded RNA (dsRNA) molecules of ~20-25 bp. These small dsRNAs associate with the pre-RISC (precursor RNA-induced silencing complex) that, upon removal of one strand of the duplex (the sense "passenger" strand), is converted to mature RISC. This complex, which contains the antisense "guide" strand, is the final effector of RNAi. Pairing of the antisense "guide" strand to specific target mRNAs may induce gene silencing by causing transcript degradation or translational inhibition depending on the degree of complementarity. The two main types of small dsRNAs that activate RNAi are microRNAs (miRNAs), which are processed from stem-loop structures present in endogenously expressed primary transcripts and short interfering RNAs (siRNAs) that are processed from longer dsRNAs.

Elucidation of miRNA biogenesis [13] has enabled the development of several strategies for harnessing RNAi pathways for therapeutic purposes [14]. Briefly, miRNAs are transcribed from the genome as larger primary miRNA transcripts (pri-miRNAs), which form intramolecular stem-loop structures. These primary transcripts are processed in the nucleus by Drosha-DGCR8, the microprocessor complex, to generate precursor miRNAs (pre-miRNAs) which are ~60-70 nucleotide stem-loop structures. Pre-miRNAs are then transported to the cytoplasm where the loop region is removed by Dicer, thus generating the mature miRNA duplex.

Based on this knowledge, the design of artificial inhibitory RNAs can be aimed at mimicking pri-miRNAs (artificial miRNAs), pre-miRNAs (shorthairpin RNAs or shRNAs), or mature miRNAs with perfect complementarity to their targets (siRNAs). Each class mediates gene silencing but enters the pathway at a different step. SiRNAs are chemically synthesized double-stranded oligoribonucleotides designed to mimic Dicer products or substrates (Dicer-ready siRNAs) which, once delivered into the cells, are loaded into the RISC, directly or following processing by Dicer, respectively [15, 16]. ShRNAs are generated within the cells upon vector-mediated delivery of their coding DNA sequences [17, 18]. They are transcribed as sense and antisense sequences connected by a loop of unpaired nucleotides, and, once exported to the cytoplasm, are converted to functional siRNAs *via* Dicer cleavage. Like shRNAs, artificial miRNAs are also expressed intracellularly from viral vectors but enter the RNAi pathway upstream of the

Drosha-DGCR8 complex [19, 20]. It is important to note that shRNAs and artificial miRNAs differ from siRNAs not only by the mode of delivery but also by the duration of gene silencing. The use of chemically synthesized molecules allows chemical modifications to be introduced in order to increase stability and efficacy and reduce off-target effects. However, gene silencing mediated by these molecules is transient whereas long-term silencing can be potentially achieved by intracellular expression of shRNAs or miRNAs.

Protein-Binding Oligonucleotides: Decoys and Aptamers

Decoys are small DNA or RNA molecules designed to provide competing binding sites for specific DNA or RNA binding proteins, respectively [21, 22]. In particular, DNA decoys are double-stranded oligodeoxynucleotides that inhibit the expression of target genes thanks to their ability to block the binding of transcriptional activators to the promoter regions of those genes. They may also be designed to provide the DNA binding site of a transcriptional repressor in order to rescue the expression of its target genes.

Aptamers are short, single stranded (ss) oligoribo- or oligodeoxyribo-nucleotides that specifically recognize and bind their targets (protein or small organic molecules) due to their stable three-dimensional structure [23, 24]. They are selected from an initial library containing 10^{13}-10^{16} random ssDNA or ssRNA sequences through an *in vitro* process termed SELEX (systematic evolution of ligands by exponential enrichment), which was developed in 1990 by two different groups [25, 26] (for a recent review see [27]). Due to their high specificity and binding affinity, aptamers are able to block the functions of specific target proteins, which make them therapeutic agents with a mode of action conceptually similar to that of antibodies. Compared to antibodies, however, aptamers show several advantages such as easier and faster production, non-immunogenicity and higher stability when chemically modified.

DELIVERY OF NUCLEIC ACIDS TO THE CENTRAL NERVOUS SYSTEM

Two distinct approaches, known as *in vivo* and *ex vivo* gene therapy, respectively, can be used for nucleic acid delivery to the CNS. In the *in vivo* approach, the therapeutic nucleic acid is directly administered to the patient, while the *ex vivo* approach involves transplantation of cells that are engineered in culture to express a therapeutic gene.

The *ex vivo* approach can only be applied to the delivery of genes encoding secretable proteins. Several types of cells have been considered to this purpose. Unfortunately, transplantation of cell lines engineered to express candidate genes entails the risk of rejection or tumor development. To overcome this problem, a polymer-encapsulated cell technology was developed [28]. Before implantation, genetically engineered cells are encapsulated in a semipermeable synthetic polymer, which allows the expressed and secreted protein to be released. The encapsulation isolates the foreign cells from the host immune cells and at the same time prevents tumor formation. Allogenic cells have also been investigated as vehicles for *ex vivo* gene delivery. Among these, bone marrow-derived cells are an interesting option since they are able to cross the blood-brain barrier (BBB) and can therefore be intravenously transplanted [29].

The *in vivo* approach based on the use of viral vectors has become the favored strategy for gene delivery to the CNS thanks to its superior efficiency in terms of both duration of transgene expression and coverage of critical target regions. Non-viral vectors are also intensely investigated for *in vivo* delivery of both coding and regulatory nucleic acid therapeutics.

Administration Routes

The development of effective *in vivo* treatments represents a considerable challenge due to the unique environment of the CNS and mainly to the presence of the BBB which limits the brain uptake of the vast majority of neurotherapeutic agents [30, 31].

The BBB is the specialized system that separates the brain from systemic circulation. It is formed by the endothelial cells surrounding the brain capillaries, together with perivascular elements such as basal lamina, pericytes and astrocyte end-feet [32]. Features that distinguish cerebral endothelial cells from other endothelial cells include the luck of *fenestrae*, the presence of tight junctions and adherens junctions between the cells, reduced vesicular transport and increased numbers of mitochondria. In the BBB, the endothelial cells are completely covered by a basal lamina in which pericytes are embedded and which is surrounded by the astrocyte end-feet. The role of BBB is to maintain chemical composition of the neuronal microenvironment for proper neuronal functions, supplement the brain with nutrients and protect it from potentially harmful substances in the blood stream. The drawback of this tightly controlled barrier is that it also limits the transport of drugs into the brain. Neurotherapeutic agents are often unable to penetrate into the brain to perform their actions. Approximately 98% of the small molecule drugs and nearly 100% of the large molecule drugs (such as peptides, proteins and nucleic acids) cannot substantially cross this barrier [33].

As a consequence of the BBB efficiency in excluding the vast majority of gene transfer vehicles from reaching the CNS *via* the vasculature, most gene therapy approaches have been based on direct administration to the CNS.

Direct Administration to the CNS

When a neurodegenerative process is restricted to a defined brain region, therapeutic agents can be delivered by localized administration. Intraparenchymal direct injection consists in the introduction of nucleic acid therapeutics inside or near the population of target cells. This is the simplest method for localized nucleic acid delivery to the brain. However, it is also the most invasive one being local trauma to the brain neuropil, toxicity, inflammation and limited therapeutic diffusion its major drawbacks. Therapeutic distribution in the brain can be improved considerably by convection-enhanced delivery (CED), a slow pressurized infusion method that exploits convective flow in the brain [34], or some variation of conventional CED, such as microfluidics-mediated CED [35]. Pressure gradients generated by CED increase interstitial flow, which in turn facilitates therapeutic distribution with little physical or functional damage. Nonetheless, therapeutic distribution after intraparenchymal injection remains mostly limited to the targeted structure.

Direct delivery of nucleic acid therapeutics to the cerebrospinal fluid (CSF) by either intraventricular or intrathecal injection allows a more global distribution throughout the CNS to be achieved. If injected into the CSF produced in the lateral ventricles of the brain, delivered nucleic acids can distribute to cells lining the ventricular and subarachnoid space. Once expressed within these cells, their products can be released into the CSF and distributed throughout the brain [36]. The alternative option is the injection of therapeutics into the intrathecal space surrounding the spinal cord [37], which is a less painful procedure and allows a larger volume of therapeutics to be delivered.

Peripheral Administration

Theoretically, peripheral injection, including intramuscular and intravenous injection, is the most attractive administration mode to deliver nucleic acid therapeutics to the CNS. Besides the obvious advantage of safer and less invasive delivery, peripheral administration offers the possibility of achieving a widespread therapeutic distribution as well as administering multiple doses. However, restricted access to the CNS as well as systemic clearance of the vectors have greatly limited the application of these approaches.

Intramuscular injection exploits retrograde transport along motor neurons which allows the BBB to be circumvented. Interesting results were obtained in proof-of-concept studies [38-40], but it is worth noticing that retrograde transport impairment occurs in various neurodegenerative disorders [41].

Intravenous injection would be an ideal administration route, provided that the problem of crossing the BBB is overcome. In the recent years, a great deal of efforts to develop strategies that aid passage across the BBB have been made, mainly in the development of non-viral delivery systems. Some progress, however, has been reported for viral vectors as well. Adeno-associated virus 9 (AAV9) vectors have been shown to reach the CNS of neonatal mice and young cats after intravenous injection and to transduce large numbers of glia and motor neurons in the spinal cord [42, 43] as well as hippocampal neurons and Purkinje cells in the cerebellum [42]. Efficient gene transfer to various regions of adult mice CNS was also observed following intravenous injection of SV40 recombinant vectors combined with intraperitoneal mannitol infusion [44]. An alternative approach based on targeting gene transfer to brain microcapillary endothelial cells is currently under investigation [45]. The basic premise is that a protein of interest expressed in and secreted from, the vascular endothelia will be endocytosed by underlying neurons and glia. Peptides able to bind the brain vascular endothelia were isolated from a phage library by *in vivo* panning. Presentation of these peptides on the capsid of the adeno-associated virus 2 (AAV2) was shown to expand the biodistribution of intravenously injected AAV2 to include the CNS. Interestingly, reconstitution of enzyme activity throughout the brain and improvement of disease phenotypes have been obtained in two distinct models by peripheral injection of the peptide modified AAV2 vectors expressing the enzymes lacking in lysosomal storage disease affected mice [45].

Delivery Systems

A prerequisite for the successful clinical application of nucleic acid-based therapeutics to the treatment of neurodegenerative diseases is the availability of safe and efficient systems for nucleic acid delivery to the CNS, or better to the relevant neuronal subpopulations depending on the specific disease. Clinical trials undertaken to date have employed viral vectors almost exclusively. However, viral vectors have several drawbacks, such as those resulting from a non-complete safety, that significantly limit their widespread clinical use. Consequently, a great deal of efforts to develop non-viral vectors for nucleic acid delivery to the CNS has been made in the recent years.

Viral Vectors

In vivo gene transfer using viral vectors is the most widely used approach for delivering therapeutic genes to the CNS in both pre-clinical and clinical investigations. This approach exploits the ability of viruses to deliver their genetic material to target cells. The ideal viral vector must be safe, *i.e.,* non-virulent, non-immunogenic, non-oncogenic. It must also be efficient, *i.e.,* able to transduce non-dividing cells and ensure long-term expression of the therapeutic gene. Furthermore, special properties (for example, axonal transport capability or ability to carry multiple genes) are, in some cases, required.

Viral vectors are designed to be replication-defective, while maintaining the ability to infect cells and transfer their genetic material into the nucleus. This manipulation not only eliminates virus pathogenicity, but also allows uncontrolled spreading of transgene delivery to be prevented.

Several types of viral vectors have been investigated in animal models of neurodegenerative diseases [46, 47], including those derived from herpes simplex virus type 1 (HSV-1), adenoviruses (AdVs), adeno-associated virus (AAV) and lentiviruses (LVs) (Table **2**) but only AAV and LV vectors are currently used in clinical trials. These vectors have emerged as the vectors of choice for gene transfer to the CNS for non-oncological applications as they mediate efficient long-term gene expression with no apparent toxicity.

Table 2: Viral vectors for nucleic acid delivery to the CNS.

Vector	Cloning Capacity	Viral Genome	Chromosomal Integration	Expression Onset	Expression Duration
AAV	~ 4 kb	ssDNA	No	1 - 2 weeks	6 months - 8 years
AdV	8 - 35 kb	dsDNA	No	1 week	~ 6 months
HSV	40 - 150 kb	dsDNA	No	1 day	~ 7 months
LV / NIL	~ 8 kb	ssRNA	Yes / No	1 - 2 weeks	~ 3 months

NIL: Non-Integrating LV vectors.

Besides efficiency in gene transfer, AAV vectors [48, 49] provide several further advantages, among which the non-pathogenicity of the virus. They do not integrate into the host cell genome, but persist in an episomal form in non-dividing cells, which minimizes risk of insertional mutagenesis. Furthermore, they do not express any viral protein, thus do not elicit an immune response against transduced cells, nor cause inflammation. As a consequence, these vectors ensure long-term transgene expression. A recent study has shown that AAV-mediated transgene expression in the primate brain persists for at least eight years with no evidence of neuroinflammation or reactive gliosis [50]. A further interesting property of AAV vectors is that they can be generated at high titers, thus allowing the simultaneous expression of different genes from the same cells or tissues. This property can be very useful when the delivery of multiple neurotrophic factors or multiple proteins involved in the same metabolic pathway is needed, or when multiple shRNAs to inhibit different proteins are to be administered. Over ten AAV serotypes have been engineered into vectors but AAV2 is to date the serotype of choice for clinical trials. The overall results so far reported have shown that direct infusion of AAV2 vectors into the human brain parenchyma is well tolerated.

Compared to AAV vectors, which accept inserts with a maximum size of just 4.5 kb, LV vectors [51, 52] can accommodate a larger transgene payload (~8 kb) which make them an attractive option for multigene treatments. Though these vectors do not naturally infect cells of the CNS, they have been pseudo-typed with envelope proteins from other viruses (*e.g.*, vesicular stomatitis virus G). The resulting pseudo-typed vectors have a broad cell tropism including neuronal and glial cells. LV vectors integrate into the host cell genome and lead to stable transgene expression. Because integration occurs at random sites, it entails the risk for insertional mutagenesis. One strategy to improve the vector safety profile was based on the introduction of self-inactivating (SIN) mutations, which knock out promoter activity of the long terminal repeats (LTRs) of the viral genome, thus reducing the risk of insertional gene activation. A further strategy was the development of non-integrating LV (NIL) vectors which carry either mutant integrase or mutations in their LTRs that inhibit integrase binding. Integration of these vectors is greatly reduced (down to 0.35-2.30%) and can be further reduced by removal of a sequence element involved in plus-strand DNA synthesis. It is worth noticing that NIL vectors show an efficiency of transduction similar to that of integrating vectors. Though LV vectors are increasingly used for gene delivery in experimental models of neurodegenerative diseases, only one of them has progressed to clinical investigation. A tri-cistronic, self-inactivating vector derived from a non-primate lentivirus, the equine infectious anemia virus (EIAV), has been safely used in a clinical trial involving patients affected by Parkinson's disease (see below).

Non-viral Vectors

Non-viral vectors offer several advantages including improved safety profiles, lower production costs and ability to target specific neuronal subpopulations, but their delivery efficiency has to be improved in order to thoroughly realize their potential in clinical settings. Several types of nano-scale nucleic acid delivery systems (nanocarriers), including lipid- and polymer-based nanoparticles (NPs) and inorganic NPs, are currently investigated for CNS targeted nucleic acid delivery [53-56]. To efficiently deliver their cargo to neurons, nanocarriers must overcome a number of hurdles, *i.e.*, internalization into the neurons, interaction with intracellular organelles such as lysosomes and transport across the nuclear membrane in the case of DNA molecules to be transcribed [54]. Furthermore, if intravenously injected, they should both avoid systemic clearance and efficiently cross the BBB to reach the target cells in the brain.

Efficacy of nanocarriers when administered by systemic route depends on their ability to circulate in the blood stream for a prolonged period of time. However, after intravenous administration, they often interact with the reticuloendothelial system (RES), leading to a rapid removal from systemic circulation. This process mainly depends on particle size, charge and surface properties of the nanocarrier. A proper modification of the characteristics of nanocarrier surface is therefore needed in order to increase their blood circulation time, thus favoring their uptake into non-RES organs. The interactions with the RES can be prevented by coating the particles with a hydrophilic or a flexible polymer and/or a surfactant. Several authors demonstrated that both polyethylene glycol (PEG) coating and direct chemical linking of PEG to the particle are able to increase circulation times as well as uptake into non-RES organs.

In order to facilitate crossing of the BBB, the addition of targeting moieties is also needed. The most commonly used strategy exploits one of the physiological transport systems, the receptor-mediated endocytosis. This strategy [57] involves the modification of nanocarrier surface by coupling to ligands that specifically bind to receptors expressed on the brain endothelial cells. Natural ligands, such as insulin and transferrin, or peptides selected from a phage-displayed peptide library can be used as targeting ligands. Monoclonal antibodies (mAbs) that mimic peptide structure also undergo receptor-mediated endocytosis and may thus be considered as useful targeting moieties. For instance, Pardridge and coworkers employed mAbs which mimic either transferrin or insulin as targeting ligands to deliver liposome-encapsulated therapeutic nucleic acids (hence the name "Trojan

Horse Liposomes") to the brain. The Trojan Horse Liposome (THL) technology developed by Pardridge's group has been recently reviewed in [58, 59].

Receptor-mediated delivery is also exploited to achieve nanocarrier internalization into neuronal cells. It is worth noticing that an appropriate choice of targeting ligands allows a more selective neuron targeting. For example, a neurotensin (NTS) polyplex has been used for targeted delivery to dopaminergic neurons. NTS-polyplex consists of NPs resulting from the electrostatic binding of the conjugate NTS-poly-L-lysine (NTS carrier) to plasmid DNA (pDNA). The NTS carrier takes advantage of the endocytosis of NTS with its high-affinity receptor (NTSR1) to transfer the pDNA into dopaminergic neurons [60]. To accomplish targeting to selected neuron populations, neurotrophin receptors may be considered as well since they are expressed by specific neuron subtypes.

NUCLEIC ACID-BASED STRATEGIES FOR THE TREATMENT OF NEURODEGENERATIVE MOVEMENT DISORDERS

Huntington's Disease

Huntington's disease (HD) is a midlife onset, progressive and fatal form of hyperkinetic disorder. It is a single gene disease with autosomal dominant inheritance pattern, which belongs to the polyglutamine (polyQ) expansion diseases [61] and its prevalence is 5-10 cases per 100,000. The majority of HD patients have inherited the disease causing mutation from one parent, whereas 1-3% of them are affected due to a *de novo* mutation [62]. The rare juvenile form of HD is found in 2% of the total population having the disease [63].

Although cognitive and psychiatric domains are also involved in HD patients, the defining phenotype of HD is generally considered to be the motor one, the chorea. Chorea consists in random, spontaneous and irrepressible dance-like movements (hence the name chorea, the Greek word for 'dance'). Furthermore, HD patients show, along with progressive weight loss, severe motor impairment such as loss of voluntary movement coordination that is later replaced by bradykinesia and rigidity and motor impersistence, *i.e.*, inability to maintain voluntary muscle contractions. Adult onset HD patients also display personality disorders and psychiatric symptoms such as depression, anxiety, mood swings, aggressiveness and paranoid psychosis, followed by a progressive decline in cognitive abilities (dementia). Symptoms observed in the juvenile form of HD (onset before the age of twenty), are somewhat different. As a matter of fact, chorea can be absent and involuntary movements can take the form of tremor. Patients affected by this form of the disease show bradykinesia, rigidity and dystonia. Furthermore, epileptic seizures are often observed in affected children. HD patients are rapidly disabled by the functional decline which early follows symptom onset. As a consequence, they require increasing care until the effects of severe physical and mental deterioration cause their death (about 15-25 years later). A correlation between the age of symptom onset and the course of the disease has been observed. In most cases, the earlier the onset occurs, the faster HD progresses. Due to the persistent and increasingly intensive multidisciplinary care made necessary by the seriousness of functional decline, HD takes up conspicuous medical, social and family resources.

HD is caused by an unstable expansion in the number of CAG trinucleotide repeats in the exon 1 of a gene located on chromosome 4, [64, 65]. This gene, named *HTT* (or *IT15*) codes for huntingtin, a protein ubiquitously expressed through the body, whose functions are still not fully understood. Normal individuals have 35 or fewer CAG repeats in this locus, while HD patients show expansions spanning from 36 to an exceptional 240 [66]. Often, the onset and severity of disease correlate with the length of the CAG expansion in the individual patient. With over 42 repeats, penetrance is complete, but this is reduced with intermediate repeat length. The trinucleotide repeat expansion is translated into an abnormally long polyglutamine (polyQ) tract close to the N-terminus of huntingtin, which induces the protein to misfold and acquire toxic properties (gain-of-function) to specific populations of vulnerable neurons. The toxicity of mutant huntingtin is, in fact, limited almost exclusively to striatal neurons, but little is known about mechanisms underlying this selectivity [67]. Mutant huntingtin also undergoes proteolytic cleavage, thus generating cytotoxic N-terminal products that are believed to be key players in the pathogenesis of the disease. Misfolded huntingtin and N-terminal fragments accumulate and aggregate in the cytoplasm and nucleus leading to the formation of insoluble inclusions, which represent the hallmark of this pathology. These inclusions also contain several other proteins including components of the ubiquitin-proteasome pathway, chaperones, synaptic proteins and transcription factors. Since their appearance often precedes the onset of symptoms, inclusions in diseased brains have been proposed to be involved in pathogenicity. Although the pathogenic mechanisms by which mutant huntingtin causes neuronal dysfunction and cell death are not yet fully understood, polyQ-huntingtin expression turned out to be involved in crucial cellular processes such as transcriptional regulation, cell survival, intracellular signaling, mitochondrial function, axonal transport and others [68, 69]. Besides the toxic gain-of-function, a loss-of-function pathogenic mechanism seems to play a role [68, 69], even though such a role remains to be elucidated.

HD is characterized by the selective neuronal loss in the striatum (caudate–putamen), being the GABAergic medium-sized spiny neurons (MSNs) projecting to the substantia nigra pars reticulata (SNpr) and globus pallidus (GP), mainly affected. In the final stages, however, cell loss in the cerebral cortex and widespread brain atrophy are also observed either by means of non-invasive brain magnetic resonance imaging or *post mortem* histological evaluation. Loss of striatal neurons leads to reduced inhibition of external segment of the globus pallidus (GPe). This, in turn, induces GPe neurons to become hyperactive in inhibiting the subthalamic nucleus (STN). As a consequence, an excessive motor activity occurs which results in chorea, the most characteristic motor symptom of the disease.

There is no cure for HD, so treatment focuses on reducing symptoms, preventing complications and helping patients and family members cope with daily challenges. On the other hand, the monogenic nature and the almost selective involvement of a specific brain area in the degenerative process make HD a potential good target for nucleic acid-based therapeutic approaches. Strategies currently under investigation in animal models of HD are mainly aimed at providing neurotrophic support or silencing the mutant allele of HD gene.

Many animal models which replicate HD symptoms and neuropathology have been developed over the years. Among them, rodents (mouse and rat) and nonhuman primates are the most widely used models to test nucleic acid-based therapeutic strategies. Two types of HD animal models have been generated, chemically induced and genetic models.

Since the initial demonstration that intrastriatal injection of kainic acid (KA), a non N-methyl-D-aspartate (NMDA) glutamate agonist, could mimic in rats the axon-sparing striatal lesion observed in the human HD [70], various HD animal models were generated by injecting neurotoxins into the striatum. Quinolinic acid (QA) has been one of the most commonly used agents to produce rodent and nonhuman primate models of HD. QA is an NMDA receptor agonist and an endogenous metabolite of tryptophan and it has been shown to be the most effective agent in stimulating the neurochemical changes observed in HD [71-73]. Intrastriatal injection of QA induces selective degeneration of striatal projection neurons but does not affect afferent fibers and some interneuron populations [71, 74, 75].

Intrastriatal injection of mitochondrial toxins, such as 3-nitropropionic acid (3-NP), was also shown to be able of replicating some of the behavioral aspects of HD in rats [76]. 3-NP is an irreversible inhibitor of succinate dehydrogenase and inhibits both Kreb's cycle and Complex II of the electron transport chain in mitochondria. 3-NP also decreases ATP and increases lactate in the striatum, indicating a similar impairment in oxidative energy metabolism to that seen in HD. The chronic, systemic administration of 3-NP produces striatal lesions in both rats and primates that resemble lesions seen in HD patients [76]. It has also been demonstrated that chronic 3-NP treatment in rats and primates replicates several of the cognitive and motor deficits seen in HD, such as bradykinesia, dystonia, gait abnormalities and frontal-type cognitive defects [76-80].

Chemically induced models replicate the regional selectivity of the neuropathology observed in HD, which makes them valuable models for testing neuroprotective and neurorestorative therapeutic strategies for this disease. However, they are unable to reproduce the pathogenetic mechanisms triggered by the HD causing mutation.

The identification of the disease-causing gene in 1993 [64] paved the way to the generation of genetic models of HD, which, compared to chemical lesion models, reproduce the molecular pathogenesis of HD more closely. Since then, several HD transgenic mouse models have been produced that differ regarding the type of mutation expressed, the portion of the protein included in the transgene, the promoter employed and the level of expression of the mutant protein (Table **3**).

The R6/1 and R6/2 transgenic mice were the first transgenic mouse models of HD that were developed [81]. They both express exon 1 of the human *HTT* gene with around 115 and 150 CAG repeats, respectively and the transgene expression is driven by the human *HTT* promoter in both of them. The phenotype and neuropathology observed in these transgenic lines replicate several features observed in humans, but these mice and in particular the R6/2 line, show an early onset of symptoms and a fast progression of the disease. The R6 transgenic lines are widely used to evaluate new therapeutic strategies. It should be noted, however, that these transgenic lines reflect more accurately the infantile/juvenile HD cases than the adult-onset form of the disease, due to the high number of CAG repeats that is expressed in these mice.

A different transgenic mouse model expressing an N-terminal fragment of human mutant huntingtin was developed by Schilling *et al.* [82]. This model, referred to as N171-82Q, expresses a cDNA encoding an N-terminal fragment (171 residues) of human huntingtin with 82 glutamine repeats, under the control of the mouse prion protein promoter. As a consequence of the lower number of glutamine repeats compared to the R6 mice, N171-82Q mice show a later onset of symptoms thus providing a better model for the evaluation of presymptomatic therapies. A limitation common to all transgenic mouse models expressing N-terminal fragments of mutant huntingtin is that they cannot be used to test therapeutic approaches that are predicted to function at the level of the full-length mutant protein.

Several transgenic mouse models expressing full-length human mutant huntingtin with different numbers of glutamine residues have been generated. These transgenic mice include the yeast artificial chromosome (YAC) and the bacterial artificial chromosome (BAC) HD mouse models.

Among the YAC HD mouse models, the YAC128 transgenic mice [83], which express the full-length human HD gene with 128 CAG repeats, constitute a model of particular interest since they replicate the slow and biphasic progression of behavioral deficits observed in the human condition and show age-dependent striatal and subsequent cortical neurodegeneration [84, 85]. For these reasons, the YAC128 mice are widely used not only for the elucidation of HD pathogenesis, but also for the screening of new therapies for HD.

BAC HD transgenic mice (BACHD mice) express a full-length human mutant huntingtin with 97 glutamine residues [86]. These mice exhibit progressive motor deficits and late-onset selective neurodegeneration in the cortex and in the striatum [86, 87], thus providing a further mouse model well suited for the evaluation of new therapeutic approaches.

In order to obtain mouse models that more faithfully replicate the human condition, knock-in mouse models of HD have been generated (reviewed in [88]). Knock-in mouse models carry the HD-causing mutation in its appropriate genomic and protein context, since they have an expanded polyQ encoding sequence inserted into the endogenous mouse *Htt* gene. The early development and slow progression of pathological and behavioral abnormalities in knock-in mice make these animals valuable models to test the ability of potential therapeutics in delaying early abnormalities onset.

Table 3: Transgenic mouse models of Huntington's Disease.

	Transgene / Promoter	CAG	Onset of symptoms / Survival (weeks)	Neuropathology	Symptoms	Refs.
R6/1	first 1.9 kb (exon 1) / human huntingtin	113	15-21 / 32-40	- overall brain atrophy - brain weight loss by 18 wk - neuronal atrophy - aberrant synaptic plasticity - nuclear inclusions and neuropil aggregates throughout the brain - reduction in dopamine levels	- clasping behavior (onset 20 wk) - rotarod deficit - gait abnormalities - progressive weight loss	[81]
R6/2		144	5-6 / 12-15	- overall brain atrophy with hyperventricular enlargement - brain weight loss by 4 wk and striatum volume reduction - neuronal atrophy and loss by 90 days - aberrant synaptic plasticity - nuclear inclusions and neuropil aggregates throughout the brain - reduction in dopamine levels	- clasping behavior (onset 8 wk) - rotarod deficit - hyperkinetic movements - resting tremors - circling behavior - increase in limb movements - decrease in grip strength - progressive weight loss - seizure - diabetes, cardiac dysfunction	
N171-82Q	first 171 aminoacids / mouse prion protein	82	10 / 10-24	- overall brain atrophy with hyperventricular enlargement - cells with degenerative morphology by 20 wk - striatal neuron atrophy and loss - nuclear inclusions in cortex, hippocampus, amygdala, striatum	- clasping behavior (onset 15 wk) - rotarod deficit - hypokinesis - resting tremors - tremors and gait abnormalities - loss of coordination - muscle weakness - progressive weight loss	[82]
YAC128	full length human *HTT* / human huntingtin	128	8-12 / normal life span	- reduction of striatum and cortex volume by 48 wk - reduction of striatum and cortex neuron number by 48 wk - inclusions in striatal cells	- clasping behavior - rotarod deficit - hyperkinesis (12 wk) and hypokinesis (24 wk) - gait abnormalities - circling behavior - ataxia - weight increase	[83]
BACHD		97	12 / normal life span	- brain atrophy - degenerating darkly stained neurons in striatum - inclusions in neuropil, cortex and striatum	- rotarod deficit - weight increase	[86]

Rat models of HD were initially developed by viral vector-mediated delivery of mutant huntingtin to the striatum [89, 90]. These animals exhibit the striatal MSN loss characteristic of human HD, but they show only certain motor deficits and no change in body weight or survival.

A transgenic rat model of HD, which carries a 1,962 bp rat HD cDNA fragment with a 51 CAG repeat expansion under the control of the endogenous rat HD promoter, has been generated by Von Horsten *et al.* [91]. This transgenic rat displays adult-onset disease with behavioral phenotypes and histological alterations in the brain.

More recently, the generation of the first transgenic nonhuman primate HD model has been reported by Yang *et al.* [92]. These authors produced transgenic monkeys (*Macacus Rhesus*) which express the exon 1 of the human *HTT* gene with 84 CAG repeats and show relevant clinical features of human HD.

Apart the *ex vivo* delivery of CNTF encoding gene (see below), none of the nucleic acid-based therapeutic strategies for HD has so far progressed towards clinical trials. In this regard, it should be noted that performing therapeutic trials in HD patients is particularly challenging. This is due to its low prevalence, which makes it difficult to recruit a sufficient number of patients, slow progression, which may lead to very long follow up periods for treatments that are hypothesized to affect disease progression, variability in the rate of progression [93] and lack of well-established biomarkers, which means that in most cases, clinical trials have to rely on subjective clinical rating scales as outcome parameters.

Neuroprotection and Neurorestoration

The neuropathology of HD patients is characterized by a wave of degeneration in the striatum, being the MSNs mainly affected, although in the final stages cortical atrophy is also observed [94, 95]. Therefore, the delivery of neurotrophic factor genes in order to protect striatal neurons against mutant huntingtin-induced toxicity is one of the major gene therapy approaches currently under investigation. Several neurotrophic factors have been tested as potential therapeutic agents for HD in pre-clinical studies and one of them has progressed towards a Phase I clinical trial [96, 97].

Brain-Derived Neurotrophic Factor (BDNF)

In consideration of its importance for MSNs, the most affected neuronal population in HD, BDNF, a member of the neurotrophin family, represents the main candidate for neuroprotective therapeutic strategies. BDNF is produced in the cerebral cortex and anterogradely transported to the striatum [98, 99] where it supports neuron survival [100-103]. Wild-type huntingtin promotes the cortical expression of BDNF [104-106] as well as its vesicular transport along the microtubules [107].

A reduction in BDNF levels has been described in *post mortem* samples from individuals affected by HD [108, 109] and in several animal models as well (reviewed in [101, 103]). Such a decrease has been shown to be due to both a reduction in BDNF gene transcription [104] and an impaired transport of BDNF vesicles along the cortico-striatal afferents [107]. The possible role of BDNF depletion in HD pathogenesis has been evaluated in animal models [100, 110, 111]. The results of these studies demonstrated that reduction in BDNF levels leads to dysfunction and loss of striatal neurons, thus supporting the conclusion that depletion of this neurotrophin is a key factor in HD onset and progression. On these basis, several efforts have been made in order to develop therapeutic strategies aimed at increasing BDNF levels.

Both direct and cell-mediated delivery of the BDNF gene have been applied in rodent models of HD. Durable expression of BDNF protein has been obtained in either case and interesting effects have been reported in terms of both neuroprotection and functional recovery. However, several problems need to be overcome before applying this approach in clinical practice.

Viral delivery of BDNF gene to the striatum of rats prior to injection of QA has been shown to be significantly neuroprotective, resulting in reduction of both cell death and lesion size [112, 113]. In a first study reported in 1999 by Bemelmans *et al.* [112], BDNF gene was administered by means of an adenoviral vector (AdV) to the striatum of rats that were submitted to QA injection two weeks later. Compared to control rats, animals treated with AdV-BDNF showed a reduction of QA-induced lesions by one half and a significant increase of MSN survival (64% *versus* 46%). Protection of striatal neurons against the same excitotoxic insult was also obtained in 2004 by Kells *et al.* [113] who used an adeno-associated viral vector (AAV) for BDNF gene delivery. Furthermore, the same authors reported that AAV-mediated delivery of BDNF gene was able to attenuate the motor function impairment induced by QA injection [114], even though side effects such as weight loss and seizure activity following long-term high level expression of BDNF were observed.

Gharami *et al.* [115] used a BDNF transgene under the control of the promoter for the alpha subunit of CaMKII (Ca^{2+} calmodulin dependent kinase II) to over-express BDNF in the forebrain of R6/1 mice, which express a fragment of mutant huntingtin with a 116-glutamine tract. Such over-expression increased TrkB signaling activity in the striatum, ameliorated motor dysfunction and reversed brain weight loss. Furthermore, a reduction of mutant huntingtin intranuclear inclusions in striatal neurons was observed as well as a normalization of the expression of DARPP-32 (dopamine- and cyclic AMP-regulated phosphoprotein), a marker of healthy MSNs and an increase of the number of enkephalinergic boutons. In a subsequent study [116], the same authors evaluated the effects of BDNF over-expression in the forebrain of YAC128 mice, a more physiological HD mouse model that expresses the whole human *HTT* gene with 128 CAG repeats and exhibits age-dependent loss of striatal neurons [83]. BDNF over-expression in this mouse model prevented loss and atrophy of striatal neurons and motor dysfunction, normalized expression of dopamine receptor D2, enkephalin and DARPP-32 and improved procedural learning. Rescue of the abnormal spine phenotype in YAC128 mice was also observed.

An interesting therapeutic strategy based on the combination of BDNF with Noggin, a soluble inhibitor of the BMPs (bone morphogenetic proteins) encoded by NOG [117] and aimed at inducing neurogenesis has been proposed by Cho *et al.* [118]. Noggin over-expression suppresses astroglial differentiation by subependymal zone (SZ) progenitor cells, thus expanding the pool of SZ cells potentially responsive to neuronal instruction by BDNF [119, 120]. Concurrent Noggin and BDNF over-expression has been shown to result in a substantial increase in the number of new neurons recruited to the neostriatum of normal adult rats [120]. On the basis of these observations, Cho *et al.* [118] investigated the feasibility of recruiting new striatal neurons in a mouse model of HD by means of BDNF/Noggin over-expression. For this purpose, R6/2 mice were intraventricularly injected with AdV-BDNF and AdV-NOG. The results of this study demonstrated that BDNF and Noggin induced striatal neuronal regeneration, delayed motor impairment and extended survival in R6/2 mice [118].

Cell-mediated delivery of BDNF gene has been employed by some authors as an alternative approach to increase neurotrophin levels in the brain of HD rodent models. In the first two studies [121, 122], BDNF-secreting grafts did not prove efficient in protecting striatum in a chemically induced rat model. Transplantation of immortalized rat fibroblasts genetically engineered to express BDNF was found by Frim *et al.* [121] not to provide any significant protection against excitotoxic damage and only slight effects were observed by Martinez-Serrano and Bjorklund [122] who used genetically modified neural stem cells (NSCs) in a similar experimental paradigm. However, more encouraging results have been obtained in the subsequent years. Perez-Navarro *et al.* [123, 124] demonstrated that BDNF-secreting fibroblasts are able to reduce the size of the lesion and to prevent the loss of striatal neurons induced by intrastriatal injection of the excitotoxin. As suggested by the authors, the absence of significant BDNF effects in previous reports could be accounted for by methodological differences, including the parameters used to assess neuronal rescue, the severity of the lesion, and/or the high doses of neurotrophin. In particular, they point out that long exposure to high doses of BDNF may reduce TrkB responsiveness to the neurotrophin.

In a recent study, the therapeutic effects of the transplantation of bone-marrow mesenchymal stem cells (MSCs), genetically engineered to over-express BDNF, on motor deficits and neurodegeneration in YAC128 transgenic mice have been evaluated [125]. The results of this study show that genetically engineered MSCs reduce behavioral deficits and increase striatal neuron survival in this transgenic model of HD.

Finally, a therapeutic strategy which takes advantage of reactive striatal astrogliosis and the consequent up-regulation of GFAP (Glial Fibrillary Acidic Protein) gene expression, has been proposed in order to obtain conditional BDNF over-expression [126, 127]. In the first study by Giralt *et al.* [126], enhanced BDNF release and higher resistance to excitotoxic damage following intrastriatal administration of QA were observed in transgenic mice over-expressing BDNF under GFAP promoter (pGFAP-BDNF mice), compared to wild-type mice. The possible protective effect of astrocyte-derived BDNF was verified by grafting astrocytes from pGFAP-BDNF mice in wild-type mice. Following QA injection, pGFAP-BDNF-derived astrocytes showed higher levels of BDNF and larger neuroprotective effects than the wild-type ones. Furthermore, significant behavioral improvements over time after QA administration were observed in mice grafted with pGFAP-BDNF astrocytes compared to those grafted with wild-type astrocytes. In a subsequent study [127], the same authors tested whether conditional and pathology-dependent delivery of BDNF could be neuroprotective in a transgenic mouse model of HD. For this purpose, they obtained double-mutant animals, R6/2:pGFAP-BDNF, to be compared with R6/2 mice. The results of this study show that BDNF over-expression from striatal astrocytes improves HD phenotype by preventing cortico-striatal synaptic dysfunction. As pointed out by the authors, conditional BDNF delivery regulated by the GFAP promoter in astrocytes could be a therapeutic strategy not only for HD, but also for other neurodegenerative diseases that are associated with reactive astrogliosis and consequent GFAP up-regulation, such as Alzheimer's disease [128, 129], Parkinson's disease [130] and amyotrophic lateral sclerosis [131].

Nerve Growth Factor (NGF)

Nerve Growth Factor (NGF), is the prototypical member of the neurotrophin family. Besides promoting neuronal survival, it is also known to down-regulate huntingtin expression in cultured striatal neurons [132]. A strong neuroprotective effect of NGF in the striatum of chemically induced rodent models of HD was observed by several authors following implantation of *ex vivo* modified cells. Frim *et al.* [133] reported that implantation of NGF-secreting fibroblasts reduces the size of adjacent striatal 3-NP lesions by an average of 64%. Implantation of polymer-encapsulated human NGF-secreting fibroblasts was shown to attenuate the behavioral and neuropathological consequences of QA injections into rodent striatum [134]. In the same model, protection of striatal neurons against excitotoxic damage by NGF-producing, genetically modified, NSCs was demonstrated by Martınez-Serrano and Bjorklund [122]. It is worth noticing that the same results have not been obtained when NGF was directly administered by intravenous injection [135] or by intrastriatal infusion [136]. More recently, transplantation of MSCs genetically engineered to over-express NGF has been reported to reduce behavioral deficits in the YAC128 mouse model of HD, while no increase in striatal neuron number or DARPP-32 expression was observed [125].

Glial Cell Line-Derived Neurotrophic Factor (GDNF)

Glial Cell Line-Derived Neurotrophic Factor (GDNF) was discovered by Lin *et al.* [137] as a factor secreted by a rat glioma cell line (B49). Although GDNF was initially characterized as a trophic factor for mesencephalic dopaminergic neurons, its ability to protect several other neuronal populations was subsequently demonstrated.

The potential protective effect of GDNF on striatal neurons in rodent models of HD was initially investigated using cell-mediated delivery approaches. Perez-Navarro *et al.* [138] demonstrated that transplantation of GDNF secreting fibroblasts into the rat striatum partially protects MSNs from QA induced damage and reported that GDNF selectively promotes the survival and regulates the phenotype of striatonigral but not striatopallidal projection neurons [139]. More recently, NSCs and neural progenitor cells (NPCs) have been employed as vehicles for GDNF gene delivery. Transplantation of NSCs engineered to over-express GDNF into the mouse striatum protects against QA toxicity, resulting in reduced loss of striatal neurons as well as behavioral improvements [140]. In a study by Ebert *et al.* [141], NPCs were isolated from developing mouse striatum and genetically engineered to over-express GDNF. Transplantation of these cells into the striatum of pre-symptomatic N171-82Q mice was shown to prevent neuronal loss and to maintain motor function to a level indistinguishable from that of wild-type control mice.

The effects of GDNF over-expression in rodent HD models have also been investigated by viral vector-mediated delivery approaches. Kells *et al.* [113] investigated whether over-expression of GDNF, achieved by AAV vector-mediated gene delivery, could protect striatal neurons in the QA rodent model of HD. Three weeks after AAV-GDNF intra-striatal injection, the rats were lesioned with QA. Similarly to those treated with AAV-BDNF, AAV-GDNF treated rats showed a significant reduction of striatal neuron loss. Significant protection of parvalbumin-immunopositive striatal interneurons was also observed while AAV-BDNF provided significant neurotrophic support to NOS (nitric oxyde synthase)-immunopositive striatal interneurons.

The effects of AAV-mediated delivery of GDNF gene were also examined in a 3-NP induced rat model [142]. AAV-GDNF treatment resulted in strong protection of the striatal neurons as well as maintenance of motor function to an almost normal level. In a subsequent study [143], the same authors demonstrated that AAV-mediated delivery of GDNF to the striatum of presymptomatic N171-82Q mice delays and attenuates behavioral deficits. Behavioral neuroprotection was found to be associated with prevention of striatal neuron death and atrophy as well as reduction of neurons containing mutant huntingtin inclusions. These findings stand in contrast to those previously reported by Popovic *et al.* [144] who delivered GDNF *via* a lentiviral vector (LV) to the striatum of R6/2 transgenic mice. In this study, GDNF over-expression failed to attenuate behavioral and neuropathological changes. This difference has been attributed to the transgenic mouse model used [143]. R6/2 mice contain a human cDNA that encodes a mutant huntingtin protein with a larger number of repeats than those of N171-82Q mice and consequently show an earlier disease onset and a more aggressive behavioral phenotype. In the study by Popovic *et al.* [144], R6/2 mice were treated with LV-GDNF after symptom onset (4-5 weeks of age), while McBride *et al.* [143] administered AAV-GDNF to presymptomatic 5 weeks old N171-82Q mice. On this basis, it has been suggested that only presymptomatic HD gene carriers might profit of this therapeutic approach.

Neurturin (NTN)

Neurturin (NTN), which is encoded by *NRTN*, is a member of the GDNF family of neurotrophic factors and shares a 42% homology with GDNF. It was found to be up-regulated in the rat striatum in response to excitotoxic insults [145], thus suggesting a role in a physiological regenerative response. Cell-mediated delivery of neurturin has been shown to protect projection neurons but not interneurons in a rat model of HD [146]. A fibroblast cell line engineered to over-express neurturin was grafted into adult rat striatum 24 h before QA injection and grafting of neurturin-secreting cells resulted in a protective effect on striatal projection neurons much more robust than that previously observed for a GDNF-secreting cell line [123]. Neurturin reduced the lesion size and rescued 50-58% of striatal calbindin-positive neurons, whereas GDNF does not modify lesion size and protects only 30-35% of these neurons [123]. In contrast, neurturin had no effect on the survival of parvalbumin- or ChAT (choline acetyltransferase)-positive neurons, while GDNF prevented the decrease in ChAT activity induced by QA. In the same HD model, it has subsequently been demonstrated that grafting of neurturin-secreting fibroblasts selectively prevents the loss of striatopallidal neurons expressing glutamic acid decarboxylase (GAD) and preproenkephalin (PPE) [147]. Neurturin also prevented the decrease in the expression of their phenotypic markers, thus suggesting a preservation of neuronal function. On the other hand, neurturin did not exert any influence on the survival or phenotype of striatonigral, preprotachykinin A (PPTA)-positive neurons. Grafting of neurturin over-expressing fibroblasts was also shown to protect rat striatal neurons against KA excitotoxicity [148].

More recently, viral delivery of *NRTN* gene has been successfully tested in both chemically induced and transgenic rodent models of HD. AAV2-*NRTN* (developed by Ceregene Inc., San Diego, CA, USA and commercially known as CERE-120) was injected into the striatum of rats who received systemic injection of 3-NP four weeks later [149]. AAV2-*NRTN* treatment partially protected striatal neurons from cell death (27.4% loss of NeuN-ir (neuronal nuclear antigen-immunoreactive) neurons relative to non-lesioned animals, *versus* 49.7% neuronal loss in the AAV2-eGFP (enhanced green fluorescent protein)-3NP group and 48.4% loss in the vehicle-3NP group, respectively) and prevented ventricular enlargement as a result of reduction in striatal volume. A significant reduction of motor function deficits induced by 3-NP was also observed in AAV2-*NRTN* treated rats. AAV2-*NRTN* delivery to the striatum of presymptomatic N171-82Q mice resulted in a significant delay of behavioral deficits onset, as well as reduction of their severity [150]. Improvements in behavioral phenotypes were found to correlate with a significant protection of striatal neurons even though their atrophy was not prevented. Since AAV2-*NRTN* had no effect on the number or percentage of neurons containing mutant huntingtin inclusions, the authors suggest that "AAV2-*NRTN* induced neuroprotection does not depend on altering inclusion number within the striatum and that significant prevention of behavioral abnormalities can be achieved despite a heavy inclusion load" [150]. Interestingly, intrastriatal AAV2-*NRTN* administration was also shown to prevent neuron loss in layers V-VI of prefrontal cortex. Even though the underlying mechanism remains unclear, these findings raises the possibility that a neurturin-based therapy may help to prevent cognitive and personality deficits seen in animal models as well as in patients.

Ciliary Neurotrophic Factor (CNTF)

Ciliary neurotrophic factor (CNTF), a member of the neuropoietic cytokine family, promotes the survival of several types of neurons, including those most susceptible to mutant huntingtin-induced toxicity [151]. The first evidence indicating CNTF as a potential therapeutic agent for HD came from a study by Anderson *et al.* [152] in which intracerebral administration of this neurocytokine to the striatum of QA-treated rats was shown to efficiently reduce cell death. Since then, CNTF has been intensely investigated and actually it is the first and, up to now, only trophic factor to enter clinical trials in HD [96, 97].

The effects of cell-mediated delivery of CNTF gene have been investigated in both rodent and nonhuman primate models of HD. In a first study, a baby hamster kidney fibroblast cell line (BHK) genetically engineered to over-express CNTF was used by Emerich *et al.* [153] in a QA rat model. Polymer-encapsulated CNTF-secreting BHK cells were transplanted into the striatum of rats that received QA injection twelve days later. CNTF-producing implants were reported to protect ChAT-ir and GAD-ir striatal neurons from excitotoxic damage and to provide a partial (task-specific) behavioral protection. The same cells were subsequently employed to deliver the CNTF gene to the striatum of a QA nonhuman primate HD model [154]. CNTF was found to exert a neuroprotective effect on several populations of striatal cells, including GABAergic, cholinergic and diaphorase-positive neurons. Interestingly, CNTF also prevented the atrophy of layer V neurons in motor cortex and exerted a significant protective effect on the two critical GABAergic efferent projections from striatum (to the GP and to the SNpr, respectively). In a further study involving a primate model of HD, the intrastriatal implant of encapsulated CNTF-secreting BHK cells was shown to restore cognitive and motor functions [155].

On the basis of these findings, a Phase I clinical trial has been carried out in France in order to evaluate the safety of cell-mediated, continuous and long-term intracerebral delivery of CNTF in HD patients [96, 97]. In this trial, BHK cells engineered to express CNTF encapsulated in a semipermeable polymer membrane were used as vehicle for the delivery. One capsule was implanted into the right lateral ventricle of each of six HD patients and the capsule was removed and replaced with a new one every 6 months, over a total period of 2 years. No side effects were observed in this study, but the retrieved capsules contained variable numbers of surviving cells and CNTF release was low in 13 out of 24 cases. Patients showed no clinical improvement, even though improvements in electrophysiological recording were observed in those patients whose capsules were shown to release the largest amount of CNTF after explantation. The absence of clinical benefits may have been due to the inadequacy of the amount of CNTF delivered to the striatum. In this respect, it is worth noticing that dose dependence of the neurochemical and functional protection provided by encapsulated CNTF-producing cells in a QA rodent model has been reported by Emerich [156]. Furthermore, the neuroprotective effects of encapsulated CNTF-producing cells have been shown, in the same animal model, to depend on the proximity of the implant to the lesion site [157]. In this study, encapsulated CNTF-producing cells were implanted into the lateral ventricle either ipsilateral or contralateral to an intrastriatal QA injection. Neurochemical and behavioral benefits were provided by implants ipsilateral to the QA injection but not by implants in contralateral ventricle, thus suggesting that the direct delivery to the striatum may be needed for clinical purposes.

Viral vectors have been employed to deliver CNTF gene in rodent models either chemically or genetically induced [158-163]. Significantly different results have been reported in these studies, depending on the typology of the animal model employed. The intrastriatal injection of a CNTF-expressing LV vector was shown to partially reverse behavioral deficits and provide a neuroprotective effect in a QA rat model [158]. Compared to control animals, the extent of the striatal damage was significantly diminished in the CNTF-treated rats as indicated by the strong reduction of the lesion volume and the sparing of DARPP-32, ChAT and NADPH-d (NADPH-diaphorase) neuronal populations. Interestingly, the use of tetracycline-regulated LV vectors allowed the same authors to demonstrate the dose dependence of CNTF neuroprotective effect [159]. Beurrier *et al.* [160] recently confirmed that CNTF protects striatal neurons against QA-induced excitotoxicity and reported evidence that an enhancement in glial glutamate uptake underlies CNTF-mediated neuroprotection. Promising results were also reported by Mittoux *et al.* [161], who investigated the effects of AdV-mediated CNTF gene delivery in a 3NP rat model. A neuroprotective efficacy of the transgene product up to 3 months after intrastriatal delivery of the viral vector was observed. Furthermore, CNTF-expressing AdV vector was shown to protect not only striatal neurons but also neurons located upstream and downstream from the striatum, and, in particular, the corticostriatopallidal pathway, a neuronal circuit severely affected in HD.

On the other hand, no beneficial effects were observed in YAC72 mice following intrastrial injection of a CNTF-expressing LV vector [162]. In a further study, AAV-mediated intrastriatal delivery of CNTF was reported to result in earlier onset of motor impairments in R6/1 mice and to cause abnormal behavior in wild-type mice [163]. Furthermore, CNTF expression caused a significant decrease in the levels of striatal-enriched transcripts in both wild-type and R6/1 mice [163]. These adverse effects may be due to the high levels of CNTF expression provided by the expression cassette used in this study.

Targeting Mutant Huntingtin

Therapeutic strategies aimed at targeting mutant huntingtin are an attractive option for treating HD since they might represent a true etiologic therapy. The first evidence supporting the therapeutic potential of knocking down the mutant allele of HD gene was obtained in a tetracycline-regulated conditional mouse model of HD [164]. In this model, the expression of a mutated huntingtin fragment resulted in formation of neuronal inclusions and progressive motor dysfunction. Blocking the expression of the mutant protein in symptomatic mice led to the disappearance of the inclusions and motor function improvement.

Both RNAi- and ASO-based approaches targeting mutant huntingtin mRNA are under pre-clinical investigation for HD gene silencing. Currently, an alternative strategy aimed at blocking the mutant huntingtin itself using intracellular antibodies is also under development.

Targeting the Mutant mRNA

Effective down-regulation of HD gene expression by an ASO-based approach was firstly reported in 2000 by Boado *et al.* [165]. In this study an oligonucleotide complementary to nt -1 to 15 surrounding the AUG initiation codon was found to inhibit huntingtin synthesis in PC12 cells genetically modified to express the exon 1 of human wild-type *HTT*. A further successful *in vitro* study was reported in the same year by Nellemann *et al.* [166] who used a phosphorothioate ASO (PS-

ASO) to down-regulate the expression of endogenous huntingtin in postmitotic neurons (NT2-N neurons) differentiated from embryonic teratocarcinoma cells (NT2 cells). In a subsequent study by the same authors [167], a PS-ASO was shown to down-regulate the expression of mutant huntingtin and reduce aggregate formation in NT2 and NT2-N cells genetically engineered to express the exon 1 of a mutant HD gene. More recently, the ASO-based approach has been further developed in order to achieve allele-specific gene silencing (see below).

RNAi-based strategies for mutant huntingtin mRNA targeting include both viral vector-mediated delivery of shRNAs and administration of chemically synthesized siRNAs. In a study by Harper *et al.* [168], AAV1 vectors expressing a shRNA directed against human mutant *HTT* mRNA were injected into the striatum of the N171-82Q transgenic mouse model which expresses an N-terminal fragment of human huntingtin with 82 CAG repeats [82]. As a result, transgene expression at both the mRNA and protein levels was found to be decreased. Notably, a reduction in mutant huntingtin inclusions and significant behavioral improvements were observed in the treated mice. In a parallel study, Rodriguez-Lebron *et al.* [169] used AAV5 to deliver two different human *HTT*-specific shRNAs to the striatum of R6/1 mice which express exon 1 of human *HTT* containing 144 CAG repeats and display a substantially more aggressive phenotype relative to N171-82Q mice [81]. One shRNA (si-Hunt1) was found to reduce both mutant huntingtin mRNA expression and inclusions and improve hind-limb clasping, an indicator of neurological impairment in mice. Furthermore, a partial normalization of striatal proenkephalin and DARPP-32 mRNA levels (which are reduced in both murine and human HD brain) was observed in the treated mice. Together, these studies provided the first evidence of RNAi potentiality as a tool for HD therapy. Similar neuropathological and/or motor improvements were obtained in subsequent shRNA-based studies using different models, RNA sequences and efficacy assessments [170-173]. In particular, delivery of huntingtin-targeting AAV-shRNAs after the onset of motor abnormalities was able to decrease neuronal aggregates and attenuate DARPP-32 down-regulation [170], thus suggesting that RNAi-based therapies could benefit HD patients even after the onset of symptoms.

Beneficial effects of *in vivo* mutant *HTT* knockdown using chemically synthesized siRNAs were initially reported by Wang *et al.* in 2005 [174]. In this study, lipid-encapsulated siRNAs specific for human *HTT* exon 1 were injected into the lateral ventricle of two day old R6/2 mice which express the same mutant human *HTT* transgene as the R6/1 mice but exhibit a more aggressive disease phenotype [81]. In treated pups, the levels of mutant huntingtin mRNA were shown to be significantly reduced up to seven days after treatment, but silencing was almost completely lost one week later. Interestingly, while silencing was only transient and preceded the disease onset, beneficial effects were later observed in terms of neuropathological and behavioral improvements as well as a modest extension in life span. In a subsequent siRNA-based study [175], an acute mouse model generated by intrastriatal injection of AAV vectors carrying a fragment of human mutant *HTT* was used. While mice receiving only AAV-m*HTT* rapidly developed HD neuropathology and motor deficits, co-injection of chemically modified and cholesterol-conjugated siRNAs targeting human mutant huntingtin mRNA resulted in a significant reduction of both huntingtin monomer levels and aggregates, improved neuronal survival and significantly delayed the onset of behavioral abnormalities.

The results of these proof-of-concept studies appear very promising. However, it is important to note that, in these studies, allele-specific silencing of the pathogenic human mRNA (while preserving the expression of endogenous wild-type huntingtin) was easily achieved thanks to sequence differences between murine and human huntingtin encoding genes. In consideration of potential deleterious effects resulting from loss of normal huntingtin functions, the application of RNAi-based therapies to HD patients might require approaches that selectively silence the mutant allele. In order to evaluate the feasibility of a non-allele-specific RNAi strategy (which implies both mutant and wild-type allele silencing) four studies investigated whether adult neurons can tolerate hypomorphic wild-type huntingtin expression [173, 176-178]. No overt histopathological changes were detected by McBride *et al.* [176] in wild-type mouse brain following AAV1-mediated delivery of a *Htt*-targeted miRNA which resulted in a reduction of normal mouse *Htt* mRNA and protein levels by 70% and 83%, respectively. In a subsequent study, a partial reduction of huntingtin levels in the normal nonhuman primate putamen was shown to be well tolerated [178]. Safety of therapeutic shRNAs and miRNAs was also assessed by the same authors [176] in the CAG140 knock-in mouse model of HD. Drouet *et al.* [173] reported the first evidence that co-silencing of mutant *HTT* and normal *Htt* in the same neurons could improve HD-related histopathology. In this study, human mutant huntingtin was expressed in rat or mouse striatum by using LV vectors. Co-injection of a LV vector carrying an shRNA that targets a region present in both transcripts prevented DARPP-32 loss and reduced ubiquitin-positive neuronal inclusions up to nine months after treatment. However, even though a substantial reduction in huntingtin mRNA levels was observed, the levels of remaining normal huntingtin protein were unclear. The first evidence that simultaneous lowering of wild-type and mutant huntingtin expression could improve HD-associated motor deficits was reported by Boudreau *et al.* [177] who used an AAV vector to deliver a miRNA to the striatum of N171-82Q mice. The overall results of these studies support the feasibility of non-allele-specific gene silencing strategies to treat HD. It is worth noticing, however, that abnormalities in molecular pathways associated with huntingtin loss of function were reported by Drouet *et al.* [173] and Boudreau *et al.* [177]. Further investigation of the safety of RNAi-mediated normal huntingtin reduction is necessary, and, in particular, the possible effects of long-term *HTT* gene silencing need to be assessed.

To avoid the potential deleterious effects of wild-type huntingtin down-regulation, allele-specific strategies are currently under development. Since siRNAs cannot distinguish between normal and CAG repeat expanded transcripts [179], targeting heterozygous single-nucleotide polymorphisms (SNPs), rather than directly targeting the CAG repeat, has been proposed as an alternative approach to distinguish between mutant and wild-type *HTT* mRNAs. The first evidence of allele-specific silencing of the HD gene using SNP-specific RNAi was obtained in human cells over-expressing two *HTT* mRNAs that differed at a SNP [180]. The first study supporting the SNP-based approach to selectively silence endogenous mutant *HTT* was reported in 2008 by van Bilsen *et al.* [181]. In this study, a siRNA targeting a SNP located several thousands bases downstream from the disease-causing mutation was shown to selectively reduce both mutant mRNA and protein in fibroblasts derived from an HD patient who was heterozygous for that SNP. Several HD-linked SNPs have subsequently been identified and characterized [182-184] and it has been suggested that the majority of HD patients are potentially eligible for allele-specific therapy [185]. A different type of polymorphism found to be associated with the mutant allele of HD gene, a microdeletion in exon 58 [186, 187] has been used by Zhang *et al.* [188] to develop siRNAs able to preferentially target the mutant mRNA in HD fibroblasts and neuroblastoma cells expressing both wild-type and mutant alleles of HD gene. More recently, Carroll *et al.* [189] developed SNP-targeted ASOs which were shown to selectively silence the mutant *HTT* allele in HD patient-derived fibroblasts and cortical neurons obtained from transgenic mouse models of HD. Modification of these ASOs with S-constrained-ethyl (cET) motifs was found to significantly improve efficiency while maintaining allele-selectivity *in vitro*. Notably, intraparenchymal bolus delivery of ASO to the transgenic mice striatum led to effective and selective silencing of the mutant *HTT* allele. It is worth noticing that, unlike siRNAs, ASOs can target SNPs anywhere in the pre-mRNA, including introns, thus substantially increasing the number of potential SNP targets.

To achieve allele-specific silencing for patients who do not exhibit either the microdeletion or an appropriate SNP, a different strategy will be obviously required. With reference to this problem, it is important to note that, as suggested by some recent studies, selective targeting of the expanded CAG repeat might be in a short time feasible. Peptide nucleic acid (PNA)-peptide conjugates and locked nucleic acid (LNA) oligomers complementary to the CAG repeat were found to selectively inhibit the expression of mutant huntingtin in HD patient-derived fibroblast cell lines [190]. The best allele-discriminating reagent was an oligomer containing 19 PNA residues and modified by addition of lysine residues at both termini, which showed a specificity of inhibition at least 10 times greater for mutant *HTT* mRNA translation than for normal mRNA. Moreover, the translation of other transcripts containing CAG repeats was not significantly inhibited in PNA- and LNA-treated cells. Oligomers with other chemical modifications such as 2',4'-constrained ethyl (cEt), carba-LNA, 2'-O-methoxyethyl (2'-MOE) or 2'-fluoro have also been shown to be promising blockers of mutant huntingtin mRNA translation [191]. An alternative approach to target expanded CAG repeats relies on switching the RNAi mechanism towards a miRNA-like RNAi mechanism [192, 193]. This is achieved by introducing one or more base substitutions into RNA duplexes composed of CAG and CUG repeat strands, thus causing mismatches with their mRNA targets. Several RNA duplexes of this type were shown to efficiently inhibit the synthesis of mutant huntingtin, without down-regulation of its transcript, in HD patient-derived fibroblasts [192, 193]. In most efficient and selective duplexes, the mismatched bases were present in the central positions of the duplex [192] or in its 3' half [193]. Evidence supporting a miRNA-like mechanism of action of anti-CAG mismatch-containing RNA duplexes was reported in a recent study by Hu *et al.* [194].

Although RNAi- and ASO-based strategies appear promising approaches to treat HD, several issues still need to be solved prior to their clinical testing, such as, for example, silencing stability, optimized delivery methods and safety. Furthermore, cellular targets of these therapeutic strategies should be carefully evaluated since several data suggest that non-cell-autonomous mechanisms (alteration of the cortico-striatal circuitry) are also operating in HD [104, 195, 196]. Besides MSNs, additional target cell types should, therefore, be considered.

Targeting the Mutant Protein

Several anti-huntingtin intrabodies (iAbs) specific for various domains of the protein have been developed and tested as potential therapeutics in both *in vitro* and *in vivo* models of HD [197-207].

The iAb C4, which recognizes the N-terminus of the protein, has been shown to reduce mutant huntingtin-induced toxicity and aggregation in cell models of HD [197-200] and in brain slices from HD mice [201]. Furthermore, the same iAb has been reported to suppress neuropathology and to prolong life span in a *Drosophila* model of HD [202, 203]. Snyder-Keller *et al.* [204] used an AAV vector to deliver the iAb C4 to the striatum of R6/1 mice before or after the development of aggregates and clinical signs. Delivery of the iAb at 5-8 weeks of age, prior to the first appearance of mutant huntingtin aggregates, was most effective, resulting in a strong reduction in the number of large aggregates within transduced striatal neurons. However, the iAb was still effective when injected at 10-12 weeks of age, *i.e.*, the time mutant huntingtin aggregates are forming. Interestingly, following injection at 16-24 weeks of age, when large aggregates are present in nearly all striatal neurons, a partial reversion of mutant huntingtin aggregation was observed. In order to enhance long-term iAb-mediated correction, the same authors engineered a fusion of scFv-C4 to a sequence that can lead to rapid and irreversible turnover of the antigen-antibody complex. Fusion to a PEST signal sequence (a peptide rich in proline, glutamic acid, serine and threonine, which is known to be associated with short-lived proteins) results in significant degradation of both soluble and insoluble mutant huntingtin in cultured ST14A striatal cells over-expressing this toxic protein [205].

Interesting results have also been obtained by Wang *et al.* [206] in a study based on the iAb EM48. This iAb recognizes an epitope in the C-terminus of huntingtin common to mutant and wild-type protein, but preferentially binds mutant huntingtin, possibly due to conformational changes induced by the polyQ expansion. EM48 has been shown to increase the ubiquitination and subsequent degradation of mutant huntingtin in a cell model of HD [206]. Intrastriatal AdV-mediated delivery of this iAb into R6/2 and N171-82Q mouse models resulted in increased mutant huntingtin degradation and reduced neuropil aggregate formation. EM48 treatment also improved motor deficits in N171-82Q mice but no significant improvement in body weight or survival was observed.

Southwell *et al.* [207] have more recently tested the effects of two different iAbs, $V_L12.3$ and Happ1, in a lentiviral mouse model of HD and in four transgenic HD mouse lines (R6/2, N171-82Q, YAC128 and BACHD) using intrastriatal AAV delivery. $V_L12.3$ recognizes the N-terminus of huntingtin, while Happ1 recognizes the proline-rich domain adjacent to the polyQ segment and selectively increases the turnover of mutant but not wild-type huntingtin in cell models of HD [208]. The selectivity of Happ1 action, which involves enhanced calpain cleavage of mutant huntingtin [209], might be due to the enhanced epitope availability in the context of the mutant protein. Different results were obtained following $V_L12.3$ treatment depending on the model used. While beneficial effects were observed in the lentiviral model, no effects were detected in YAC128 HD mouse model and an increase in the severity of the phenotype was observed in the R6/2 HD model. On the other hand, Happ1 significantly ameliorated the neuropathology as well as motor and cognitive deficits in all five HD mouse models. Importantly, Happ1 increased the body weight and prolonged the life span by 30% in the N171-82Q model, while no benefit in these outcomes was observed in the more severe and earlier onset R6/2 model [207].

Considered as a whole, the results so far obtained in iAb based studies support the idea that therapies directed at the specific degradation of mutant huntingtin represent a selective and effective strategy for the treatment of HD, although further improvements are still needed.

A different strategy for reducing mutant huntingtin levels involves induction of autophagy. Selective degradation of the mutant protein by chaperone-mediated autophagy has been recently reported by Bauer *et al.* [210]. This study was based on the expression of a construct comprising the polyglutamine binding peptide 1 (QBP1), which binds mutant but not wild-type huntingtin, fused to two different heat shock cognate protein 70 (HSC70) binding motifs, which target bound proteins to lysosomes for degradation. The expression of this construct in a cell model of HD was shown to result in the specific degradation of mutant huntingtin and its intrastriatal AAV-mediated delivery ameliorated the disease phenotype in the R6/2 mouse model of HD. Reduced levels of both aggregated and soluble mutant huntingtin in brain lysates from the treated animals and decreased inclusions in the striatum were observed. Moreover, improvements in motor deficits and in body weight were reported as well as a significant increase in the survival of this very severe model [210].

A few other therapeutic strategies based on mutant huntingtin targeting have also been suggested. Wang *et al.* [211] used both *in vitro* and *in vivo* models to investigate the effects of over-expression and under-expression of ubiquilin-1 [212] on mutant huntingtin-induced cell death and/or toxicity. Knockdown of ubiquilin expression by RNAi was shown to increase both aggregation and cytotoxicity of GFP-huntingtin fusion proteins containing expanded polyQ repeats in HeLA cells and in primary cortical neurons. On the contrary, a reduction of both aggregation and cytotoxicity was observed as a consequence of ubiquilin over-expression in the same cells. In a *Caenorhabditis elegans* model of HD, RNAi of the ubiquilin gene exacerbated the motility defect, whereas over-expression of ubiquilin prevented and could rescue, loss of worm movement induced by over-expression of GFP-huntingtin (Q55). Another therapeutic strategy has been suggested by Perrin *et al.* [213] who investigated the possibility of harnessing heat-shock proteins which refold denatured proteins. The results of this study demonstrated that over-expression of Hsp104 and Hsp27 rescue striatal dysfunction in primary neuronal cultures and in rat models based on LV-mediated over-expression of a mutated huntingtin fragment.

Finally, short G-rich oligonucleotides able to inhibit mutant huntingtin aggregation have also been proposed by Skogen *et al.* [214] as potential therapeutics for HD. Using a cell-based assay, these authors demonstrated that a 20-mer, all G-oligonucleotide (HDG) capable of adopting a G-quartet conformation, effectively inhibits the aggregation of a fusion protein comprising a mutant huntingtin N-fragment. They also reported that this oligonucleotide is able to improve cell survival in PC12 cells over-expressing a mutant *HTT* fragment fusion gene.

Further Approaches

Advances in understanding the mechanisms of HD pathogenesis have led to devise mechanism-based strategies that might be of therapeutic value. The proposed approaches mainly focus on the normalization of either mitochondrial function and calcium signaling pathway and on the restoration of transcriptional balance.

Normalization of Mitochondrial Function and Calcium Signaling

Strong evidence indicates that mitochondrial defects may play a central role in HD pathogenesis [215]. Studies investigating the toxic effects of mutant huntingtin *in vivo* and *in vitro* models have demonstrated mitochondrial alterations including reduction of Ca^{2+} buffering capacity, loss of membrane potential and decreased expression of oxidative phosphorylation enzymes. Several approaches aimed at restoring mithocondrial function and normalizing calcium signaling have been reported in recent years.

In a study by Benchoua *et al.* [216] the over-expression of either Ip or Fp subunit of succinate dehydrogenase (SDH), the main component of mitochondrial complex II, was shown to block mitochondrial dysfunction and cell death induced by mutant huntingtin in primary rat striatal neurons. This study suggests that regulating complex II / SDH expression may be of therapeutic interest to slow down striatal degeneration in HD.

Cui *et al.* [217] demonstrated that one of the key mechanisms involved in mutant huntingtin-induced mithocondrial dysfunction is the down-regulation of the expression of peroxisome proliferator-activated receptor gamma coactivator-1α (PGC-1α), a transcriptional coactivator that regulates several metabolic processes, including mitochondrial biogenesis and oxidative phosphorilation [218, 219]. In this study of Cui and coworkers, the possible protective role of PGC-1α over-expression against mutant huntingtin-induced mitochondrial dysfunction and striatal toxicity was also investigated. PGC-1α over-expression was shown to significantly reverse the mitochondrial defect in STHdhQ111 cells and to abrogate mutant huntingtin toxicity in primary striatal neurons. In agreement with *in vitro* results, LV vector-mediated delivery of PGC-1α to the striatum of R6/2 mice completely prevented neuronal atrophy. Even though the effects of PGC-1α over-expression on motor deficits and life span of R6/2 mice have not been investigated, the results reported by Cui and coworkers suggest that this approach could provide potential clinical benefit at early stages of HD.

Dai *et al.* [220] investigated the therapeutic potential of the expression of a calmodulin (CaM) fragment in the R6/2 mouse model of HD. Huntingtin has been shown to interact with many proteins, among which CaM [221] and mutant huntingtin has been reported to interact with CaM with a higher affinity compared to wild-type huntingtin [221]. CaM is a calcium binding protein that activates many enzymes including transglutaminase (TGase), an enzyme which modifies mutant huntingtin, thus possibly contributing to its toxicity. AAV-mediated delivery of a CaM fragment (containing aminoacids 76-121 of CaM) to the striatum of R6/2 mice resulted in a reduction of TGase-modified mutant huntingtin and in a decrease of the number and size of nuclear aggregates. A significant reduction of body weight loss and improvements in motor function were also reported in the treated animals, while no significant increase in survival was observed [220].

The dysregulation of type 1 inositol 1,4,5-trisphosphate receptor (InsP$_3$R1)-mediated calcium signaling in MSNs, which plays an important role in HD pathogenesis, is due to the interaction between mutant huntingtin and the C-terminal cytosolic tail of InsP$_3$R1 (IC10 fragment) [222]. On this basis, Tang *et al.* [223] investigated whether the expression of IC10 fragment in HD MSNs could disrupt InsP$_3$R1-mutant huntingtin association and consequently prevent Ca^{2+} overload and apoptosis. Infection of cultured YAC128 MSNs with a LV vector expressing a GFP-IC10 fusion protein stabilized Ca^{2+} signaling and protected against glutamate-induced apoptosis. Furthermore, intrastriatal AAV-mediated delivery of the same fusion protein significantly alleviated motor deficits and reduced MSN loss and shrinkage in YAC128 mice.

Restoration of Transcriptional Balance

A large number of studies have shown the involvement of a widespread transcriptional dysregulation in the molecular pathology of HD [224, 225]. Although many transcriptional regulatory molecules have been implicated in HD pathogenesis, increasing evidence supports a key role of RE1 Silencing Transcription Factor (REST) [226]. The target genes of this repressor include protein encoding genes, among which BDNF, as well as genes coding for microRNAs and long non-coding RNAs. Rescuing REST target gene expression might offer a new therapeutic approach for the treatment of HD.

In 2007, Zuccato *et al.* [227] reported a strategy based on a dominant-negative mutant form of REST comprising only the eight zinc finger DNA-binding domain and lacking any co-repressor interaction domain. The expression of the cDNA construct encoding this dominant-negative REST in a cell model of HD was shown to reduce the binding of REST to its binding sites in genomic DNA and consequently increase the transcription of BDNF and other REST-regulated genes.

More recently, a decoy strategy to rescue REST target gene expression has been developed by the same authors [228]. Phosphorothioate double-stranded oligonucleotides corresponding to the DNA-binding element of REST were delivered to cells expressing mutant huntingtin. Following its specific interaction with REST, the decoy acted to sequester it, resulting in reduced REST occupancy of target sites in the genome and rescue of the expression of target genes, including BDNF.

Parkinson's Disease

Parkinson's disease (PD) is a progressive neurodegenerative disease with a mean age of onset of 55 years old and the probability of occurrence increasing markedly with age. It is the second most common neurodegenerative disorder and the most common neurodegenerative movement disorder. The cardinal motor symptoms observed in PD patients include tremor at rest, rigidity, akinesia and bradykinesia (paucity and slowness of movement, respectively) and postural instability. Gait disturbances, swallowing and speech problems, autonomic dysfunctions and cognitive decline can also be present. PD motor symptoms can be attributed to the selective gradual loss of dopaminergic (DAergic) neurons in the substantia nigra pars compacta (SNpc). This leads to a progressive reduction of dopamine (DA) levels in the striatum and consequent dysfunction of the basal ganglia, a cluster of nuclei involved in the initiation and execution of movement. Apart from neuron loss, one major pathological feature of PD is the presence of cytoplasmic proteinaceous inclusions, known as Lewy bodies (LBs), in surviving DAergic neurons. LBs contain a variety of proteins with α-synuclein as a major component.

PD appears essentially as a sporadic condition with unknown etiology, in which, however, a combination of environmental and genetic risk factors is believed to be involved. A few rare cases of the disease are caused by several different gene mutations and are inherited as mendelian characters [229, 230]. Such inherited cases usually have an earlier age of onset than sporadic forms and often show atypical clinical features. The mutations identified in PD families include autosomal dominant mutations in the α-synuclein [231-233] and LRRK-2 encoding genes [234, 235] and autosomal recessive mutations in several genes such as, for example, those encoding for Parkin [236], DJ-1 [237] and PINK-1 [238]. Studying the proteins encoded by these genes and their pathogenic mutations has provided important insight about the mechanisms underlying neurodegeneration in PD. Although PD pathogenesis is not yet clearly understood, these studies have confirmed the involvement of oxidative stress, mitochondrial dysfunction and reduced ability to degrade abnormal proteins in the progressive degeneration of the DAergic neurons. Moreover, a crucial role of neuroinflammation in the disease progression has been supported by clinical, epidemiological and experimental evidence [239, 240].

To date, no effective therapies have been developed to cure PD. Current pharmacological and surgical treatments offer symptomatic improvements to PD patients, but none of these has convincingly shown effects on disease progression [241]. The most commonly used treatment is administration of the DA precursor, L-dopa, which replaces lost DA in the denervated striatum and relieves motor symptoms. While PD patients generally respond very well in the earlier stages of the disease, the clinical benefit of L-dopa tends to be overbalanced by substantial side effects as the disease progresses. Surgical methods such as deep brain stimulation (DBS) of the subthalamic nucleus (STN) or globus pallidus (GP) have also been used with success, but these procedures are not widely available or applicable for all patients.

Table 4: Neurotoxin-induced animal models of Parkinson's Disease.

	Neuropathology	**Behavioral Symptoms**
6-OHDA	- massive loss of DAergic neurons - no intracellular aggregates - reduced DA levels in the striatum	- rotational behavior after unilateral lesion - motor impairments after bilateral lesion
MPTP	- dose-dependent loss of DAergic neurons (95% in acute high-dose conditions) - few cases of α-synuclein aggregates in nonhuman primates - increased α-synuclein immunoreactivity in rodents - formation of intracellular aggregates in case of chronic administration - reduced DA levels in the striatum	- motor impairments in nonhuman primates and, less obvious, in rodents
Rotenone	- loss of DAergic neurons - reduced striatal DA innervation - formation of intracellular aggregates	- decreased motor activity in rodents
Paraquat	- loss of DAergic neurons - decreased TH immunoreactivity in the striatum - increased α-synuclein immunoreactivity - formation of LB-like structures	- motor deficits in most but not all cases

Since the available treatments do not provide safe and long-lasting relief from the symptoms and do not have any effect on the progression of the disease, much effort is currently made to develop novel therapeutic strategies able to slow or even halt the neurodegenerative process, rather than simply treating the symptoms of PD. Among these, nucleic acid-based strategies may have the potential for substantial advancements in the treatment of PD, which presents several features that make it a good target for such strategies. A main favorable feature is a relatively selective localization of the pathology to a brain area, which can be targeted using standard stereotactic procedures. Furthermore, although the actual cause of the disease is still unclear, a lot of information about the pathogenetic mechanisms has been acquired and several genes responsible for inherited forms of PD have been identified.

The three main nucleic acid-based strategies currently under investigation are aimed, respectively, at protecting and restoring neuronal function, improving availability of DA to the striatum and reducing activity in the STN. In this context, various approaches have proven successful in pre-clinical experimentation. The availability of both rodent and nonhuman primate models for proof-of-principle pre-clinical testing has obviously been of paramount importance. Several models of PD have been developed in various species by administration of toxins (Table **4**) or by genetic manipulations (for a recent review on classic and new animal models of PD see [242]).

The earliest and most widely used neurotoxin-induced models are those produced by administration of the DA system-specific neurotoxins 6-hydroxydopamine (6-OHDA) and 1-methyl-4-phenyl-1,2,3,6-tetrahydropyridine (MPTP). Only these two animal models of PD are briefly described in this section since most of the studies here reported employed just these models.

The neurotoxin 6-OHDA does not efficiently cross the BBB and therefore it must be injected directly into the nigro-striatal tract where it is taken up into the DAergic neurons *via* the DA transporter (DAT). The injection of 6-OHDA into the substantia nigra (SN) induces a fast and massive loss of tyrosine hydroxylase (TH)-positive neurons in this area of the rodent brain [243], with subsequent loss of TH-positive terminals in the striatum. When the toxin is injected into the striatum, TH-positive terminals in this area degenerate prior to the TH-positive neurons in the SN [244]. The ability of 6-OHDA to induce DA degeneration has been linked to oxidative stress mechanisms and mitochondrial dysfunction [245]. Although 6-OHDA does not induce the formation of LB-like inclusions, it has been reported to inhibit α-synuclein proteasomal degradation and promote its aggregation [246].

The 6-OHDA model displays many of the pathological and biochemical features of PD. Reduced levels of striatal DA and TH are associated to the decrease of TH-positive neurons in the SN. Furthermore, features of neuroinflammation have also been reported in 6-OHDA lesioned mice [247]. This neurotoxin is best used as a unilateral model since the bilateral injection produces severe adipsia, aphagia, seizures and frequently death. Following unilateral application of 6-OHDA, a quantifiable rotational behavior can be induced by injection of DA receptor agonists or amphetamine. This behavior, which measures the extent of the induced SN or striatal lesion, is often used to test the efficacy of potential PD therapeutics. Other behavioral assessments have also been developed to measure striatal DA loss [248]. In addition, rodent models for dyskinesia are mostly based on unilateral intracerebral injections of 6-OHDA followed by chronic L-dopa treatment [249-251].

Generation of a further experimental model of PD was made possible by the accidental discovery of the ability of MPTP to produce selective nigral cell degeneration and parkinsonism in humans after systemic administration [252, 253]. MPTP is a lipophilic pro-toxin that, following systemic administration, easily crosses the BBB and is converted into the active toxic metabolite 1-methyl-4-phenyl-2,3-dihydropyridium ion (MPP$^+$) by the enzyme monoamine oxidase B (MAO-B), mainly located in serotoninergic neurons and astrocytes [254, 255]. Once released into the extracellular space, MPP$^+$ is selectively taken up into DAergic neurons by DAT and interferes with complex I of the mitochondrial respiratory chain. As a consequence, ATP production is decreased, ROS (reactive oxygen species) levels are increased and eventually cell death pathways are activated. Although many species, including rats [256], are insensitive to MPTP, various strains of mice, such as C57Bl, are sensitive to MPTP-induced toxicity, displaying many features of the human disease. In mice, systemic MPTP administration induces bradykinesia, rigidity and posture abnormalities combined with bilateral degeneration of the nigro-striatal tract with reductions in striatal DA and TH [257]. Levels of extracellular glutamate [258] and glutathione [259] are increased and decreased, respectively, in the SN of MPTP-treated mice. Furthermore, a glial response peaking prior to DAergic neuron loss has been shown to be induced by MPTP in mice [260].

Importantly, the toxicity of MPTP to DAergic cells in nonhuman primates provided the first effective primate model of PD [261, 262]. Besides nigro-striatal degeneration, MPTP-treated primates show the major motor symptoms of PD (with the exception of tremor) and respond to all currently used drugs that are effective in the treatment of PD. Another important feature is that these animals rapidly develop dyskinesia in response to the repeated administration of L-dopa [263-265], thus providing an effective model of dyskinesia as it occurs in PD.

More recently, toxin-induced animal models of PD were developed by using agents with general toxicity, such as the mitochondrial function inhibitors paraquat and rotenone [266, 267]. These models, however, have not so far been employed, with very few exceptions, in testing nucleic acid-based therapeutic strategies.

All neurotoxin-induced models share the ability to generate oxidative stress and cause DAergic neurons death. This feature replicates what is seen in PD, but with a substantial difference. What is invariably induced in these models is an acute loss of DAergic neurons in the SN and not a progressive cell degeneration as occurs in the human condition. In spite of this drawback, neurotoxin-based animal models provide a valued tool in the study of PD as well as the development of novel treatment strategies.

Some of the nucleic acid-based therapeutic approaches successfully tested in animal models of PD have recently progressed towards clinical trials (see Table **5**).

Neuroprotection and Neurorestoration

In PD, the main pathological feature is the loss of a specific neuron population, namely the DAergic neurons in the SN that project to the striatum (nigrostriatal pathway). The well established link between the selective and progressive degeneration of nigrostriatal neurons and the neurological deficits in PD patients provides defined targets for disease-modifying strategies to prevent, slow down or even reverse the degenerative process. From several years ago, intense efforts are, thus, made to develop neurotrophic factor-based gene therapies.

Glial Cell Line-Derived Neurotrophic Factor (GDNF)

Glial Cell Line-Derived Neurotrophic Factor (GDNF) was identified thanks to its potent neurotrophic activity on embryonic rodent midbrain DAergic neurons [137], promoting their survival *in vitro*. The potential of this neurotrophic factor as a therapeutic agent for PD was, therefore, obvious, and, since then, GDNF has become the most studied molecule in the pursuit of neuroprotection in this disease. Importantly, as shown more recently in a conditional knock-out mouse, GDNF is indispensable for the survival of DAergic neurons in the adult brain [268].

A few *in vitro* studies demonstrated the ability of GDNF to protect DAergic neurons against MPP^+ and 6-OHDA toxicity [269-271]. The *in vivo* neuroprotective potential of GDNF was intensely investigated in both rodent [272-296] and primate [286, 297-307] models by administration of the recombinant protein either into intracerebroventricular (ICV) space or into the brain parenchyma. Considered as a whole, the results of these studies indicated that the efficacy of GDNF in protecting DAergic neurons varied depending on the site of delivery, time and duration of treatment, as well as the toxin used to generate the DAergic lesion.

Although GDNF was found to be very promising in animal models of PD, the results from clinical trials have been inconclusive. While the first study based on infusion of recombinant GDNF protein into the ICV space [308] reported no clinical improvement (probably due to poor penetration of GDNF into the brain parenchyma [309]) and several adverse effects, encouraging results emerged from two small open-label trials, which used intraputamenal infusion of GDNF [310-313]. However, a randomized, double-blind, placebo-controlled trial [314] failed to confirm the preliminary observations of therapeutic efficacy reported in the open-label trials. Furthermore, subsequent studies found that the clinical improvements in one of the open-label trials were lost a year after treatment withdrawal [315]. In consideration of these findings, as well as safety issues regarding antibody formation in some of the treated patients [313, 314, 316] and cerebellar toxicity in a primate model [317], the company that licensed the use of recombinant GDNF (Amgen) decided to stop clinical work on GDNF [318-320].

While several possible explanations for the different outcomes in the above mentioned trials have been proposed, a major factor contributing to the negative results was suggested to be the insufficient distribution of GDNF throughout the striatal parenchyma [318, 321-325]. As supported by the results of a recent study performed in nonhuman primates [326], gene delivery-based approaches might provide a more efficacious methodology for GDNF administration.

A large number of *in vivo* studies have been conducted in rodent and nonhuman primate models of PD using Adenovirus (AdV)- [327-339], Adeno-Associated Virus (AAV)- [340-352], Lentivirus (LV)- [353-367] and Herpes Simplex Virus (HSV)- [368-371] based vectors to express GDNF in the striatum or the SNpc. Overall, promising results in terms of GDNF diffusion within the brain parenchyma, reduction of DAergic neuron loss and behavioral recovery were obtained in toxin-based models as well as aged monkeys. On the other hand, viral vector-mediated delivery of the GDNF gene failed to provide any neuroprotective effect in genetic rat models of PD [352, 363].

A clinical trial aimed at testing the safety and effectiveness of AAV2-GDNF gene transfer for advanced PD (ClinicalTrials.gov identifier: NCT01621581) was launched in May 2012 and is expected to be completed by January 2018. In this Phase I, open-label, dose escalation study, an AAV2 vector containing the human GDNF cDNA will be bilaterally delivered by CED (convection-enhanced delivery) to the putamen of twenty-four patients with advanced PD. Assessing the safety and tolerability of four different dose levels of AAV2-GDNF is the primary outcome measure of this trial. Secondary outcome measures are preliminary data regarding the potential efficacy of the treatment by assessing its effects *via* clinical, laboratory and neuroimaging studies.

In a recent study by Drinkut *et al.* [372], the efficacy of viral vector-mediated expression of GDNF in striatal astrocytes rather than neurons was evaluated in the mouse MPTP and rat 6-OHDA models of PD. The authors engineered an AAV5 vector for mutually exclusive neuronal or astrocytic transgene expression through the use of appropriate transcriptional control elements (hSYN and hGFAP promoters, respectively). GDNF expression in astrocytes demonstrated the same efficacy as neuron-derived GDNF in terms of protection of DAergic cell bodies and projections, DA synthesis and behavior. On the other hand, unilateral striatal GDNF expression in astrocytes did not result in delivery of bio-active GDNF to the contralateral hemispheres as happened when GDNF was expressed in neurons. Based on these results, the authors suggest that neurotrophic factor expression in astrocytes could represent a safer therapeutic strategy able to avoid potential off-target effects.

As an alternative to the use of viral vectors, nanocarrier-based approaches for GDNF gene delivery are being investigated in pre-clinical studies [373-376].

Neurotensin polyplexes [377] were used by Gonzalez-Barrios *et al.* [373] for delivering the human GDNF gene into nigral DAergic neurons of hemiparkinsonian rats. This polyplex is a nanoparticle carrier system able to target genes in nigral DAergic neurons *in vivo* [60]. Delivery of the GDNF gene to the SN of rats one week after a 6-OHDA injection was shown to ensure transgene expression and to produce biochemical, anatomical and functional recovery from hemiparkinsonism.

The ability of THLs (Trojan horse liposomes) to deliver a therapeutic plasmid DNA across the BBB [378] was exploited in a study by Xia *et al.* [374]. In this study, THLs targeted with a monoclonal antibody to the rat transferrin receptor (TfR) were used to deliver a plasmid DNA in which the GDNF gene is placed downstream of the rat TH promoter in order to confine the expression of the transgene to catecholaminergic cells, thus avoiding possible side-effects of widespread expression of GDNF throughout the brain. A single intravenous injection was given to rats at two weeks after experimental PD had been induced by intra-cerebral 6-OHDA administration. Sustained therapeutic effects were observed following the treatment, including behavioral improvements and a significant increase in striatal TH enzyme activity. Nevertheless, the rescue of neurons of the nigra-striatal tract is not complete and the TH enzyme activity in the striatum is still only 10% of the contralateral striatum. As suggested by the authors, the therapeutic effect of intravenous, THL-mediated delivery of the GDNF gene could be augmented in various ways, such as earlier treatment, more frequent dosing, or combination with other neurotrophin gene(s). Furthermore, the TfRMAb used by Xia and colleagues [374] is only active in rats. However, THLs could be delivered to human brain with humanized forms of the human insulin receptor MAb (HIRMAb) [379].

Huang and coworkers [380] reported the development of lactoferrin-modified nanoparticles (Lf-modified NPs) for brain-targeted gene delivery and their use in two different rat models of PD [375, 376]. Lactoferrin was conjugated to a polyamidoamine (PAMAM)-based gene vector as a brain-targeting ligand to bind specifically with Lf receptors on endothelial cells of the CNS. The effects of Lf-modified NPs loading the human GDNF gene were examined in the 6-OHDA rat model *via* a regimen of multiple dosing intravenous administrations [375]. The results showed improvement in locomotor activity, reduction in DAergic neuronal loss and enhancement in monoamine neurotransmitter levels in PD rats. It is worth noticing that five injections of Lf-modified NPs loading GDNF exerted much more powerful neuroprotective effects than a single injection. Similar results were also obtained in the rotenone-induced chronic model of PD [376]. Development and application of a different nanocarrier was reported in a recent paper by the same authors [381]. In this case, a synthetic peptide (angiopep) was employed as a ligand specifically binding to low-density lipoprotein receptor-related protein (LRP) which is over-expressed on BBB and conjugated to biodegradable dendrigraft poly-L-lysine (DGL) *via* hydrophilic polyethyleneglycol (PEG), yielding DGL-PEG-angiopep (DPA). Multiple dosing intravenous administrations of DPA/hGDNF NPs was shown to induce locomotor activity improvements as well as recovery of DAergic neurons in rotenone-induced parkinsonian rats.

Ex vivo approaches for the delivery of GDNF gene, based on either naked or encapsulated cells, have been reported by several authors. [382-396].

A variety of GDNF over-expressing cells, such as neural progenitor cells (NPCs) [384, 393, 395], astrocytes [387, 391] and fibroblasts [390], were shown to have beneficial effects when grafted into the brain of rodent models of PD. Interestingly, bone marrow-derived cells have also been used as vehicles for GDNF gene delivery by intravenous transplantation in MPTP-treated mice [386, 396]. This therapeutic strategy exploits the ability of bone marrow-derived cells to cross the BBB and preferentially home to sites of neuronal degeneration. In particular, Biju *et al.* [396] used bone marrow stem cells transduced *ex vivo* with a LV vector expressing the GDNF gene under the control of a macrophage-specific promoter in order to restrict transgene expression to microglial precursor cells. Macrophage-mediated GDNF delivery was found to protect against MPTP-induced DAergic neurodegeneration.

Beneficial effects have also been observed following implantation of encapsulated GDNF over-expressing cells, such as BHK [382, 383, 385, 388, 392], PC12 [382] and fibroblasts [394], in rodent models of PD. However, intraventricular implantation of encapsulated GDNF-producing C2C12 cells was found to only induce a transient recovery of motor deficits in MPTP-treated nonhuman primates [389].

Besides its possible role as a therapeutic protein to ameliorate DAergic neuron degeneration, the potential use of GDNF to support DAergic neuron replacement strategies has also been investigated. Since the limited success of neural transplants observed in clinical trials [397-403] seems to be due to the significant cell loss occurring after transplantation into the brain, attempts have been made to enhance the survival of the grafted DAergic neurons by supplementing neural grafts with GDNF [348, 404-409].

In a study by Bauer *et al.* [405] fetal ventral mesencephalic cultures genetically modified to produce GDNF were found to induce earlier functional recovery after transplantation in 6-OHDA-lesioned rats compared to non-transfected cultures. Implantation of polymer-encapsulated cells genetically engineered to secrete GDNF was shown to improve DAergic graft survival and function [404] and to enhance fiber outgrowth from the grafted DAergic neurons [406] in 6-OHDA-lesioned rats. GDNF-over-expressing neural precursor cells also significantly increased the survival of co-grafted primary DAergic neurons, even though lacking of long term effects on fiber outgrowth and behavior was reported [407].

A different strategy based on direct delivery of the GDNF gene by means of viral vectors has been adopted by Georgievska *et al.* [408] and Elsworth *et al.* [348]. In the former study [408], lentiviral vector (LV vector)-mediated transfer of the GDNF gene to the striatum resulted in a short term increase of the graft survival in 6-OHDA-lesioned rats, while no improvements in long term graft survival and motor behavior compared to animals with control grafts were observed. In the study by Elsworth *et al.* [348], AAV2-mediated delivery of GDNF gene to the striatum of MPTP-treated monkeys was found to enhance the survival and outgrowth of co-implanted fetal DAergic neurons. More recently, Yurek *et al.*

[409] used DNA plasmids encoding GDNF and compacted into nanoparticles (pGDNF DNPs) to transfect the striatum of 6-OHDA-lesioned rats prior to graft implantation. Pretreatment with pGDNF DNPs was found to enhance the survival of the grafted neurons and to improve functional recovery compared to grafted animals receiving saline control pretreatment.

Neurturin (NTN)

Neurturin (NTN), which belongs to the GDNF family of neurotrophic factors, was originally identified through its ability to promote the survival of sympathetic neurons *in vitro* [410]. NTN gained attention as a potential therapeutic agent for PD due to the similarities in brain-expression patterns [411] and signaling pathways between this factor and GDNF [412]. Soon after its discovery, NTN was indeed found to promote the survival of midbrain DAergic neurons both *in vitro* and *in vivo* [411, 413] and intraparenchymal delivery of recombinant NTN protein was shown to provide neuroprotective activity comparable to that of GDNF in animal models of PD [289, 414, 415].

The effects of *ex vivo* delivery of the NTN gene have been investigated in the 6-OHDA rat model of PD. Liu *et al.* [416] transplanted a mouse neural stem cell line genetically engineered to over-express NTN into the striatum of adult rats which were then subjected to 6-OHDA lesioning. The transplanted cells differentiated into neurons, astrocytes and oligodendrocytes and maintained stable, high-level NTN expression. Protection of SN DAergic neurons against 6-OHDA toxicity and significant reduction of behavioral abnormalities were observed. Ye *et al.* [417] used bone marrow stromal cells to deliver the NTN gene to the striatum of 6-OHDA lesioned rats and observed an increase in striatal DA content for up to three months post-transplantation.

Extending their previous work on bone marrow-derived microglia (BMDM)-based delivery of GDNF [396], Biju *et al.* [29] have recently tested the efficacy of BMDM-mediated delivery of NTN in the murine MPTP model. Bone marrow cells were transduced *ex vivo* with a LV-vector expressing the NTN gene under the control of a macrophage-specific promoter and transplanted into recipient animals by intravenous injection. Eight weeks after transplantation, the mice were treated with the neurotoxin MPTP for seven days to induce DAergic neurodegeneration. Microglia-mediated NTN delivery was found to significantly reduce degeneration of DAergic neurons of the SN and their terminals in the striatum and to improve motor function, without any major side effect. The neuroprotective effects of BMDM-mediated delivery of NTN were similar to those of GDNF. These findings provide further support for bone marrow transplantation-based neurotrophic factor gene therapies for PD.

The first study based on *in vivo* delivery of the NTN gene was reported in 2005 by Fjord-Larsen *et al.* [418], who used a LV vector to transfer a modified NTN construct into the striatum of 6-OHDA lesioned rats. Initially, the results of viral vector-mediated expression of wild-type human NTN were disappointing because the protein was insufficiently processed and therefore secreted as inactive pro-NTN. By replacing the NTN signal peptide with the immunoglobulin heavy-chain signal peptide (IgSP), the secretion properties and biological activity of NTN could be regained and a protective effect on lesioned nigral DAergic neurons comparable to that of GDNF was observed.

A similar approach led to the development of a recombinant AAV2-based vector containing a human NTN cDNA in which the pre-pro domain has been exchanged to that of the human NGF (CERE-120, Ceregene Inc.). The therapeutic potential of CERE-120 has been widely investigated in both rodent and nonhuman primate models. Several studies demonstrated that NTN expression following intrastriatal injection of CERE-120 is stable and substantially restricted to the targeted region [419-424]. The neurotrophic activity of the NTN protein expressed following CERE-120 administration has been investigated in a variety of animal models. Delivery of CERE-120 to the striatum of young healthy monkeys was shown to induce appropriate trophic signaling in the SN (*e.g.*, enhanced nigrostriatal TH and activation of pERK) [422, 423]. Similarly appropriate responses to CERE-120 were observed in aged rats and monkeys, including hypertrophy of DAergic neurons in aged rats [424] and increased numbers of TH-positive neurons and enhanced ^{18}F-fluorodopa PET uptake in aged monkeys [421]. In the 6-OHDA rat model of PD, protection of nigral neurons and behavioral improvements were observed following CERE-120 administration [419, 420]. Delivery of CERE-120 to the nigrostriatal system of MPTP-treated monkeys virtually eliminated parkinsonian symptoms for up to 10 months, protected DAergic nigral neurons and partially preserved striatal DAergic innervation [425]. Considered as a whole, the results of these studies consistently demonstrated that the NTN protein expressed following CERE-120 administration is able to induce a neurotrophic response in nigrostriatal neurons in terms of enhancement of their function and improvement of their status, as well as protection from degeneration and death. Furthermore, the studies conducted in rats and monkeys to evaluate safety and toxicity of CERE-120 up to one year following administration did not reveal any toxicological finding or safety concern even at doses far exceeding those intended for the clinic [419, 421-424].

Based on the promising results of the pre-clinical studies, an open label Phase I trail (ClinicalTrials.gov identifier: NCT00252850) was conducted in moderately advanced PD patients by Ceregene Inc. [426]. Purposes of this study were to test the safety and tolerability of bilateral, stereotactic intraputamenal injections of CERE-120 and to gain preliminary evidence of possible efficacy. This study was conducted at the University of California, San Francisco and at Rush University Medical Center, Chicago and involved twelve patients grouped into two cohorts of six subjects each, who received two different dose levels of CERE-120 and were monitored for twelve months. No serious adverse events occurred during this period thus suggesting that CERE-120 could be safely used in humans. Furthermore, an improvement on various motor and quality-of-life endpoints was observed at twelve months post-dosing, with no significant differences between the two groups for any endpoint.

The efficacy of CERE-120 was more extensively evaluated in a Phase II trial initiated in 2006 (ClinicalTrials.gov identifier: NCT00400634), which also enabled to further evaluate the safety of this therapeutic approach [427]. This study was designed as a multicenter, double-blind, sham-surgery-controlled trial in patients with moderately advanced PD. Fifty-eight patients were randomly assigned (2:1) to receive either CERE-120 injected bilaterally into the putamen or sham surgery. While providing further support to the safety of CERE-120, the trial did not result in any benefit compared to sham surgery on the primary endpoint, the Unified Parkinson's Disease Rating Scale (UPDRS) motor-off at twelve months. However, eighteen months follow-up of a subgroup of patients showed a moderate but significant reduction in UPDRS score. In addition, a modest clinical benefit was suggested by several secondary endpoints either at twelve and eighteen months. Considered as a whole, the results of this study led to the conclusion that CERE-120 indeed provided some reliable clinical benefit, but, at the same time, they clearly indicated that the neurotrophic effect still needed to be optimized.

Two patients enrolled in the Phase II trial died of events unrelated to CERE-120 treatment and histological analysis of their brains not only supported the potential benefit of NTN in PD, but also provided crucial insight for optimizing CERE-120 targeting and dosing in moderately advanced PD patients [428]. The effects of delivering CERE-120 to the putamen in these patients were investigated comparing the expression of NTN in putamen and nigra and TH-induction in nigrostriatal neurons to those following the same treatment in young, aged and MPTP-treated monkeys. In this study, expression of NTN was clearly detected in the targeted PD putamen following CERE-120 treatment, as well as a modest increase in TH immunoreactivity, thus providing evidence of improved DAergic function. In sharp contrast to the findings in nonhuman primate models, however, NTN was almost undetectable in the cell bodies of SNpc and no increase in TH immunoreactivity was observed in these cells. This discrepancy is thought to be due to a defect in axonal retrograde transport of NTN occurring in moderately advanced PD but not in currently available animal models of the disease [428-431], even though species-specific differences in the capacity for anterograde and retrograde transport of NTN cannot be excluded [432]. In any case, these findings provided a plausible explanation for the unsatisfactory results of the Phase II trial and led to the conclusion that delivery of CERE-120 directly to the degenerating cell bodies of the SN, as well as their terminals in the putamen, would be necessary in order to optimize the neurotrophic effects of NTN [428-431]. Further support for this conclusion came from the analysis of two additional, longer-term post CERE-120 autopsy brains [433].

To evaluate the feasibility, safety and effectiveness of targeting the SN with CERE-120, additional pre-clinical experiments were conducted [429]. No serious side effects or toxicity, over a range of doses, were observed in this study and support for the potential usefulness of targeting the SN with CERE-120 in advanced PD patients was obtained. Furthermore, the study provided useful information to define an appropriate dose of CERE-120 for testing in PD patients, taking into account the volume of human SN.

A Phase I/II clinical trial was started in 2009 to test the safety and efficacy of nigral and putaminal administration of CERE-120 in PD patients (ClinicalTrials.gov identifier: NCT00985517). The revised clinical protocol applied in this study optimized not only the targeting but also the dosing of CERE-120 and planned a later endpoint. The Phase I safety component of the trial, which involved six patients grouped into two dose cohorts (the first using the same putaminal dose as in the prior Phase II trial and adding the SN dose; the second retaining that SN dose but increasing the putaminal dose by 4-fold), has been successfully completed. Absence of any serious safety issue during a period of several months after treatment allowed the Phase II component of the protocol to be started. Data for twenty-four months post-dosing with no significant or unexpected safety issues have also been recently reported [434, 435]. The Phase II efficacy/safety component of the trial was a multicenter, double-blind, sham-surgery-controlled study and involved fifty-one patients with moderately advanced PD. Approximately half of the enrolled subjects received the treatment while the other half received a sham surgery procedure. To evaluate possible clinical improvements, besides treatment safety, the patients were monitored for 15-24 months. Multiple endpoints, among which UPDRS, Daily Diaries that assess motor function throughout the day and PDQ-39 (a measure of quality of life), were used. In April 2013, Ceregene announced [436] that "The trial did not demonstrate statistically significant efficacy on the primary endpoint (UPDRS-motor off). However, one of the "key secondary endpoints" (Diary-off score), as defined and pre-specified in the Statistical Analysis Plan, did produce statistically significant benefit. The trial also provided further evidence for the safety of CERE-120 and the dosing methods employed. A marked placebo effect was observed in this trial in that both the sham-surgery-controlled patients and the CERE-120 treated patients showed significant improvement following their surgery". In conclusion, the promising pre-clinical findings have not yet been replicated in a clinical setting. One possible explanation is that, despite the efforts made to find the most promising method for CERE-120 delivering and the brain areas to be targeted, the right delivery method to provide therapeutic benefit has not yet been found. It is also possible that successful CERE-120 and possibly other neurotrophic factor, therapy might require patients to be treated in an earlier stage of the disease, when a greater number of DAergic neurons able to respond to the treatment is still available. In more advanced PD patients, like those investigated in the trial, there may not be enough healthy neurons left in the brain for the trophic factor to take effect.

Other Neuroprotective/Neurorestorative Factors

Two other members of the GDNF family, Artemin (ART) [437] and Persephin (PSP) [438], have been shown to have neurotrophic effects on midbrain DAergic neurons.

To investigate the neuroprotective effects of ART *in vivo*, Rosenblad *et al.* [439] injected LV vectors carrying the ART encoding cDNA into the striatum and ventral midbrain of adult rats that were subjected to an axon terminal lesion of the nigrostriatal DAergic neurons by intrastriatal 6-OHDA injection one week later. Three weeks after the lesion, the amount of rescued cells was about 90% (compared to 20% of the control group), similar to that obtained with GDNF and the striatal TH-immunoreactive innervation was partly spared. As pointed out by the authors, the ability of ART to protect not only the cell bodies but also the striatal DAergic innervation against the 6-OHDA-induced toxic insult supports its potential as a powerful neuroprotective agent for nigrostriatal DAergic neurons.

The therapeutic potential of PSP, delivered by neural stem cells (NSCs), in a rodent model of PD was investigated by Åkerud *et al.* [440]. Intrastriatal grafting of NSCs engineered to over-express PSP prevented the degeneration of DAergic neurons by 68% and significantly prevented motor deficits in mice injected with 6-OHDA 16 days after grafting. These findings suggest that the delivery of PSP by NSCs may constitute a useful strategy in the treatment of PD.

In the early '90s, several observations led to a significant interest in the neurotrophin BDNF as potential therapeutic agent for the treatment of PD. *In vitro*, BDNF was shown to promote the survival and differentiation of rat fetal DAergic neurons [441, 442] and to exert partial protection against the selective DAergic neurotoxins MPP$^+$, 6-hydroxy-DOPA (TOPA or 2,4,5-trihydroxyphenylalanine) and 6-OHDA [443-445]. *In vivo*, BDNF was found to modulate DAergic neurotransmission in nigrostriatal neurons of unlesioned rats, as evidenced by elevated contralateral rotational behavior and postural bias, in combination with increased turnover of DA in the striatum [446-448]. It was also reported that radiolabelled BDNF, when injected into the striatum, is specifically taken up and retrogradely transported by the DAergic neurons in the SNpc [449]. Furthermore, BDNF administration increased the spontaneous activity and firing rate of SNpc DAergic neurons in unlesioned rats [450] and prevented motor deficits in 6-OHDA-treated animals at least for periods of one-two weeks [451]. In subsequent years, further support to the role of BDNF in the etiology and potential treatment of PD came from a number of studies. One of these studies was conducted by Hagg [452] who reported a protective effect of this neurotrophin on nigrostriatal neurons of 6-OHDA lesioned rats. While BDNF is expressed widely throughout the SN [453] and its receptor TrkB is expressed within the striatum and the SN [454], a significant reduction of BDNF levels was observed in the SN of PD patients [455, 456] suggesting a role for this factor in the survival of DAergic neurons. Additionally, Guillin *et al.* [457] demonstrated that BDNF controls the expression of DA receptor D3, which is abnormally expressed in PD. Finally, BDNF up-regulation has been implicated as the mechanism of action of various PD drugs like L-dopa [458], selegiline [459] and omega-3 fatty acids [460].

For delivering the BDNF gene in animal models of PD, both *ex vivo* and *in vivo* approaches have been used. In a rat model of PD, *ex vivo* delivery of the BDNF gene to the SN was shown to protect nigral DAergic neurons from degeneration [461]. In this study, immortalized rat fibroblasts genetically engineered to produce human BDNF were implanted near the SN of adult rats that received striatal MPP$^+$ infusion seven days after cell implantation. BDNF-secreting fibroblasts were found to markedly increase nigral DAergic neuron survival when compared to control fibroblast implants. In a later study, the same authors demonstrated that cell-mediated delivery of BDNF enhances DA levels in the same rat model [462]. BDNF-secreting fibroblasts were also shown by Levivier et al. [463] to protect DAergic neurons against 6-OHDA-induced toxicity when implanted into the striatum. These authors implanted BDNF-secreting fibroblasts into the striatum of adult rats two weeks before an intrastriatal injection of 6-OHDA, and, three weeks after the lesioning, they observed a complete protection of DAergic neurons within the SN and a partial protection of the striatal DAergic nerve terminals. In a study reported by Yoshimoto et al. [464], genetically engineered astrocytes were used to deliver the BDNF gene. Intrastriatal injection of the modified astrocytes fifteen days after lesioning with 6-OHDA resulted in improvement of motor behavior, while no effect on the density of TH-positive fibers in the striatum was observed.

In vivo AAV-mediated delivery of the BDNF gene to the SN of adult rats was reported to significantly reduce motor deficits induced by a 6-OHDA lesion without significantly affecting the number of TH-positive neurons in the SNpc [465]. More recently, the effects of BDNF were compared to those of GDNF in a 6-OHDA rat model [371]. In this study, rats received unilateral intrastriatal injection of HSV-1 vectors expressing either GDNF or BDNF, or both vectors, followed by intrastriatal injection of 6-OHDA. Persistent expression of GDNF or BDNF was detected in striatal neurons but with a different outcome. In fact, a significantly higher effectiveness of GDNF compared to BDNF was detected in terms of both correction of behavioral deficits and protection of nigrostriatal DAergic neurons. The combination of the two neurotrophic factors was no more effective than GDNF alone. With regard to BDNF and GDNF relative effectiveness, it is worth noticing that Stahl et al. [466] have come up to a different conclusion in a study performed on an organotypic culture model of PD. They applied the neurotoxin 6-OHDA unilaterally in slices of rat mesencephalon and evaluated the cytoprotective and regenerative effects exerted by BDNF, GDNF and the combination of these. Pre-, co-, or post-treatment with neurotrophic factors clearly protected DAergic neurons from cell death. Pre-treatment with BDNF was particularly effective in promoting cell survival and the combination with GDNF did not result in any further increase of this effect. Furthermore, BDNF, but neither GDNF nor the combination of the two factors, was shown to up-regulate the DAergic phenotype at the transcriptional level. On the basis of these findings, the authors concluded that BDNF could be a more promising agent than GDNF for SN neuron protection.

Non-viral delivery of the BDNF gene to the brain of 6-OHDA mice was recently reported by Fu *et al.* [467]. In this study, a plasmid containing the BDNF gene was complexed with the brain-targeting peptide RDP (Rabies virus glycoprotein-derived peptide) [468] and intravenously injected in 6-OHDA-lesioned mice. The results demonstrated that the DNA delivered by RDP was able to cross the BBB and induce specific gene expression in neuronal cells, resulting in a significant improvement of motor behavior.

In a study reported by Ebert *et al.* [395], the effects of *ex vivo* delivery of the IGF-1 gene in a rat model of PD were tested and compared to those of GDNF. Intrastriatal transplantation of human NPCs (hNPCs) over-expressing either IGF-1 or GDNF seven days after a 6-OHDA striatal lesion was shown to reduce to a similar extent DAergic neurons loss as well as behavioral abnormalities. Interestingly, GDNF, but not IGF-1, was able to preserve TH-positive DAergic projection fibers in the striatum. On the other hand, IGF-1, but not GDNF, significantly increased hNPC survival both *in vitro* and following transplantation. On the basis of these findings, the authors suggest that IGF-1 might be of therapeutic interest and therefore deserves to be further investigated in animal models of PD. The authors also suggest that a combined treatment with GDNF and IGF-1 might prove to be even more effective than with either factor alone.

Growth/Differentiation Factor 5 (GDF5), a member of the Transforming Growth Factor β superfamily of proteins, is also being investigated for its therapeutic potential in PD. *In vitro* studies have shown that GDF5 has trophic actions on DAergic neurons, comparable to those of GDNF and protects them against the neurotoxin MPP^+ and against free radical-induced damage [469-473]. Intracerebral infusion of recombinant human GDF5 protein has been reported to protect and restore nigrostriatal DAergic neurons in 6-OHDA lesioned adult rats [474-476]. The effects of *ex vivo* delivery of the GDF5 gene in the same rat model of PD have recently been reported by O'Sullivan *et al.* [477]. In this study, intrastriatal transplantation of GDF5 over-expressing embryonic rat DAergic neurons was found to significantly improve motor behavior compared to the untreated 6-OHDA lesioned animals.

Cerebral Dopamine Neurotrophic Factor (CDNF) and Mesencephalic Astrocyte-derived Neurotrophic Factor (MANF) are members of a recently described family of evolutionarily conserved proteins showing neurotrophic activity [478]. Since CDNF and MANF are structurally unrelated to the classical NTF families, it is possible to hypothesize a novel mechanism of action and, therefore, a novel therapeutic potential for the treatment of neurodegenerative diseases. Intracerebral injection of recombinant CDNF or MANF protein has been shown to be both neuroprotective and neurorestorative in animal models of PD [479-482]. On the basis of the promising results of these studies, pre-clinical evaluation of CDNF gene delivery has been undertaken. A first study by Back *et al.* [483] investigated the neuroprotective effects of viral vector-mediated delivery of the CDNF gene in a rat 6-OHDA model of PD. AAV2 vectors encoding CDNF were injected into the striatum of rats which were treated with 6-OHDA two weeks later. A single AAV2-CDNF injection was shown to lead to long lasting expression of the neurotrophic factor and functional recovery of neural circuits controlling movements. On the other hand, only partial protection of TH-positive neurons in the SN and TH-positive fibers in the striatum was observed, probably due to the low viral vector titers used as well as restricted and mainly intracellular expression of CDNF protein. Long-term neurorestorative and potential therapeutic effects of AAV2-mediated CDNF delivery in 6-OHDA-lesioned rats have been subsequently reported by Ren *et al.* [484]. In this study, AAV2-CDNF was injected into the rat striatum after 6-OHDA treatment and was found to rescue 6-OHDA-induced TH-positive neuron pathology and behavior deficits. The results emerging from these studies indicate that CDNF gene delivery is worth to be further investigated in view of future clinical applications in PD therapy.

A recent study by Xue *et al.* [485] investigated the use of erythropoietin (Epo) as a therapeutic agent for PD. In this study, an AAV9 vector was used to deliver the human Epo encoding gene by direct injection into the striatum of 6-OHDA lesioned rats. Protection of nigral DAergic neurons against 6-OHDA toxicity, as well as improvements in behavioural outcomes, were observed in the treated animals. Anyway, Epo as a potential neuroprotective agent has a drawback represented by the induction of polycythaemia, which entails the risk of complications such as ischaemic heart disease or stroke.

Enhancement of Dopamine Levels

In PD, the selective degeneration of SNpc DAergic neurons combined with dystrophy of the striatal projection fibers result in a loss of DA in the striatum, which becomes more severe as the disease progresses. In consideration of the inherent limitations of L-dopa based therapy, considerable efforts have been made towards the development of strategies aimed at achieving ectopic DA synthesis in the striatum. These approaches are based on the delivery of genes required for DA synthesis to the striatal GABAergic neurons. Three major strategies have been developed which differ from one another in the enzymes considered to be necessary and sufficient to express ectopically in the striatum in order to confer the DA synthesis capacity.

Dopamine Synthesis in Striatal Neurons

A possible strategy to enhance DA levels in the parkinsonian striatum is to convert striatal neurons into DA synthesizing cells by the co-delivery of the genes encoding all the enzymes required for the synthesis of DA from the dietary tyrosine. DA is synthesized in a two-step reaction by the enzymes tyrosine hydroxylase (TH), which catalyzes the synthesis of L-dopa from tyrosine and aromatic L-amino acid decarboxylase (AADC), which converts L-dopa to DA. Additionally, TH requires a cofactor, tetrahydrobiopterin (BH4), the biosynthesis of which is rate-limited by GTP cyclohydrolase 1 (GCH1).

A successful application of this strategy to 6-OHDA-lesioned rats was reported by Shen *et al.* [486]. In this study, co-delivery of three different AAV-vectors encoding TH, GCH1 and AADC, respectively, was shown to increase DA levels in the DA denervated striatum and reduce motor deficits. Using the same AAV vectors in MPTP treated monkeys, Muramatsu *et al.* [487] observed a marked behavioral improvement that was sustained for up to ten months.

As a further progress in the development of this therapeutic approach, a tri-cistronic LV vector (ProSavin®) based on the equine infectious anemia virus (EIAV), which encodes all the three enzymes required for DA biosynthesis, was developed by Oxford BioMedica [488]. In this vector, a truncated form of TH enzyme, lacking the 160 aminoacid long N-terminal fragment, is encoded in order to prevent the negative feedback induced by DA [489]. Injection of ProSavin® into the striatum of 6-OHDA lesioned rats resulted in sustained expression of each enzyme and was able to induce DA production as well as functional improvement [488]. In a subsequent study performed in MPTP-treated monkeys, intrastriatal injection of ProSavin® was shown to restore extracellular DA concentration and correct motor deficits for twelve months without associated dyskinesia [490].

On the basis of these results, an open-label Phase I/II clinical trial sponsored by Oxford BioMedica in affiliation with the French social security health care system, was started in 2008 (ClinicalTrials.gov identifier: NCT00627588) in order to assess the safety, efficacy and dose evaluation of ProSavin® in patients with mid-stage PD. The successful completion of this trial was announced by Oxford BioMedica in April 2012 [491]. The study, which evaluated three ascending dose levels (1x, 2x and 5x) in a total of fifteen patients, met both its primary and secondary endpoints (safety and efficacy, respectively). More in detail, ProSavin® demonstrated a long-term safety profile, up to forty-eight months post-treatment for the first two patients treated with the lowest dose. An improvement in motor function, as measured by the UPDRS in off-state, was observed in all fifteen patients at six months post-treatment and in the first nine patients (1x and 2x doses) the improvements were found to remain statistically significant up to twelve months. In the six patients treated with the highest dose, an average motor function improvement of 30% was observed, even reaching 41% in one patient. Furthermore, oral DAergic therapy was reduced in all six patients, when normally it increases as PD progresses. The results of this trial were indeed encouraging, particularly at the 5x dose and provided a solid basis for further clinical development of ProSavin®.

A further clinical study is currently ongoing (ClinicalTrials.gov identifier: NCT01856439). This trial is aimed at evaluating long-term safety and efficacy of ProSavin® in PD patients from the previous study. Safety and tolerability of ProSavin® as well as patients' responses to the treatment will be assessed in a time frame of ten years.

An alternative strategy based on a poly-cistronic vector has been pre-clinically tested by Sun and colleagues [492] who developed a vector which also carries the VMAT2 (vesicular monoamine transporter 2) encoding gene, in order to improve the ability of the transduced neurons to store and release the *de novo* synthesized DA. These authors compared the effects of HSV vectors co-expressing either the three DA biosynthetic enzymes (3-gene-vector) or these three enzymes plus VMAT2 (4-gene-vector). Following intrastriatal injection in 6-OHDA-lesioned rats, both vectors induced long-term transgene expression (fourteen months) and supported long-term behavioral correction (six months). However, a more pronounced behavioral correction was induced by the 4-gene-vector compared to the 3-gene-vector and only the 4-gene-vector supported significant K^+-dependent release of DA. Based on these findings, the authors suggest that co-expression of VMAT2 with DA biosynthetic enzymes could provide a useful strategy for restoring striatal DA levels in PD patients.

L-dopa Conversion in Striatal Neurons

An alternative strategy, also known as "pro-drug approach", is the delivery of the AADC encoding gene to the striatum in order to increase the conversion efficiency of exogenously administered L-dopa into DA. This approach is based on the observation that AADC levels are lower in the striatum of PD patients compared to controls [493-495], possibly contributing to the reduced efficacy of L-dopa administration associated with disease progression. Hopefully, restoration of striatal AADC levels could lower the effective dose of L-dopa, thus limiting the side effects associated with higher doses.

The pro-drug approach was pre-clinically tested both in rodent and non-human primate models of PD. Long-term restoration of striatal AADC activity was reported following AAV vector-mediated delivery of the AADC encoding gene to the striatum of 6-OHDA-lesioned rats [496, 497]. In MPTP-lesioned monkeys, convection-enhanced delivery of AAV-AADC resulted in increased conversion efficiency of peripheral L-dopa to DA [498]. The long-term effects of this therapeutic approach in the same nonhuman primate model were evaluated in a subsequent study [499]. AADC activity was shown to persist for at least six years and long-term motor improvements were observed with significantly lowered L-dopa requirements and a reduction in L-dopa-induced side effects. These findings were also confirmed at the 8-year time point with no evidence of neuroinflammation or reactive gliosis [50], although the number of monkeys used in this specific component of the overall study precluded statistical analysis.

The promising results of the pre-clinical studies led to a Phase I clinical trial [500, 501] sponsored by Genzyme Inc. (ClinicalTrials.gov identifier: NCT00229736). In this open-label, dose-escalation study, ten patients with moderately advanced PD received bilateral intraputamenal infusion of AAV-AADC vector. The therapy was found to be safe and well tolerated in both dose cohorts (five patients in each) with no significant adverse effects related to the viral vector itself (even though intracranial hemorrhages occurred in some patients, which were most likely related to the method of vector administration). A dose-dependent increase in striatal AADC activity was detected by PET imaging at six months and preliminary results on the efficacy of the treatment were also reported. Significant improvements in both the off- and on-state UPDRS scores and in motor diaries were observed and the majority of patients were able to reduce their total dose of L-dopa.

An independent confirmation of the safety, tolerability and potential efficacy of AAV vector-mediated delivery of AADC gene was provided by a Phase I clinical study reported by Muramatsu *et al.* [502]. In this study, based on the same vector preparation as the study reported by Christine *et al.* [501], six patients with mid- to late-stage PD received intraputamenal infusion of AAV-AADC. The procedure was well tolerated and PET analysis revealed a significant increase in AADC activity which persisted up to ninety-six weeks. A significant improvement in motor function in the off-state was shown by the UPDRS scores six months after surgery.

Although both trials [501, 502] are open-label studies involving a small number of patients and the non-blinded, uncontrolled analysis limits the interpretation, these preliminary clinical data are encouraging. Thus AAV vector-mediated delivery of AADC gene warrants further evaluation in a randomized, controlled, Phase II setting.

L-dopa Synthesis in Striatal Neurons

A third strategy relies on endogenous AADC activity to convert L-dopa ectopically synthesized in the striatum following combined intrastriatal delivery of TH and GCH1 encoding genes. This approach is based on the assumption that sufficient AADC activity remains available in the parkinsonian striatum despite DA terminal degeneration, which is known to increase as the disease progresses. Besides the DA terminals, serotoninergic terminals do indeed provide AADC activity to the striatum and degenerate to a lesser extent compared to DA terminals in PD patients [503], thus providing a reliable long-term AADC source. Moreover, the existence of an additional source of AADC activity in this brain region, possibly including neuronal as well as non-neuronal cells, has been indicated by several reports [504-508].

The initial efforts to establish a continuous production of L-dopa in the 6-OHDA denervated rat striatum by AAV vector-mediated delivery of TH and GCH1 genes did not allow to reach therapeutic levels of L-dopa synthesis [509], most likely due to insufficient infectivity of the first generation AAV vectors. In a subsequent study by Kirik *et al.* [510] based on a new generation of AAV vectors, co-transduction of TH and GCH1 genes was shown to induce functional recovery in rats with a partial 6-OHDA-induced lesion. Furthermore, this strategy was found to strongly reduce or even abolish dyskinesia caused by intermittent administration of L-dopa in 6-OHDA lesioned rats [511]. The same authors recently reported an investigation based on rats with a complete 6-OHDA-induced lesion [512]. In this study, AAV5-mediated TH and GCH1 delivery resulted in motor improvements and was also able to prevent L-dopa-induced dyskinesia. Furthermore, the results demonstrated that even in animals with complete combined DAergic and serotoninergic denervation AADC activity as well as functional benefits of TH plus GCH1 gene delivery are partially maintained. These findings support the existence of an additional source of AADC activity and suggest that this therapeutic approach could produce significant motor functional restoration even in patients with advanced PD where the serotoninergic terminals may be seriously affected. It is important to notice that behavioral recovery promoted by combined TH and GCH1 gene delivery correlates with the normalization of DA neurotransmission [513].

A single AAV5 vector including both TH and GCH1 genes was recently developed and found to result in very efficient L-dopa synthesis, superior to what had been achieved with two separate vector constructs of the same serotype [514]. Injection of this vector into the striatum of 6-OHDA lesioned rats was shown to restore extracellular DA levels and provide motor function recovery [514, 515]. Safety and efficacy of TH-GCH1 co-expressing vector were also investigated in MPTP-treated monkeys [515]. In this animal model, however, no increase in L-dopa, DA or DA metabolites was detected and no functional improvement was observed. *Post mortem* analysis of transgene expression demonstrated that GCH1 expression was robust but TH was not detectable.

Overall, the results so far achieved encourage further investigations on this gene therapy strategy, which might offer some advantage compared to the other enzyme replacement strategies. In particular, according to this approach, the conversion of L-dopa to DA does not take place in the transduced striatal neurons, thus avoiding the risks associated with intracytoplasmic accumulation of DA in those neurons.

Inhibition of Subthalamic Nucleus

In PD, the degeneration of DAergic neurons of the SN with consequent loss of DA input to the striatum affects the activity of several deep brain nuclei. In particular, the inhibitory control exerted by the external segment of the globus pallidus (GPe) on the subthalamic nucleus (STN) is decreased (*via* the "indirect" pathway) as a consequence of striatal DA depletion. The lack of inhibition of the STN does, in its turn, alter the signals emanating from the basal ganglia circuitry, which leads to pathological excitation of the internal segment of the globus pallidus (GPi) and substantia nigra pars reticulata (SNpr). Major symptoms of PD, *i.e.*, tremor, rigidity, bradykinesia and gait disturbance, are thought to be due to the increased GPi/SNpr outflow. As a matter of fact, inhibition of the overactive STN, by either ablation [516, 517], or electrical inhibition by high-frequency stimulation [518-520], or pharmacological silencing by local GABA-agonist infusion [521], is known to improve the motor symptoms of PD. In particular, high-frequency deep brain stimulation (DBS) of the subthalamic nucleus (STN-HFS) is currently considered as the "gold-standard" surgical treatment for PD. However, this procedure suffers from some drawbacks, *i.e.*, adverse neurocognitive effects and side-effects consequent to spread of stimulation to surrounding structures [522].

An alternative, nucleic acid-based approach to inhibit the overactive STN was proposed by During *et al.* [523] who developed a strategy based on the delivery of the GAD (glutamic acid decarboxylase) encoding gene to the STN. GAD is the limiting enzyme in the synthesis of GABA, the main inhibitory neurotransmitter in the brain. Over-expression of GABA in the STN is, therefore, expected to convert the excitatory neurons to an inhibitory phenotype, thus normalizing the output of the basal ganglia circuitry.

Pre-clinical testing of this strategy was initially performed in the 6-OHDA rat model of PD [524]. Injection of an AAV2 vector encoding GAD into the STN of unlesioned and 6-OHDA lesioned animals led to increased GABA release in the SNpr in response to STN stimulation and suppressed the firing activity of the innervated SN. The effects of the phenotypic shift of the transduced neurons were investigated in rats that received intra-STN injection of AAV-GAD three weeks before 6-OHDA lesioning. A strong neuroprotection of nigral DAergic neurons and rescue of the parkinsonian behavioral phenotype were observed in these animals. Significant behavioral improvements in 6-OHDA lesioned rats following AAV-GAD intra-STN delivery were also reported by Lee *et al.* [525].

In a subsequent study, the same therapeutic approach was applied to MPTP-treated monkeys [526]. In this nonhuman primate model of PD, the treatment was found to be safe and improvements in clinical rating scores were observed. Furthermore, PET scans showed an increase of glucose metabolism in the motor cortex ipsilateral to the AAV-GAD injection, which correlated with the improvement in clinical ratings. The neuroprotective effects of GAD gene delivery were not confirmed in this study.

In 2003, a Phase I clinical trial [527, 528] (ClinicalTrials.gov registry number NCT00195143), sponsored by Neurologix Inc., was initiated in order to evaluate the safety, tolerability and potential efficacy of subthalamic GAD gene transfer in PD patients. In this open-label, dose-escalation study, three groups of four moderate-advanced PD patients each received unilateral subthalamic injection of low, medium and high dose of AAV-GAD, respectively. As reported by Kaplitt *et al.* [527], no adverse events related to the intervention occurred for at least one year after the treatment. Statistically significant improvements in motor UPDRS scores, in both on- and off-states, were observed three months after surgery and persisted up to twelve months. These improvements were localized predominantly to the side of the body contralateral to the treatment. Functional imaging studies were also performed to evaluate the changes in regional metabolism and network activity induced by the treatment [527, 528]. PET scans revealed a substantial reduction in thalamic metabolism that was restricted to the treated hemisphere side, as well as concurrent metabolic increases in ipsilateral motor and pre-motor cortical regions. Furthermore, evidence of significant modulation of PDRP (abnormal Parkinson's Disease-Related covariance Pattern) network activity was obtained and these network changes were found to correlate with motor benefit. Despite objective improvement in motor function and functional imaging, it was not possible to draw any firm conclusion about the efficacy of the treatment. A placebo effect could not be ruled out because this trial, being a pilot study to assess safety, did not include a sham-operated control group.

Based on the results of the Phase I trial, a Phase II randomized, double-blind, sham-surgery-controlled study (ClinicalTrials.gov registry number NCT00643890) was initiated in 2008. This trial, sponsored by Neurologix Inc., was aimed at assessing the effect of bilateral delivery of AAV2-GAD into the STN compared with bilateral sham surgery in patients with advanced PD. As reported by LeWitt *et al.* [529], twenty-three patients were randomly assigned to sham surgery and twenty-two to AAV2-GAD delivery; of those, 21 and 16, respectively, were analyzed. No major effects related to the treatment occurred during the six month course of the study. At the end of the trial, the AAV2-GAD group showed a greater improvement of the UPDRS motor scores compared with the sham group, with a modest but statistically significant difference between the two groups. This trial supports the further development of AAV2-GAD treatment for PD and provides useful information for designing a larger clinical trial aimed at assessing whether this approach may be considered for more widespread clinical application. Importantly, a comparison with DBS in terms of safety and efficacy will be needed, since the clinical benefit obtained so far was smaller than that reported previously for DBS [519, 530].

A long term follow-up study of AAV-GAD-treated patients (ClinicalTrials.gov registry number NCT01301573) was initiated in 2011 in order to evaluate the long term effects of AAV-GAD and provide long term safety information. Unfortunately, this trial, sponsored by Neurologix Inc., was terminated due to financial reasons.

The intra-STN delivery of GAD gene has been combined with intrastriatal delivery of TH gene in a recent pre-clinical study by Zheng *et al.* [531]. In this study, an AAV vector was used to deliver GAD gene into the STN while genetically engineered fibroblasts were used to transfer the TH gene into the striatum of 6-OHDA lesioned rats. In this model of PD, the combination treatment was found to be more efficacious than each single treatment in alleviating motor symptoms.

Table 5: Clinical trials of nucleic acid-based therapeutics for Parkinson's disease.

Therapeutic Approach	Transgene(s)	Vector	Study Phase	Study Design	ClinicalTrials.gov Identifier
Neurotrophic support	GDNF	AAV2	I	OL, DE	NCT01621581
	NTN	AAV2	I	MC, OL, DE	NCT00252850
	(CERE-120)				
	NTN	AAV2	II	MC, R, DB, SSC	NCT00400634
	(CERE-120)				
	NTN	AAV2	I/II	(I) OL, DE (II) MC, R, DB, SSC	NCT00985517
	(CERE-120)				
Enhancement of dopamine levels	TH, GCH1, AADC	EIAV	I/II	MC, R, OL, DE	NCT00627588
	(ProSavin®)				
	TH, GCH1, AADC	EIAV	I/II	MC, OL, LTFU	NCT01856439
	(ProSavin®)				
	AADC	AAV2	I	OL, DE	NCT00229736
	AADC	AAV2	I	OL	*
Inhibition of subthalamic nucleus	GAD	AAV2	I	OL, DE	NCT00195143
	GAD	AAV2	II	MC, R, DB, SSC	NCT00643890
	GAD	AAV2	II	MC, LTFU	NCT01301573

Study design abbreviations: DB, double-blind; DE, dose escalation; LTFU, long-term follow-up; MC, multicenter; OL, open-label; R, randomized; SSC, sham-surgery-controlled.
* Trial performed in Japan and reported in [502].

Further Approaches

The identification of genes responsible for the inherited forms of PD as well as advances in the study of PD pathogenesis have allowed further therapeutic strategies to be proposed. Approaches based on disease genes are aimed at delivering a wild-type copy of a recessively inherited gene, such as *PARK2*, or silencing a dominantly inherited gene, such as *SNCA*. Anti-oxidative and mitochondrial restorative strategies are also being investigated. Finally, as the crucial role of miRNAs in PD is increasingly recognized [532], nucleic acid-based strategies aimed at modulating the expression of specific miRNAs can be expected to be developed in a short time frame.

Targeting Alpha-Synuclein

The main neuropathological feature of PD is the intracytoplasmic occurring of the so-called Lewy bodies (LBs) predominantly in the SN that is the primary site of neurodegeneration [533]. LBs are inclusions resulting from the accumulation of cellular proteins as a consequence of abnormal protein clearance and are primarily composed of α-synuclein [534], a protein prone to aggregate due to its hydrophobic non-Aβ component domain.

Based on numerous lines of evidence [535-537] α-synuclein is considered to be a key player in the pathogenesis of PD. Several missense mutations in the α-synuclein gene, as well as gene duplication and triplication, have been linked to familial PD [231-233, 538-540]. Moreover, genetic variability of the α-synuclein gene promoter is associated with sporadic PD [541] and increased levels of α-synuclein mRNA have been observed in the midbrain of sporadic PD patients [542]. Thus, both over-expression of wild-type α-synuclein and α-synuclein gain-of-function mutations are linked to neurodegenerative pathology. Accordingly, α-synuclein over-expression has been shown to be toxic both *in vitro* [543-547] and *in vivo* [548, 549]. Overall, these findings suggest that down-regulation of α-synuclein expression or increase of its degradation by the ubiquitin ligase parkin (see below) may provide therapeutic benefit for patients with either familial or sporadic PD.

Effective silencing of human α-synuclein gene (*SNCA*), both *in vitro* and *in vivo*, by a LV vector-mediated RNAi mechanism was reported by Sapru *et al.* in 2006 [550]. At first, the authors identified an effective target for RNAi-mediated gene silencing in the coding region of the human α-synuclein gene. Then, they demonstrated the effectiveness of a LV vector expressing a human α-synuclein-targeting shRNA in silencing the expression of the α-synuclein gene in the human DAergic cell line SH-SY5Y as well as the expression of a human α-synuclein transgene in the rat brain. In the same study, the authors also designed a synthetic A53T-specific siRNA that was able to mediate *in vitro* allele-specific silencing of the mutant A53T allele of the human α-synuclein gene. In a more recent study by Junn *et al.* [551], miR-7-induced down-regulation of the same mutant form of α-synuclein was found to protect NS20Y cells against α-synuclein-mediated proteasome impairment and susceptibility to oxidative stress.

RNAi-mediated knock-down of α-synuclein was also reported by Fountaine and Wade-Martins [552] who achieved 80% protein reduction in SH-SY5Y cells by using siRNA molecules targeted to endogenous α-synuclein. The first evidence of successful anti-α-synuclein intervention in nonhuman primate SN was obtained by McCormack *et al.* in 2010 [553]. In this study, feasibility and safety of α-synuclein suppression were evaluated by treating monkeys with a targeted siRNA that was directly infused into the left SN. A significant reduction of α-synuclein mRNA and protein was observed in the infused hemisphere compared to the untreated one, with no evident adverse consequences.

An alternative approach based on the AAV-mediated delivery of an anti-α-synuclein ribozyme was investigated in a rodent model of PD [554]. Injection of this construct into the SN of MPP^+-treated rats resulted in a significant protection of TH-positive cells against apoptotic death.

The above findings seem to support a protective role of α-synuclein down-regulation. However, possible undesired effects of uncontrolled α-synuclein silencing should be taken into account since the physiological role of this protein has not yet been thoroughly elucidated. As a matter of fact, while no overt abnormalities were observed in α-synuclein knock-out mice [555], a recent study reported that RNAi-mediated silencing of endogenous α-synuclein in the adult rat SN leads to the degeneration of nigral DAergic neurons and motor deficits [556].

Targeting α-synuclein at the protein level by the intrabody approach has also been investigated. Zhou *et al.* [557] selected a human single-chain intrabody, D10, from a phage library specific for the recombinant monomeric α-synuclein. By transfecting the D10 encoding gene into an HEK 293 cell line that over-expresses wild-type α-synuclein, the authors demonstrated that the D10 intrabody stabilizes the monomeric α-synuclein and inhibits the formation of high-molecular-weight α-synuclein species. Furthermore, the D10 intrabody was shown to rescue the cell adhesion defect that characterizes the α-synuclein over-expressing cells [557]. Emadi *et al.* [558] reported the isolation of a human single chain antibody fragment (scFv) specific for α-synuclein oligomers which are thought to be the most toxic species [559]. This scFv was isolated from a phage displayed antibody library against the target antigen morphology using a novel biopanning technique that utilizes Atomic Force Microscopy (AFM) to image and immobilize specific morphologies of α-synuclein. The selected scFv was shown to bind only to an oligomeric form of α-synuclein and to inhibit both aggregation and toxicity of α-synuclein in SH-SY5Y cells.

A different approach for the selection of anti-α-synuclein intrabodies has been adopted by Lynch *et al.* [560]. The authors used a yeast surface display library of an entire naïve repertoire of human scFv antibodies to select for binding to the non-amyloid component (NAC) region of α-synuclein. This region has been shown to be necessary for aggregation *in vitro* [561] and toxicity to DAergic neurons in a *Drosophila* model [562], thus providing a potential therapeutic target. The most effective of the selected intrabodies, NAC32, was able to reduce α-synuclein intracellular aggregation and toxicity [560]. Cytoplasmic solubility as well as efficacy of various anti-α-synuclein intrabodies have been recently improved by their fusion to a highly charged proteasome-targeting sequence [563].

Parkin Over-Expression

Loss-of-function mutations in the parkin encoding gene (*PARK2*) are responsible for autosomal recessive juvenile parkinsonism (AR-JP), an early-onset form of PD with typical symptoms and pathology and very slow disease progression [236]. Parkin is an E3 ubiquitin ligase that polyubiquitylates unfolded or short-lived proteins, directing the substrates to proteasomal degradation [564]. Loss of this activity leads to the accumulation of potentially toxic substrate proteins and consequent DAergic neuron degeneration. Therefore, it is reasonable to think that *PARK2*-type PD patients would profit by the delivery of wild-type *PARK2* gene. Furthermore, it was reported that oxidative post-translational modifications of parkin, *e.g., S*-nitrosylation [565, 566] or covalent binding of DA [567], may be responsible for impairment of parkin activity in sporadic cases of PD. Such processes could lead to accumulation of substrate proteins with subsequent DAergic neuron degeneration similar to what occurs in *PARK2* patients. In addition, evidence for beneficial effects of Parkin expression on oxidative stress levels [568] and more recently, on mitochondrial homeostasis [569] has been reported. Overall, these findings suggest that parkin over-expression may represent a useful therapeutic approach not only for *PARK2*-type but also for sporadic PD patients.

Parkin over-expression has been shown to provide neuroprotection in genetic animal models of PD, including *Drosophila* and rat models. In transgenic *Drosophila* models, expression of the neuronal parkin substrate Pael-R leads to loss of DAergic neurons, but co-expression of parkin, that catalyzes proteasomal degradation, is able to almost completely block neurodegeneration [570]. Mitochondrial dysfunction and DAergic neurodegeneration caused by *PINK1* defect in *Drosophila* models were also reported to be rescued by parkin [571-573]. Viral vector-mediated delivery of wild-type parkin encoding gene to the rat SN was shown to protect against α-synuclein mediated neurodegeneration [574, 575]. In particular, Lo Bianco and colleagues [574] used LV vectors to co-express parkin and the disease-causing mutant (A30P) human α-synuclein in the rat SN. Parkin over-expression resulted in a significant reduction in α-synuclein-induced neuropathology. Both TH-positive cell bodies in the SN and TH-positive nerve terminals in the striatum were found to be protected. Interestingly, a concomitant increase in the number of hyperphosphorylated α-synuclein inclusions was observed, thus suggesting that the protective mechanism is not through α-synuclein degradation, as previously hypothesized [576]. On the basis of their results, Lo Bianco and colleagues suggested that parkin might protect DAergic neurons by promoting the sequestration of toxic prefibrillar oligomers in mature hyperphosphorylated inclusions [574]. In a similar study, Yamada *et al.* [575] examined the effect of parkin by using AAV vectors to co-express α-synuclein and parkin in the rat SN and showed that parkin over-expression protects against α-synuclein induced DAergic neurodegeneration and consequent motor dysfunction.

The neuroprotective function of parkin has also been investigated in the classic 6-OHDA rat [577, 578] and MPTP mouse models of PD [579]. In a study by Vercammen *et al.* [577], a LV vector encoding human wild-type parkin was injected into the rat SN two weeks prior to a striatal 6-OHDA lesion and a significant preservation of DAergic cell bodies and nerve terminals was observed at three weeks after lesioning. Furthermore, lesioned rats over-expressing parkin displayed milder motor deficits as compared with the control group. The effects on motor function persisted up to twenty weeks after lesioning, while the difference between the control and parkin expressing rats was found to be greatly reduced at forty weeks in terms of numbers of TH-positive cells in the SN. The neuroprotective effect of parkin was not confirmed in a subsequent study based on a more severe 6-OHDA lesioning [578]. In this study, an AAV vector encoding human parkin was injected into the rat SN six weeks prior to a four-site striatal 6-OHDA lesioning. Vector-mediated parkin over-expression was found to be partially protective against 6-OHDA-induced functional impairments, but this effect was not associated to an increase in striatal TH-positive innervation or nigral TH-positive survival neurons compared to control lesioned animals. On the other hand, parkin over-expression was shown to increase both TH and DA levels. These findings led the authors to suggest that parkin, rather than exerting any protective effect on the nigrostriatal tract, could improve motor function by enhancing nigral DA neurotransmission in surviving cells [578]. In the MPTP mouse model, AAV vector-mediated parkin expression induced a modest but statistically significant protection of DAergic neurons, which was associated to behavioral benefits, even though no protection against striatal DA depletion was observed [579]. Overall, the results of these studies indicate that even though parkin over-expression alleviates motor symptoms, the protection of DAergic neurons observed at one-three weeks undergoes a time-dependent decrease.

More recently, the neuroprotective effects of parkin over-expression have been investigated in a modified long-term mouse model of PD using osmotic minipump administration of MPTP [580] AAV vector-mediated delivery of parkin into the SN was shown to prevent motor deficits and DAergic neuron loss and also alleviate the MPTP-induced decrease of the active phosphorylated form of Akt. However, parkin was not effective in reducing up-regulation of p53 and mitochondrial alterations induced by chronic MPTP administration. On the basis of these findings, the authors suggested that the neuroprotective actions of parkin may be impaired in severe PD.

The first study involving a nonhuman primate model of PD was reported by Yasuda *et al.* in 2007 [581]. In order to investigate the effect of parkin co-expression in α-synuclein over-expressing monkeys, a cocktail of AAV1-α-synuclein and AAV1-parkin was unilaterally injected into the striatum while AAV1-α-synuclein alone was injected into the contralateral side. Even though parkin protected against α-synuclein-induced degeneration of DAergic axon terminals in the striatum, no protective effect against the α-synuclein-induced loss of DAergic neurons in the SN was observed. These results have been attributed to insufficient retrograde transport of delivered genes to the SN [581]. Therefore, an alternative strategy aimed at introducing AAV1 vectors directly into the SN is currently in progress [582].

Taken together, the results so far obtained seem to indicate that the effects of parkin are different from those observed with neurotrophic factors. On this basis, it has been suggested that a combined treatment strategy could be more effective than either factor alone as it would interfere with different pathways operating in the disease process [583].

Glutathione Peroxidase Over-Expression

Since oxidative stress is believed to be a key player in the pathogenesis of PD, viral vector-mediated over-expression of antioxidant enzymes could represent an interesting therapeutic approach. Ridet *et al.* [584] investigated, in both *in vitro* and *in vivo* models of PD, the potential neuroprotective effects of the antioxidant enzyme glutathione peroxidase (GPX), which is known to be decreased in parkinsonian brains [585]. LV vector-mediated over-expression of GPX was shown to significantly protect SH-SY5Y cells against 6-OHDA-induced toxicity and a small, but significant protection of nigral DAergic neurons was also observed with LV-GPX in mice injected with 6-OHDA. The authors underline that the actual neuroprotective potential of GPX over-expression *in vivo* may have been underestimated since only 30% of nigral DAergic neurons were transduced with the LV vector.

VMAT2 Over-Expression

Increasing attention has been given to a potential dysregulation of cytoplasmic DA levels and DA-dependent oxidative stress [586-590] as an important contribution to the selective degeneration of nigral DAergic neurons in PD. The neuronal-specific vesicular monoamine transporter 2 (VMAT2), which transports monoamines from the cytosol into secretory vesicles [591], plays a prominent role in intraneuronal DA homeostasis and vesicular turnover. Therefore it is reasonable to think that VMAT2 over-expression could promote DAergic neuron survival and restore neuronal function in conditions of dysregulated DA homeostasis, such as PD, by increasing vesicular DA sequestration. Evidence supporting this hypothesis has been reported by Vergo *et al.* [592]. In this study, the authors used two different *in vitro* models, the DA-producing PC12 cell line and primary postnatal DAergic neurons, to investigate the effects of VMAT2 over-expression on DA compartmentalization and cell viability. Transfection of VMAT2 in PC12 cells was found to increase both intracellular DA content and potassium-induced DA release and to attenuate cell death induced by the cytosolic DA enhancer, methamphetamine. In rat ventral mesencephalic cultures highly enriched for DAergic neurons, LV vector-mediated delivery of VMAT2 also resulted in elevated intracellular DA content and neurotransmitter release after depolarization. Furthermore, silencing endogenously expressed VMAT2 by virally delivered shRNAs was shown to produce the opposite effects in the same primary cultures. These data indicate that DA dependent oxidative stress decrease promoted by VMAT2 over-expression could protect DAergic neurons. This suggests that the course of the disease might be altered by targeting an early pathogenic event such as dysregulation of cytoplasmic DA storage.

Complex I Over-Expression

Several lines of evidence support the involvement of mitochondrial dysfunction in PD pathogenesis [593, 594] and provide the basis for the development of a mitochondrial restorative gene therapy.

Dysfunction of the mitochondrial respiratory chain, specifically the nicotinamide adenine dinucleotide (NADH)-quinone oxidoreductase or complex I, leads to increased ROS levels that are believed to contribute to neuronal damage and death [595-597]. Enhancing complex I activity might thus be a useful approach to restore mitochondrial function. A neuroprotective effect of long-term NDI1 gene expression in a chronic mouse model of PD has been demonstrated in a study by Barber-Singh *et al.* [598]. NDI1 is a *Saccharomyces cerevisiae* gene coding for the internal rotenone-insensitive NADH–quinone oxidoreductase, which was previously shown to be fully functional when delivered to the rodent CNS [599, 600]. Eight months after AAV vector-mediated delivery into the unilateral SN of normal mice, a chronic PD model was created by administration of MPTP with probenecid and neurochemical and behavioral responses were evaluated one-four weeks post-MPTP/probenecid injection. Expression of NDI1 enzyme was shown to significantly prevent the loss of DA and TH as well as the DAergic transporters in the striatum of the chronic parkinsonian mice. Neurological preservation in the NDI1-treated parkinsonian mice was also supported by behavioral assessment.

CONCLUSION AND PERSPECTIVES

The development of nucleic acid-based therapeutics has recently aroused an increasing interest due to the great promise they hold for the treatment of a wide range of both inherited and acquired disorders, including neurodegenerative diseases. As a matter of fact, expectations about gene therapy have followed a cyclical course over the years, with periods of over-optimism followed by bouts of excessive pessimism as a consequence of significant setbacks. Many hurdles have been met in developing safe and effective gene therapy tools. Besides difficulties in scaling up from animal models to humans, vector toxicity, immune rejections and cancer risk inherent to insertional mutagenesis were the major problems to be faced out. In the recent years, however, substantial advances in gene therapy technology, such as improved viral vectors for nucleic acid delivery, have given a new strong impulse to gene therapy research programs. Moreover, the results from several proof-of-concept studies have highlighted the potential of RNAi therapeutics, although their transition to the clinic requires significant problems to be overcome.

Nucleic acid-based therapeutic strategies are under investigation from several years ago and continue to evolve as a viable treatment for neurodegenerative diseases. Among these disorders, PD has been the main field of application for such strategies, and, after extensive pre-clinical experimentation, a number of clinical trials have been performed. The results of these studies are very encouraging as far as the safety of the therapeutic tools used is concerned. With respect to the efficacy, even though data from Phase I trials appeared promising for all the four tested strategies, data from Phase II trials, where available, significantly differ depending on the specific strategy. In particular, the Phase II trial of AAV-GAD in PD patients was the first successful randomized, controlled trial of gene therapy for a neurological disorder. More recently, the successful completion of an open-label Phase I/II trial of ProSavin® has been announced. On the other hand, the promising findings from both pre-clinical and Phase I clinical studies of CERE-120 have not been replicated in Phase II trials. These disappointing results, however, should not lead to give up NTF-based therapeutic strategies, but they should rather induce to increased efforts towards improving theoretical and practical approaches to such strategies. For example, delivery methods as well as diffusion of NTFs into the brain parenchyma should be improved. Importantly, the design of clinical trials evaluating neuroprotective NTF-based therapies would greatly benefit from the development of biomarkers able to detect the initial stages of nigral dysfunction and more sensitive clinical indicators might be needed to enable the detection of disease-modifying effects. Finally, PD gene therapy might benefit from combination treatment approaches, which might be more effective in improving both neuronal survival and function compared to each treatment alone. In particular, such approaches might be based either on the combination of more NTFs or the combination of NTF use with other therapies (*e.g.*, AAV-NTN combined with AAV-AADC and peripheral L-dopa).

The success of gene therapy approaches for neurodegenerative diseases will be facilitated by further improvements in viral vector technology that enable enhanced delivery and diffusion as well as regulated spatial and temporal transgene expression. A further contribution is also expected to come from non-invasive imaging techniques to monitor diffusion of viral vectors and transgene expression. Moreover, a uniform study design with respect to patient selection criteria, vector serotype, injection paradigm and outcome measurements, would be highly desirable in order to allow a more reliable analysis of clinical trial results.

Finally, the routine administration of therapeutic nucleic acids to the CNS will most likely involve nanomedicine and, more specifically, CNS-targeted nanocarriers. Further technological advancements in nanocarrier composition as well as their functionalization and targeting can be expected in the near future. The use of effective *in vitro* models of the BBB will play a central role in testing several properties of novel nanocarriers, including targeting ability and efficiency of the delivery across the BBB, thus providing useful pilot data. However, for a successful translation of brain targeted nanocarriers from bench to clinics, efforts will also have to be focused on other challenges. In consideration of the complexity of the CNS and the safety concerns related to the use of nanomaterials, an accurate evaluation of the interactions of nanocarriers and their hosts in terms of biodistribution, accumulation, degradation and potential toxicity is needed.

REFERENCES

[1] Giacca M, Ed. Gene Therapy. Milan: Springer-Verlag Italia 2010.
[2] Friedmann T, Roblin R. Gene therapy for human genetic disease? Science 1972; 175: 949-55.
[3] Cardinale A, Biocca S. The potential of intracellular antibodies for therapeutic targeting of protein-misfolding diseases. Trends Mol Med 2008; 14: 373-80.
[4] Zhou C, Przedborski S. Intrabody and Parkinson's disease. Biochim Biophys Acta 2009; 1792: 634-42.
[5] Gambari R, Fabbri E, Borgatti M *et al.* Targeting microRNAs involved in human diseases: A novel approach for modification of gene expression and drug development. Biochem Pharmacol 2011; 82: 1416-29.
[6] Ruberti F, Barbato C, Cogoni C. Targeting microRNAs in neurons: Tools and perspectives. Exp Neurol 2012; 235: 419-26.
[7] Stenvang J, Petri A, Lindow M, Obad S, Kauppinen S. Inhibition of microRNA function by antimiR oligonucleotides. Silence 2012, 3:1
[8] Fichou Y, Férec C. The potential of oligonucleotides for therapeutic applications. Trends Biotechnol 2006; 24: 563-70.
[9] Watts JK, Corey DR. Silencing disease genes in the laboratory and the clinic. J Pathol 2012; 226: 365-79.
[10] Bhindi R, Fahmy RG, Lowe HC, *et al.* DNA enzymes, short interfering RNA and the emerging wave of small-molecule nucleic acid-based gene-silencing strategies. Am J Pathol 2007; 171: 1079-88.
[11] Grimpe B. Deoxyribozymes: new therapeutics to treat central nervous system disorders. Front Mol Neurosci 2011; 4: 25.
[12] Meister G, Tuschl T. Mechanisms of gene silencing by double-stranded RNA. Nature 2004; 431: 343-9.
[13] Kim VN. MicroRNA biogenesis: coordinated cropping and dicing. Nat Rev Mol Cell Biol 2005; 6, 376-385.
[14] Davidson BL, McCray PB Jr. Current prospects for RNA interference-based therapies. Nat Rev Genet 2011; 12: 329-40.
[15] Elbashir SM, Harborth J, Lendeckel W, Yalcin A, Weber K, Tuschl T. Duplexes of 21-nucleotide RNAs mediate RNA interference in cultured mammalian cells. Nature 2001; 411: 494-8.
[16] Kim DH, Behlke MA, Rose SD, Chang MS, Choi S, Rossi JJ. Synthetic dsRNA Dicer substrates enhance RNAi potency and efficacy. Nat Biotechnol 2005; 23: 222-6.
[17] Paul CP, Good PD, Winer I, Engelke DR. Effective expression of small interfering RNA in human cells. Nat Biotechnol 2002; 20: 505-8.
[18] Sui G, Soohoo C, Affar EB, Gay F, Shi Y, ForresterWC. A DNA vector-based RNAi technology to suppress gene expression in mammalian cells. Proc Natl Acad Sci USA 2002; 99: 5515-20.
[19] Zeng Y, Wagner EJ, Cullen BR. Both natural and designed micro RNAs can inhibit the expression of cognate mRNAs when expressed in human cells. Mol Cell 2002; 9: 1327-33.
[20] Chung KH, Hart CC, Al-Bassam S, *et al.* Polycistronic RNA polymerase II expression vectors for RNA interference based on BIC/miR-155. Nucleic Acids Res 2006; 34: e53.
[21] Mann MJ, Dzau VJ. Therapeutic applications of transcription factor decoy oligonucleotides. J Clin Invest 2000; 106: 1071-5.

[22]   Osako MK, Nakagami H, Morishita R. Development and modification of decoy oligodeoxynucleotides for clinical application. Adv Polym Sci 2012; 249: 49-60.

[23]   Zhou J, Bobbin ML, Burnett JC, Rossi JJ. Current progress of RNA aptamer-based therapeutics. Front Genet 2012; 3: 234.

[24]   Radom F, Jurek PM, Mazurek MP, Otlewski J, Jeleń F. Aptamers: Molecules of great potential. Biotechnol Adv 2013; doi: 10.1016/j.biotechadv.2013.04.007.

[25]   Ellington AD, Szostak JW. *In vitro* selection of RNA molecules that bind specific ligands. Nature 1990; 346: 818-22.

[26]   Tuerk C, Gold L. Systematic evolution of ligands by exponential enrichment: RNA ligands to bacteriophage T4 DNA polymerase. Science 1990; 249: 505-10.

[27]   Yang X, Li N, Gorenstein DG. Strategies for the discovery of therapeutic Aptamers. Expert Opin Drug Discov 2011; 6: 75-87.

[28]   Sagot Y, Tan SA, Baetge E, Schmalbruch H, Kato AC, Aebischer P. Polymer encapsulated cell lines genetically engineered to release ciliary neurotrophic factor can slow down progressive motor neuronopathy in the mouse. Eur J Neurosci 1995; 7: 1313-22.

[29]   Biju KC, Santacruz RA, Chen C *et al.* Bone marrow-derived microglia-based neurturin delivery protects against dopaminergic neurodegeneration in a mouse model of Parkinson's disease. Neurosci Lett 2013; 535: 24-9.

[30]   Pardridge WM. The blood-brain barrier: bottleneck in brain drug development. NeuroRx 2005; 2: 3-14.

[31]   de Boer AG, Gaillard PJ. Drug targeting to the brain. Annu Rev Pharmacol Toxicol 2007; 47: 323-55.

[32]   Nag S. Morphology and molecular properties of cellular components of normal cerebral vessels. In: Nag S, Ed. The blood-brain barrier: biology and research protocols. Totowa (NJ): Humana Press 2003; pp. 3-36.

[33]   Pardridge WM. Blood-brain barrier delivery. Drug Discov Today 2007; 12: 54-61.

[34]   Bobo RH, Laske DW, Akbasak A, Morrison PF, Dedrick RL, Oldfield EH. Convection-enhanced delivery of macromolecules in the brain. Proc Natl Acad Sci USA 1994; 91: 2076-80.

[35]   Neeves KB, Lo CT, Foley CP, Saltzman WM, Olbricht WL. Fabrication and characterization of microfluidic probes for convection enhanced drug delivery. J Control Release 2006; 111: 252-62.

[36]   Heistad DD, Faraci FM. Gene therapy for cerebral vascular disease. Stroke 1996; 27: 1688-93.

[37]   Watson G, Bastacky J, Belichenko P, *et al.* Intrathecal administration of AAV vectors for the treatment of lysosomal storage in the brains of MPS I mice. Gene Ther 2006; 13: 917-25.

[38]   Schneider H, Groves M, Muhle C, *et al.* Retargeting of adenoviral vectors to neurons using the Hc fragment of tetanus toxin. Gene Ther 2000; 7: 1584-92.

[39]   Wang S, Ma N, Gao SJ, Yu H, Leong KW. Transgene expression in the brain stem effected by intramuscular injection of polyethylenimine/DNA complexes. Mol Ther 2001; 3: 658-64.

[40]   Barati S, Hurtado PR, Zhang SH, Tinsley R, Ferguson IA, Rush RA. GDNF gene delivery *via* the p75(NTR) receptor rescues injured motor neurons. Exp Neurol 2006; 202:179-88.

[41]   Devon RS, Orban PC, Gerrow K, *et al.* Als2-deficient mice exhibit disturbances in endosome trafficking associated with motor behavioral abnormalities. Proc Natl Acad Sci USA 2006; 103: 9595-600.

[42]   Foust KD, Nurre E, Montgomery CL, Hernandez A, Chan CM, Kaspar BK. Intravascular AAV9 preferentially targets neonatal neurons and adult astrocytes. Nat Biotechnol 2009; 27: 59-65.

[43]   Duque S, Joussemet B, Riviere C, *et al.* Intravenous administration of self-complementary AAV9 enables transgene delivery to adult motor neurons. Mol Ther 2009; 17: 1187-96.

[44]   Louboutin JP, Chekmasova AA, Marusich E, Chowdhury JR, Strayer DS. Efficient CNS gene delivery by intravenous injection. Nat Methods 2010; 7: 905-7.

[45]   Chen YH, Chang M, Davidson BL. Molecular signatures of disease brain endothelia provide new sites for CNS-directed enzyme therapy. Nat Med 2009; 15: 1215-8.

[46]   Lim ST, Airavaara M, Harvey BK. Viral vectors for neurotrophic factor delivery: A gene therapy approach for neurodegenerative diseases of the CNS. Pharmacol Res 2010; 61: 14-26.

[47]   Lentz TB, Gray SJ, Samulski RJ. Viral vectors for gene delivery to the central nervous system. Neurobiol Dis 2012; 48: 179-88.

[48]   Gray SJ, Woodard KT, Samulski RJ. Viral vectors and delivery strategies for CNS gene therapy. Ther Deliv 2010; 1: 517-34.

[49]   McCown TJ. Adeno-Associated Virus (AAV) Vectors in the CNS. Curr Gene Ther 2011; 11; 181-8.
[50]   Hadaczek P, Eberling JL, Pivirotto P, Bringas J, Forsayeth J, Bankiewicz KS. Eight years of clinical improvement in MPTP-lesioned primates after gene therapy with AAV2-hAADC. Mol Ther 2010; 18: 1458-61.
[51]   Jakobsson J, Lundberg C. Lentiviral vectors for use in the central nervous system. Mol Ther 2006; 13: 484-93.
[52]   Dreyer JL. Lentiviral vector-mediated gene transfer and RNA silencing technology in neuronal dysfunctions. Mol Biotechnol 2011; 47: 169-87.
[53]   Jeong JH, Park TG, Kim SH. Self-assembled and nanostructured siRNA delivery systems. Pharm Res 2011; 28: 2072-85.
[54]   Pérez-Martínez FC, Guerra J, Posadas I, Ceña V. Barriers to non-viral vector-mediated gene delivery in the nervous system. Pharm Res 2011; 28: 1843-58.
[55]   Petkar KC, Chavhan SS, Agatonovik-Kustrin S, Sawant KK. Nanostructured materials in drug and gene delivery: a review of the state of the art. Crit Rev Ther Drug Carrier Syst 2011; 28: 101-64.
[56]   Jafari M, Soltani M, Naahidi S, Karunaratne DN, Chen P. Nonviral approach for targeted nucleic acid delivery. Curr Med Chem 2012; 19: 197-208.
[57]   Rogers ML, Rush RA. Non-viral gene therapy for neurological diseases, with an emphasis on targeted gene delivery. J Control Release 2012; 157: 183-9.
[58]   Pardridge WM. Preparation of Trojan horse liposomes (THLs) for gene transfer across the blood-brain barrier. Cold Spring Harb Protoc 2010; doi: 10.1101/pdb.prot5407.
[59]   Boado RJ, Pardridge WM. The trojan horse liposome technology for nonviral gene transfer across the blood-brain barrier. J Drug Deliv 2011; 2011: 296151.
[60]   Martinez-Fong D, Bannon MJ, Trudeau LE, *et al.* NTS-Polyplex: a potential nanocarrier for neurotrophic therapy of Parkinson's disease. Nanomedicine 2012; 8: 1052-69.
[61]   He XH, Lin F, Qin ZH. Current understanding on the pathogenesis of polyglutamine diseases. Neurosci Bull 2010; 26: 247-56.
[62]   Myers RH, Macdonald ME, Koroshetz WJ, *et al. De novo* expansion of a (CAG)n repeat in sporadic Huntington's disease. Nat Genet 1993; 5: 168-73.
[63]   Jervis GA. Huntington's chorea in childhood. Arch Neurol 1963; 9: 244-57.
[64]   Huntington's Disease Collaborative Research Group. A novel gene containing a trinucleotide repeat that is expanded and unstable on Huntington's disease chromosomes. Cell 1993; 72: 971-83.
[65]   Gusella JF, MacDonald ME, Ambrose CM, Duyao MP. Molecular genetics of Huntington's disease. Arch Neurol 1993; 50: 1157-63.
[66]   Nance MA, Mathias-Hagen V, Breningstall G, Wick MJ, McGlennen RC. Analysis of a very large trinucleotide repeat in a patient with juvenile Huntington's disease. Neurology 1999; 52: 392-4.
[67]   Han I, You YM, Kordower JH, Brady S, Morfini GA. Differential vulnerability of neurons in Huntington's disease: the role of cell type-specific features. J Neurochem 2010; 113: 1073-91.
[68]   Zuccato C, Valenza M, Cattaneo E. Molecular mechanisms and potential therapeutical targets in huntington's disease. Physiol Rev 2010; 90: 905-81.
[69]   Schulte J, Littleton JT. The biological function of the Huntingtin protein and its relevance to Huntington's Disease pathology. Curr Trends Neurol 2011; 5: 65-78.
[70]   McGeer EG, McGeer PL. Duplication of biochemical changes of Huntington's chorea by intrastriatal injections of glutamic and kainic acids. Nature 1976; 263: 517-9.
[71]   Beal MF, Kowall NW, Ellison DW, Mazurek MF, Swartz KJ, Martin JB. Replication of the neurochemical characteristics of Huntington's disease by quinolinic acid. Nature 1986; 321: 168-71.
[72]   Ferrante RJ, Kowall NW, Cipolloni PB, Storey E, Beal MF. Excitotoxin lesions in primates as a model for Huntington's disease: Histopathologic and neurochemical characterization. Exp Neurol 1993; 119: 46-71.
[73]   Hantraye P, Riche D, Maziere M, Isacson O. A primate model of Huntington's disease: Behavioral and anatomical studies of unilateral excitotoxic lesions of the caudate-putamen in the baboon. Exp Neurol 1990; 108: 91-104.
[74]   Beal MF, Ferrante RJ, Swartz KJ, Kowall NW. Chronic quinolinic acid lesions in rats closely resemble Huntington's disease. J Neurosci 1991; 11: 1649-59.
[75]   DiFiglia M. Excitotoxic injury of the neostriatum: A model for Huntington's disease. Trends Neurosci 1990; 13: 286-9.

[76]   Brouillet E, Conde F, Beal MF, Hantraye P. Replicating Huntington's disease phenotype in experimental animals. Prog Neurobiol 1999; 59: 427-68.

[77]   Guyot MC, Hantraye P, Dolan R, Palfi S, Maziere M, Brouillet E. Quantifiable bradykinesia, gait abnormalities and Huntington's disease-like striatal lesions in rats chronically treated with 3-nitropropionic acid. Neuroscience 1997; 79: 45-56.

[78]   Guyot MC, Palfi S, Stutzmann JM, Maziere M, Hantraye P, Brouillet E. Riluzole protects from motor deficits and striatal degeneration produced by systemic 3-nitropropionic acid intoxication in rats. Neuroscience 1997; 81: 141-9.

[79]   Palfi S, Leventhal L, Goetz CG, *et al.* Delayed onset of progressive dystonia following subacute 3-nitropropionic acid treatment in *Cebus apella* monkeys. Mov Disord 2000; 15: 524-30.

[80]   El Massioui N, Ouary S, Cheruel F, Hantraye P, Brouillet E. Perseverative behavior underlying attentional set-shifting deficits in rats chronically treated with the neurotoxin 3-nitroproprionic acid. Exp Neurol 2001; 172: 172-81.

[81]   Mangiarini L, Sathasivam K, Seller M, *et al.* Exon 1 of the HD gene with an expanded CAG repeat is sufficient to cause a progressive neurological phenotype in transgenic mice. Cell 1996; 87: 493-506.

[82]   Schilling G, Becher MW, Sharp AH, *et al.* Intranuclear inclusions and neuritic aggregates in transgenic mice expressing a mutant N-terminal fragment of huntingtin. Hum Mol Genet 1999; 8: 397-407.

[83]   Slow EJ, van Raamsdonk J, Rogers D, *et al.* Selective striatal neuronal loss in a YAC128 mouse model of Huntington disease. Hum Mol Genet 2003; 12: 1555-67.

[84]   Van Raamsdonk JM, Murphy Z, Slow EJ, Leavitt BR, Hayden MR. Selective degeneration and nuclear localization of mutant huntingtin in the YAC128 mouse model of Huntington disease. Hum Mol Genet 2005a; 14: 3823-35.

[85]   Van Raamsdonk JM, Pearson J, Slow EJ, Hossain SM, Leavitt BR, Hayden MR. Cognitive dysfunction precedes neuropathology and motor abnormalities in the YAC128 mouse model of Huntington's disease. J Neurosci 2005b; 25: 4169-80.

[86]   Gray M, Shirasaki DI, Cepeda C, *et al.* Full-length human mutant huntingtin with a stable polyglutamine repeat can elicit progressive and selective neuropathogenesis in BACHD mice. J Neurosci 2008; 28: 6182-95.

[87]   Spampanato J, Gu X, Yang XW, Mody I. Progressive synaptic pathology of motor cortical neurons in a BAC transgenic mouse model of Huntington's disease. Neuroscience 2008;157: 606-20.

[88]   Menalled LB. Knock-in mouse models of Huntington's Disease. NeuroRx 2005; 2: 465-470.

[89]   Pereira de Almeida L, Ross CA, Zala D, Aebischer P, Déglon N. Lentiviral-mediated delivery of mutant huntingtin in the striatum of rats induces a selective neuropathology modulated by polyglutamine repeat size, huntingtin expression levels and protein length. J Neurosci 2002; 22: 3473-83.

[90]   Régulier E, Trottier Y, Perrin V, Aebischer P, Déglon N. Early and reversible neuropathology induced by tetracycline-regulated lentiviral overexpression of mutant huntingtin in rat striatum. Hum Mol Genet 2003; 12: 2827-36.

[91]   Von Horsten S, Schmitt I, Nguyen HP, *et al.* Transgenic rat model of Huntington's disease. Hum Mol Genet 2003; 12: 617-24.

[92]   Yang SH, Cheng PH, Banta H, *et al.* Towards a transgenic model of Huntington's disease in a non-human primate. Nature 2008; 453: 921-4.

[93]   Squitieri F, Cannella M, Simonelli M. CAG mutation effect on rate of progression in Huntington's disease. Neurol Sci 2002; 23: S107–S108.

[94]   Ellison DW, Beal MF, Mazurek MF, Malloy JR, Bird ED, Martin JB. Amino acid neurotransmitter abnormalities in Huntington's disease and the quinolinic acid animal model of Huntington's disease. Brain 1987; 110: 1657-73.

[95]   Heinsen H, Strik M, Bauer M, *et al.* Cortical and striatal neurone number in Huntington's disease. Acta Neuropathol 1994; 88: 320-33.

[96]   Bachoud-Lévi AC, Déglon N, Nguyen JP, *et al.* Neuroprotective gene therapy for Huntington's disease using a polymer encapsulated BHK cell line engineered to secrete human CNTF. Hum Gene Ther 2000; 11(12): 1723-9.

[97]   Bloch J, Bachoud-Levi AC, Deglon N, *et al.* Neuroprotective gene therapy for Huntington's disease, using polymer encapsulated cells engineered to secrete human ciliary neurotrophic factor: results of a phase I study. Hum Gene Ther 2004; 15(10): 968-75.

[98]   Altar CA, Cai N, Bliven T, *et al.* Anterograde transport of brain-derived neurotrophic factor and its role in the brain. Nature 1997; 389: 856-60.

[99]   Canals JM, Checa N, Marco S, *et al.* Expression of brain-derived neurotrophic factor in cortical neurons is regulated by striatal target area. J Neurosci 2001; 21: 117-24.

[100]   Baquet ZC, Gorski JA, Jones KR. Early striatal dendrite deficits followed by neuron loss with advanced age in the absence of anterograde cortical brain-derived neurotrophic factor. J Neurosci 2004; 24: 4250-8.

[101]   Zuccato C, Cattaneo E. Role of brain-derived neurotrophic factor in Huntington's disease. Prog Neurobiol 2007; 81: 294-330.

[102]   De March Z, Zuccato C, Giampa C, *et al.* Cortical expression of brain derived neurotrophic factor and type-1 cannabinoid receptor after striatal excitotoxic lesions. Neuroscience 2008; 152: 734-40.

[103]   Zuccato C, Cattaneo E. Brain-derived neurotrophic factor in neurodegenerative diseases. Nat Rev Neurol 2009; 5: 311-22.

[104]   Zuccato C, Ciammola A, Rigamonti D, *et al.* Loss of huntingtin mediated BDNF gene transcription in Huntington's disease. Science 2001; 293: 493-8.

[105]   Zuccato C, Tartari M, Crotti A, *et al.* Huntingtin interacts with REST/NRSF to modulate the transcription of NRSE controlled neuronal genes. Nat Genet 2003; 35: 76-83.

[106]   Shimojo M. Huntingtin regulates RE1-silencing transcription factor/ neuron restrictive silencer factor (REST/NRSF) nuclear trafficking indirectly through a complex with REST/NRSF-interacting LIM domain protein (RILP) and dynactin p150 Glued. J Biol Chem 2008; 283: 34880-6.

[107]   Gauthier LR, Charrin BC, Borrell-Pages M, *et al.* Huntingtin controls neurotrophic support and survival of neurons by enhancing BDNF vesicular transport along microtubules. Cell 2004; 118: 127-38.

[108]   Ferrer I, Goutan E, Marin C, Rey MJ, Ritalta T. Brain-derived neurotrophic factor in Huntington disease. Brain Res 2000; 866: 257-61.

[109]   Zuccato C, Marullo M, Conforti P, MacDonald ME, Tartari M, Cattaneo E. Systematic assessment of BDNF and its receptor levels in human cortices affected by Huntington's disease. Brain Pathol 2008; 182: 225-38.

[110]   Canals JM, Pineda JR, Torres-Peraza JF, *et al.* Brain derived neurotrophic factor regulates the onset and severity of motor dysfunction associated with enkephalinergic neuronal degeneration in Huntington's disease. J Neurosci 2004; 24: 7727-39.

[111]   Strand AD, Baquet ZC, Aragaki AK, *et al.* Expression profiling of Huntington's disease models suggests that brain-derived neurotrophic factor depletion plays a major role in striatal degeneration. J Neurosci 2007; 27: 11758-68.

[112]   Bemelmans AP, Horellou P, Pradier L, Brunet I, Colin P, Mallet J. Brain-derived neurotrophic factor-mediated protection of striatal neurons in an excitotoxic rat model of Huntington's disease, as demonstrated by adenoviral gene transfer. Hum Gene Ther 1999; 10: 2987-97.

[113]   Kells AP, Fong DM, Dragunow M, During MJ, Young D, Connor B. AAV-mediated gene delivery of BDNF or GDNF is neuroprotective in a model of Huntington disease. Mol Ther 2004 May; 9(5): 682-8.

[114]   Kells AP, Henry RA, Connor B. AAV-BDNF mediated attenuation of quinolinic acid-induced neuropathology and motor function impairment. Gene Ther 2008; 15(13): 966-77.

[115]   Gharami K, Xie Y, An JJ, Tonegawa S, Xu B. Brain-derived neurotrophic factor over-expression in the forebrain ameliorates Huntington's disease phenotypes in mice. J Neurochem 2008; 105: 369-79.

[116]   Xie Y, Hayden MR, Xu B. BDNF overexpression in the forebrain rescues Huntington's disease phenotypes in YAC128 mice. J Neurosci 2010; 30: 14708-18.

[117]   Zimmerman LB, De Jesus-Escobar JM, Harland RM. The Spemann organizer signal noggin binds and inactivates bone morphogenetic protein 4. Cell 1996; 86: 599-606.

[118]   Cho SR, Benraiss A, Chmielnicki E, Samdani A, Economides A, Goldman SA. Induction of neostriatal neurogenesis slows disease progression in a transgenic murine model of Huntington disease. J Clin Invest 2007; 117: 2889-902.

[119]   Chmielnicki E, Goldman SA. Induced neurogenesis by endogenous progenitor cells in the adult mammalian brain. Prog Brain Res 2002; 138: 451-64.

[120]   Chmielnicki E, Benraiss A, Economides AN, Goldman SA. Adenovirally expressed noggin and brain-derived neurotrophic factor cooperate to induce new medium spiny neurons from resident progenitor cells in the adult striatal ventricular zone. J Neurosci 2004; 24: 2133-42.

[121]   Frim DM, Uhler TA, Short MP, *et al.* Effects of biologically delivered NGF, BDNF and bFGF on striatal excitotoxic lesions. Neuroreport 1993; 4(4): 367-70.

[122]   Martinez-Serrano A, Bjorklund A. Protection of the neostriatum against excitotoxic damage by neurotrophin producing, genetically modified neural stem cells. J Neurosci 1996; 16(15): 4604-16.

[123]   Perez-Navarro E, Alberch J, Neveu I, Arenas E. Brain-Derived Neurotrophic Factor, Neurotrophin-3 and Neurotrophin-4/5 differentially regulate the phenotype and prevent degenerative changes in striatal projection neurons after excitotoxicity *in vivo*. Neuroscience 1999; 91(4): 1257-64.

[124]   Perez-Navarro E, Canudas AM, Akerund P, Alberch J, Arenas E. Brain-derived neurotrophic factor, neurotrophin-3 and neurotrophin-4/5 prevent the death of striatal projection neurons in a rodent model of Huntington's disease. J Neurochem 2000; 75: 2190-9.

[125]   Dey ND, Bombard MC, Roland BP, *et al.* Genetically engineered mesenchymal stem cells reduce behavioral deficits in the YAC 128 mouse model of Huntington's disease. Behav Brain Res 2010; 214(2): 193-200.

[126]   Giralt A, Friedman HC, Caneda-Ferron B, *et al.* BDNF regulation under GFAP promoter provides engineered astrocytes as a new approach for long-term protection in Huntington's disease. Gene Ther 2010; 17: 1294-308.

[127]   Giralt A, Carretón O, Lao-Peregrin C, Martín ED, Alberch J. Conditional BDNF release under pathological conditions improves Huntington's disease pathology by delaying neuronal dysfunction. Mol Neurodegener 2011; 6: 71.

[128]   Ingelsson M, Fukumoto H, Newell KL, *et al.* Early Abeta accumulation and progressive synaptic loss, gliosis and tangle formation in AD brain. Neurology 2004; 62: 925-31.

[129]   Verkhratsky A, Olabarria M, Noristani HN, Yeh CY, Rodriguez JJ. Astrocytes in Alzheimer's disease. Neurotherapeutics 2010; 7: 399-412.

[130]   Braak H, Sastre M, Del TK. Development of alpha synuclein immunoreactive astrocytes in the forebrain parallels stages of intraneuronal pathology in sporadic Parkinson's disease. Acta Neuropathol 2007; 114: 231-241.

[131]   Vargas MR, Johnson JA. Astrogliosis in amyotrophic lateral sclerosis: role and therapeutic potential of astrocytes. Neurotherapeutics 2010; 7: 471-81.

[132]   Haque N, Isacson O. Neurotrophic factors NGF and FGF-2 alter levels of huntingtin (IT15) in striatal neuronal cell cultures. Cell Transplant 2000; 9: 623-7.

[133]   Frim DM, Simpson J, Uhler TA, *et al.* Striatal degeneration induced by mitochondrial blockade is prevented by biologically delivered NGF. J Neurosci Res 1993; 35(4): 452-8.

[134]   Emerich DF, Hammang JP, Baetge EE, Winn SR. Implantation of polymer-encapsulated human nerve growth factor-secreting fibroblasts attenuates the behavioral and neuropathological consequences of quinolinic acid injections into rodent striatum. Exp Neurol 1994; 130(1): 141-50.

[135]   Kordower JH, Charles V, Bayer R, *et al.* Intravenous administration of a transferrin receptor antibody-nerve growth factor conjugate prevents the degeneration of cholinergic striatal neurons in a model of Huntington disease. Proc Natl Acad Sci USA 1994; 91(19): 9077-80.

[136]   Venero JL, Beck KD, Hefti F. Intrastriatal infusion of nerve growth factor after quinolinic acid prevents reduction of cellular expression of choline acetyltransferase messenger RNA and trkA messenger RNA, but not glutamate decarboxylase messenger RNA. Neuroscience 1994; 61(2): 257-68.

[137]   Lin LF, Doherty DH, Lile JD, Bektesh S, Collins F. GDNF: a glial cell line-derived neurotrophic factor for midbrain dopaminergic neurons. Science 1993; 260: 1130-2.

[138]   Pérez-Navarro E, Arenas E, Reiriz J, Calvo N, Alberch J. Glial cell line-derived neurotrophic factor protects striatal calbindin-immunoreactive neurons from excitotoxic damage. Neuroscience 1996; 75(2): 345-52.

[139]   Perez-Navarro E, Arenas E, Marco S, Alberch J. Intrastriatal grafting of a GDNF-producing cell line protects striatonigral neurons from quinolinic acid excitotoxicity *in vivo*. Eur J Neurosci 1999; 11: 241-9.

[140]   Pineda JR, Canals JM, Bosch M, *et al.* Brain-derived neurotrophic factormodulates dopaminergic deficits in a transgenic mouse model of Huntington's disease. J Neurochem 2005; 93: 1057-68.

[141]   Ebert AD, Barber AE, Heins BM, Svendsen CN. *Ex vivo* delivery of GDNF maintains motor function and prevents neuronal loss in a transgenic mouse model of Huntington's disease. Exp Neurol 2010; 224(1): 155-62.

[142]   McBride JL, During MJ, Wuu J, Chen EY, Leurgans SE, Kordower JH. Structural and functional neuroprotection in a rat model of Huntington's disease by viral gene transfer of GDNF. Exp Neurol 2003; 181: 213-23.

[143]   McBride JL, Ramaswamy S, Gasmi M, *et al.* Viral delivery of glial cell linederived neurotrophic factor improves behavior and protects striatal neurons in a mouse model of Huntington's disease. Proc Natl Acad Sci USA 2006; 103: 9345-50.

[144]   Popovic N, Maingay M, Kirik D, Brundin P. Lentiviral gene delivery of GDNF into the striatum of R6/2 Huntington mice fails to attenuate behavioral and neuropathological changes. Exp Neurol 2005; 193: 65-74.

[145]   Marco S, Canudas AM, Canals JM, Perez-Navarro E, Alberch J. Intrastriatal quinolinate or kainate injection differentially regulate GDNF, neurturin and their receptors in adult rat striatum. Soc Neurosci Abstr 2000; 26: 377.20.

[146]   Pérez-Navarro E, Åkerud P, Marco S, *et al.* Neurturin protects striatal projection neurons but not interneurons in a rat model of Huntington's disease. Neuroscience 2000; 98(1): 89-96.

[147]   Marco S, Perez-Navarro E, Tolosa E, *et al.* Striatopallidal neurons are selectively protected by neurturin in an excitotoxic model of Huntington's disease. J Neurobiol 2002; 50(4): 323-32.

[148]   Gratacos E, Perez-Navarro E, Tolosa E, Arenas E, Alberch J. Neuroprotection of striatal neurons against kainate excitotoxicity by neurotrophins and GDNF family members. J Neurochem 2001; 78: 1287-96.

[149]   Ramaswamy S, McBride JL, Herzog CD, *et al.* Neurturin gene therapy improves motor function and prevents death of striatal neurons in a 3-nitropropionic acid rat model of Huntington's disease. Neurobiol Dis 2007; 26: 375-84.

[150]   Ramaswamy S, McBride JL, Han I, *et al.* Intrastriatal CERE-120 (AAV-Neurturin) protects striatal and cortical neurons and delays motor deficits in a transgenic mouse model of Huntington's disease. Neurobiol Dis 2009; 34: 40-50.

[151]   Ip NY, Yancopoulos GD. The neurotrophins and CNTF: two families of collaborative neurotrophic factors. Ann Rev Neurosci 1996; 19: 491-515.

[152]   Anderson KD, Panayotatos N, Corcoran TL, Lindsay RM, Wiegand SJ. Ciliary neurotrophic factor protects striatal output neurons in an animal model of Huntington disease. Proc Natl Acad Sci USA 1996; 93: 7346-51.

[153]   Emerich DF, Lindner MD, Winn SR, Chen EY, Frydel BR, Kordower JH. Implants of encapsulated human CNTF-producing fibroblasts prevent behavioral deficits and striatal degeneration in a rodent model of Huntington's disease. J Neurosci 1996; 16: 5168-81.

[154]   Emerich DF, Winn SR, Hantraye PM, *et al.* Protective effect of encapsulated cells producine neurotrophic factor CNTF in a monkey model of Huntington's disease. Nature 1997; 386: 395-9.

[155]   Mittoux V, Joseph JM, Conde F, *et al.* Restoration of cognitive and motor functions by ciliary neurotrophic factor in a primate model of Huntington's disease. Hum Gene Ther 2000; 11: 117787.

[156]   Emerich DF. Dose-dependent neurochemical and functional protection afforded by encapsulated CNTF-producing cells. Cell Transplant 2004; 13(7-8): 839-44.

[157]   Emerich DF, Winn SR. Neuroprotective effects of encapsulated CNTF-producing cells in a rodent model of Huntington's disease are dependent on the proximity of the implant to the lesioned striatum. Cell Transplant 2004; 13(3): 253-9.

[158]   Pereira de Almeida L, Zala D, Aebischer P, Déglon N. Neuroprotective effect of a CNTF-expressing lentiviral vector in the quinolinic acid rat model of Huntington's disease. Neurobiol Dis 2001; 8(3): 433-46.

[159]   Regulier E, Pereira de Almeida L, Sommer B, Aebischer P, Deglon N. Dose-dependent neuroprotective effect of ciliary neurotrophic factor delivered *via* tetracyclineregulated lentiviral vectors in the quinolinic acid rat model of Huntington's disease. Hum Gene Ther 2002; 13: 1981-90.

[160]   Beurrier C, Faideau M, Bennouar K-E, *et al.* Ciliary neurotrophic factor protects striatal neurons against excitotoxicity by enhancing glial glutamate uptake. PLoS One 2010; 5: e8550.

[161]   Mittoux V, Ouary S, Monville C, *et al.* Corticostriatopallidal neuroprotection by adenovirus-mediated ciliary neurotrophic factor gene transfer in a rat model of progressive striatal degeneration. J Neurosci 2002; 22: 4478-86.

[162]   Zala D, Bensadoun JC, Pereira de Almeida L, *et al.* Long-term lentiviral-mediated expression of ciliary neurotrophic factor in the striatum of Huntington's disease transgenic mice. Exp Neurol 2004; 185: 26-35.

[163]   Denovan-Wright EM, Attis M, Rodriguez-Lebron E, Mandel RJ. Sustained striatal ciliary neurotrophic factor expression negatively affects behavior and gene expression in normal and R6/1 mice. J Neurosci Res 2008; 86: 1748-57.

[164]   Yamamoto A, Lucas JJ, Hen R. Reversal of neuropathology and motor dysfunction in a conditional model of Huntington's disease. Cell 2000; 101: 57-66.

[165]   Boado RJ, Kazantsev A, Apostol BL, Thompson LM, Pardridge WM. Antisense-mediated down-regulation of the human huntingtin gene. J Pharmacol Exp Ther 2000; 295: 239-43.

[166]   Nellemann C, Abell K, Norremolle A, *et al.* Inhibition of Huntington synthesis by antisense oligodeoxynucleotides. Mol Cell Neurosci 2000; 16: 313-23.

[167]   Hasholt L, Abell K, Nørremølle A, Nellemann C, Fenger K, Sørensen SA. Antisense downregulation of mutant huntingtin in a cell model. J Gene Med 2003; 5: 528-38.

[168]   Harper SQ, Staber PD, He X, *et al.* RNA interference improves motor and neuropathological abnormalities in a Huntington's disease mouse model. Proc Natl Acad Sci USA 2005; 102: 5820-5.

[169]   Rodriguez-Lebron E, Denovan-Wright EM, Nash K, Lewin AS, Mandel RJ. Intrastriatal rAAV-mediated delivery of anti-huntingtin shRNAs induces partial reversal of disease progression in R6/1 Huntington's disease transgenic mice. Mol Ther 2005; 12: 618-33.

[170]   achida Y, Okada T, Kurosawa M, Oyama F, Ozawa K, Nukina N. rAAV-mediated shRNA ameliorated neuropathology in Huntington disease model mouse. Biochem Biophys Res Commun 2006; 343: 190-197.

[171]   Huang B, Schiefer J, Sass C, Landwehrmeyer GB, Kosinski CM, Kochanek S. High-capacity adenoviral vector-mediated reduction of huntingtin aggregate load *in vitro* and *in vivo*. Hum Gene Ther 2007; 18: 303-11.

[172]   Franich NR, Fitzsimons HL, Fong DM, Klugmann M, During MJ, Young D. AAV vector-mediated RNAi of mutant huntingtin expression is neuroprotective in a novel genetic rat model of Huntington's disease. Mol Ther 2008; 16: 947-56.

[173]   Drouet V, Perrin V, Hassig R, *et al.* Sustained effects of nonallele-specific Huntingtin silencing. Ann Neurol 2009; 65: 276-85.

[174]   Wang Y-L, Liu W, Wada E, Murata M, Wada K, Kanazawa I. Clinico-pathological rescue of a model mouse of Huntington's disease by siRNA. Neurosci Res 2005; 53: 241-9.

[175]   DiFiglia M, Sena-Esteves M, Chase K, *et al.* Therapeutic silencing of mutant huntingtin with siRNA attenuates striatal and cortical neuropathology and behavioral deficits. Proc Natl Acad Sci USA 2007; 104: 17204-9.

[176]   McBride JL, Boudreau RL, Harper SQ, *et al.* Artificial miRNAs mitigate shRNA-mediated toxicity in the brain: implications for the therapeutic development of RNAi. Proc Natl Acad Sci USA 2008; 105: 5868-73.

[177]   Boudreau R, McBride J, Martins I, *et al.* Nonallele-specific silencing of mutant and wild-type huntingtin demonstrates therapeutic efficacy in Huntington. Mol Ther 2009; 17: 1053-63.

[178]   McBride JL, Pitzer MR, Boudreau RL, *et al.* Preclinical safety of RNAi-mediated HTT suppression in the rhesus macaque as a potential therapy for Huntington's disease. Mol Ther 2011; 19: 2152-62.

[179]   Caplen NJ, Taylor JP, Statham VS, Tanaka F, Fire A, Morgan RA. Rescue of polyglutamine-mediated cytotoxicity by double-stranded RNA-mediated RNA interference. Hum Mol Genet 2002; 11: 175-84.

[180]   Schwarz DS, Ding H, Kennington L, *et al.* Designing siRNA that distinguish between genes that differ by a single nucleotide. PLoS Genet 2006; 2: 1307-18.

[181]   Van Bilsen PH, Jaspers L, Lombardi MS, Odekerken JC, Burright EN, Kaemmerer WF. Identification and allele-specific silencing of the mutant huntingtin allele in Huntington's disease patient-derived fibroblasts. Hum Gene Ther 2008; 19: 710-9.

[182]   Warby SC, Montpetit A, Hayden AR, *et al.* CAG expansion in the Huntington disease gene is associated with a specific and targetable predisposing haplogroup. Am J Hum Genet 2009; 84: 351-66.

[183]   Lombardi MS, Jaspers L, Spronkmans C, *et al.* A majority of Huntington's disease patients may be treatable by individualized allele-specific RNA interference. Exp Neurol 2009; 217: 312-9.

[184]   Pfister EL, Kennington L, Straubhaar J, *et al.* Five siRNAs targeting three SNPs may provide therapy for three-quarters of Huntington's disease patients. Curr Biol 2009; 19: 774-8.

[185]   Pfistera EL, Zamore PD. Huntington's disease: Silencing a brutal killer. Exp Neurol 2009; 220: 226-9.

[186]   Ambrose CM, Duyao MP, Barnes G, *et al.* Structure and expression of the Huntington's disease gene: evidence against simple inactivation due to an expanded CAG repeat. Somat Cell Mol Genet 1994; 20: 27-38.

[187]   Novelletto A, Persichetti F, Sabbadini G, *et al.* Polymorphism analysis of the huntingtin gene in Italian families affected with Huntington disease. Hum Mol Genet 1994; 3: 1129-32.

[188]   Zhang Y, Engelman J, Friedlander RM. Allele-specific silencing of mutant Huntington's disease gene. J Neurochem 2009; 108: 82-90.

[189]   Carroll JB, Warby SC, Southwell AL. Potent and selective antisense oligonucleotides targeting single-nucleotide polymorphisms in the huntington disease gene / allele-specific silencing of mutant huntingtin. Mol Ther 2011; 19: 2178-85.

[190]   Hu J, Matsui M, Gagnon KT, *et al.* Allele-specific silencing of mutant huntingtin and ataxin-3 genes by targeting expanded CAG repeats in mRNAs. Nat Biotechnol 2009; 27: 478-84.

[191]   Gagnon KT, Pendergraff HM, Deleavey GF, *et al.* Allele-selective inhibition of mutant huntingtin expression with antisense oligonucleotides targeting the expanded CAG repeat. Biochemistry 2010; 49: 10166-78.

[192]   Hu J, Liu J, Corey DR. Allele-selective inhibition of huntingtin expression by switching to an miRNA-like RNAi mechanism. Chem Biol 2010; 17: 1183-8.

[193]   Fiszer A, Mykowska A, Krzyzosiak WJ. Inhibition of mutant huntingtin expression by RNA duplex targeting expanded CAG repeats. Nucleic Acids Res 2011; 39: 5578-85.

[194]   Hu J, Liu J, Yu D, Chu Y, Corey DR. Mechanism of allele-selective inhibition of huntingtin expression by duplex RNAs that target CAG repeats: function through the RNAi pathway. Nucleic Acids Res 2012; 40: 11270-80.

[195]   Shin JY, Fang ZH, Yu ZX, Wang CE, Li SH, Li XJ. Expression of mutant huntingtin in glial cells contributes to neuronal excitotoxicity. J Cell Biol 2005; 171: 1001-12.

[196]   Imarisio S, Carmichael J, Korolchuk V, *et al.* Huntington's disease: from pathology and genetics to potential therapies. Biochem J 2008; 412: 191-209.

[197]   Lecerf JM, Shirley TL, Zhu Q, *et al.* Human single-chain Fv intrabodies counteract *in situ* huntingtin aggregation in cellular models of Huntington's disease. Proc Natl Acad Sci USA 2001; 98: 4764-9.

[198]   Khoshnan A, Ko J, Watkin EE, Paige LA, Reinhart PH, Patterson PH. Activation of the IkappaB kinase complex and nuclear factor-kappaB contributes to mutant huntingtin neurotoxicity. J Neurosci 2004; 24: 7999-8008.

[199]   Colby DW, Chu Y, Cassady JP, *et al.* Potent inhibition of huntingtin aggregation and cytotoxicity by a disulfide bond-free single-domain intracellular antibody. Proc Natl Acad Sci USA 2004; 101: 17616-21.

[200]   Miller VM, Nelson RF, Gouvion CM, *et al.* CHIP suppresses polyglutamine aggregation and toxicity *in vitro* and *in vivo*. J Neurosci 2005; 25: 9152-61.

[201]   Murphy RC, Messer A. A single-chain Fv intrabody provides functional protection against the effects of mutant protein in an organotypic slice culture model of Huntington's disease. Brain Res 2004; 121: 141-5.

[202]   Wolfgang WJ, Miller TW, Webster JM, *et al.* Suppression of Huntington's disease pathology in *Drosophila* by human single-chain Fv antibodies. Proc Natl Acad Sci USA 2005; 102: 11563-8.

[203]   McLear JA, Lebrecht D, Messer A, Wolfgang WJ. Combinational approach of intrabody with enhanced Hsp70 expression addresses multiple pathologies in a fly model of Huntington's disease. FASEB J 2008; 22: 2003-11.

[204]   Snyder-Keller A, McLear JA, Hathorn T, Messer A. Early or late-stage anti-N-terminal huntingtin intrabody gene therapy reduces pathological features in B6.HDR6/1 mice. J Neuropath Exp Neurol 2010; 69(10): 1078-85.

[205]   Snyder-Keller A, Butler D, Messer A. Engineered antibody fragments can reduce Huntington's disease phenotypes *in vivo* and *in situ*. Society for Neuroscience 2010; 859.21.

[206]   Wang CE, Zhou H, McGuire JR, *et al.* Suppression of neuropil aggregates and neurological symptoms by an intracellular antibody implicates the cytoplasmic toxicity of mutant huntingtin. J Cell Biol 2008; 181(5): 803-16.

[207]   Southwell AL, Ko J, Patterson PH. Intrabody gene therapy ameliorates motor, cognitive and neuropathological symptoms in multiple mouse models of Huntington's Disease. J Neurosci 2009; 29(43): 13589-602.

[208]   Southwell AL, Khoshnan A, Dunn DE, Bugg CW, Lo DC, Patterson PH. Intrabodies binding the proline-rich domains of mutant huntingtin increase its turnover and reduce neurotoxicity. J Neurosci 2008; 28(36): 9013-20.

[209]   Southwell AL, Bugg CW, Kaltenbach LS, *et al.* Perturbation with intrabodies reveals that calpain cleavage is required for degradation of huntingtin exon 1. PLoS One 2011; 6: e16676.

[210]   Bauer PO, Goswami A, Wong HK, *et al.* Harnessing chaperone-mediated autophagy for the selective degradation of mutant huntingtin protein. Nat Biotechnol 2010; 28(3): 256-63.

[211]   Wang H, Lim PJ, Yin C, Rieckher M, Vogel BE, Monteiro MJ. Suppression of polyglutamine-induced toxicity in cell and animal models of Huntington's disease by ubiquilin. Hum Mol Genet 2006; 15(6): 1025-41.

[212]   Mah AL, Perry G, Smith MA, Monteiro MJ. Identification of ubiquilin, a novel presenilin interactor that increases presenilin protein accumulation. J Cell Biol 2000; 151: 847-62.

[213]   Perrin V, Régulier E, Abbas-Terki T, *et al.* Neuroprotection by Hsp104 and Hsp27 in lentiviral-based rat models of Huntington's disease. Mol Ther 2007; 15(5): 903-11.

[214]   Skogen M, Roth J, Yerkes S, Parekh-Olmedo H, Kmiec E. Short G-rich oligonucleotides as a potential therapeutic for Huntington's Disease. BMC Neurosci 2006; 7: 65-80.

[215]   Damiano M, Galvan L, Déglon N, Brouillet E. Mitochondria in Huntington's disease. Biochim Biophys Acta 2010; 1802: 52-61.

[216]   Benchoua A, Trioulier Y, Zala D, *et al.* Involvement of Mitochondrial Complex II Defects in Neuronal Death Produced by N-Terminus Fragment of Mutated Huntingtin. Mol Biol Cell 2006; 17: 1652-63.

[217]   Cui L, Jeong HY, Borovecki F, Parkhurst CN, Naoko Tanese N, Krainc D. Transcriptional Repression of PGC-1a by Mutant Huntingtin Leads to Mitochondrial Dysfunction and Neurodegeneration. Cell 2006; 127: 59-69.

[218]   Puigserver P, Spiegelman BM. Peroxisome proliferator-activated receptor-gamma coactivator 1 alpha (PGC-1 alpha): transcriptional coactivator and metabolic regulator. Endocr Rev 2003; 24: 78-90.

[219]   Finck BN, Kelly DP. PGC-1 coactivators: inducible regulators of energy metabolism in health and disease. J Clin Invest 2006; 116: 615-22.

[220]   Dai Y, Dudek NL, Li Q, Fowler SC, Muma NA. Striatal expression of a calmodulin fragment improved motor function, weight loss and neuropathology in the R6/2 mouse model of huntington's disease. J Neurosci 2009; 29(37): 11550-9.

[221]   Bao J, Sharp AH, Wagster MV, *et al.* Expansion of polyglutamine repeat in huntingtin leads to abnormal protein interactions involving calmodulin. Proc Natl Acad Sci USA 1996; 93: 5037-42.

[222]   Tang TS, Tu H, Chan EY, *et al.* Huntingtin and huntingtin-associated protein 1 influence neuronal calcium signaling mediated by inositol-(1,4,5) triphosphate receptor type 1. Neuron 2003 39: 227-39.

[223]   Tang TS, Guo CX, Wang HY, Chen X, Bezprozvanny I. Neuroprotective Effects of Inositol 1,4,5 Trisphosphate Receptor C-Terminal Fragment in a Huntington's Disease Mouse Model. J Neurosci 2009; 29(5): 1257-66.

[224]   Sugars KL, Rubinsztein DC. Transcriptional abnormalities in Huntington disease. Trends Genet 2003; 19: 233-8.

[225]   Cha JH. Transcriptional signatures in Huntington's disease. Prog Neurobiol 2007. 83: 228-48.

[226]   Buckley NJ, Johnson R, Zuccato C, Bithell A, Cattaneo E. The role of REST in transcriptional and epigenetic dysregulation in Huntington's disease. Neurobiol Dis 2010; 39: 28-39.

[227] Zuccato C, Belyaev N, Conforti P, *et al.* Widespread disruption of repressor element-1 silencing transcription factor/neuron-restrictive silencer factor occupancy at its target genes in Huntington's disease. J Neurosci 2007; 27: 6972-83.

[228] Soldati C, Bithell A, Conforti P, Cattaneo E, Buckley NJ. Rescue of gene expression by modified REST decoy oligonucleotides in a cellular model of Huntington's disease. J Neurochem 2011; 116: 415-25.

[229] Martin I, Dawson VL, Dawson TM. The impact of genetic research on our understanding of Parkinson's disease. Prog Brain Res 2010; 183: 21-41.

[230] Shulman JM, De Jager PL, Feany MB. Parkinson's disease: genetics and pathogenesis. Annu Rev Pathol 2011; 6: 193-222.

[231] Kruger R, Kuhn W, Müller T, *et al.* Ala30Pro mutation in the gene encoding alpha-synuclein in Parkinson's disease. Nat Genet 1998; 18: 106-8.

[232] Polymeropoulos MH, Lavedan C, Leroy E, *et al.* Mutation in the alpha-synuclein gene identified in families with Parkinson's disease. Science 1997; 276: 2045-7.

[233] Zarranz JJ, Alegre J, Gómez-Esteban JC *et al.* The new mutation, E46K, of alpha-synuclein causes Parkinson and Lewy body dementia. Ann Neurol 2004; 55: 164-73.

[234] Paisan-Ruiz C, Jain S, Evans EW *et al.* Cloning of the gene containing mutations that cause PARK8-linked Parkinson's disease. Neuron 2004; 44: 595-600.

[235] Zimprich A, Biskup S, Leitner P *et al.* Mutations in LRRK2 cause autosomal-dominant parkinsonism with pleomorphic pathology. Neuron 2004a; 44: 601-7.

[236] Kitada T, Asakawa S, Hattori N *et al.* Mutations in the parkin gene cause autosomal recessive juvenile parkinsonism. Nature 1998. 392; 605-8.

[237] Bonifati V, Rizzu P, van Baren MJ *et al.* Mutations in the DJ-1 gene associated with autosomal recessive early-onset parkinsonism. Science 2003; 299: 256-9.

[238] Valente EM, Abou-Sleiman PM, Caputo V *et al.* Hereditary early-onset Parkinson's disease caused by mutations in PINK1. Science 2004; 304: 1158-60.

[239] Nolan YM, Sullivan AM, Toulouse A. Parkinson's disease in the nuclear age of neuroinflammation. Trends Mol Med 2013; 19: 187-96.

[240] Taylor JM, Main BS, Crack PJ. Neuroinflammation and oxidative stress: Co-conspirators in the pathology of Parkinson's disease. Neurochem Int 2013; 62: 803-19.

[241] Toulouse A, Sullivan AM. Progress in Parkinson's disease – where do we stand? Prog Neurobiol 2008; 85: 376-92.

[242] Blesa J, Phani S, Jackson-Lewis V, Przedborski S. Classic and new animal models of parkinson's disease. J Biomed Biotechnol 2012; Volume 2012, Article ID 845618, 10 pages doi:10.1155/2012/845618.

[243] Ungerstedt U. 6-Hydroxy-dopamine induced degeneration of central monoamine neurons. Eur J Pharmacol 1968; 5: 107-10.

[244] Lee CS, Sauer H, Bjorklund A. Dopaminergic neuronal degeneration and motor impairments following axon terminal lesion by instrastriatal 6-hydroxydopamine in the rat. Neuroscience 1996; 72: 641-53.

[245] Barnum CJ, Tansey MG. Modeling neuroinflammatory pathogenesis of Parkinson's disease. Prog Brain Res 2010; 184: 113-32.

[246] Alves da Costa C, Dunys J, Brau F, Wilk S, Cappai R, Checler F. 6-Hydroxydopamine but not 1-methyl-4-phenylpyridinium abolishes alpha-synuclein anti-apoptotic phenotype by inhibiting its proteasomal degradation and by promoting its aggregation. J Biol Chem 2006; 281: 9824-31.

[247] Cicchetti F, Brownell AL, Williams K, Chen YI, Livni E, Isacson O. Neuroinflammation of the nigrostriatal pathway during progressive 6-OHDA dopamine degeneration in rats monitored by immunohistochemistry and PET imaging. Eur J Neurosci 2002; 15: 991-8.

[248] Emborg ME. Evaluation of animal models of Parkinson's disease for neuroprotective strategies. J Neurosci Methods 2004; 139: 121-43.

[249] Cenci MA, Whishaw IQ, Schallert T. Animal models of neurological deficits: how relevant is the rat? Nat Rev Neurosci 2002; 3: 574-9.

[250] Lundblad M, Usiello A, Carta M, Hakansson K, Fisone G, Cenci MA. Pharmacological validation of a mouse model of l-DOPA-induced dyskinesia. Exp Neurol 2005; 194: 66-75.

[251] Santini E, Valjent E, Fisone G. Parkinson's disease: Levodopa-induced dyskinesia and signal transduction. FEBS J 2008; 275: 1392-9.

[252] Davis GC, Williams AC, Markey SP, *et al.* Chronic Parkinsonism secondary to intravenous injection of meperidine analogues. Psychiatry Res 1979; 1: 249-54.

[253] Langston JW, Ballard P, Tetrud JW, *et al.* Chronic Parkinsonism in humans due to a product of meperidine-analog synthesis. Science 1983; 219: 979-80.

[254] Singer TP, Ramsay RR, McKeown K, *et al.* Mechanism of the neurotoxicity of 1-methyl-4-phenylpyridinium (MPP+), the toxic bioactivation product of 1-methyl-4-phenyl-1,2,3,6-tetrahydropyridine (MPTP). Toxicology 1988; 49: 17-23.

[255] Riachi NJ, LaManna JC, Harik SI. Entry of 1-methyl-4-phenyl-1,2,3,6-tetrahydropyridine into the rat brain. J Pharmacol Exp Ther 1989; 249: 744-8.

[256] Chiueh CC, Markey SP, Burns RS, Johannessen JN, Jacobowitz DM, Kopin IJ. Neurochemical and behavioral effects of 1-methyl-4-phenyl-1,2,3,6- tetrahydropyridine (MPTP) in rat, guinea pig and monkey. Psychopharmacol Bull 1984; 20: 548-53.

[257] Sedelis M, Hofele K, Auburger GW, Morgan S, Huston JP, Schwarting RK. MPTP susceptibility in the mouse: behavioral, neurochemical and histological analysis of gender and strain differences. Behav Genet 2000; 30: 171-82.

[258] Meredith GE, Totterdell S, Beales M, Meshul CK. Impaired glutamate homeostasis and programmed cell death in a chronic MPTP mouse model of Parkinson's disease. Exp Neurol 2009; 219: 334-0.

[259] Ferraro TN, Golden GT, DeMattei M, Hare TA, Fariello RG. Effect of 1-methyl-4-phenyl-1,2,3,6-tetrahydropyridine (MPTP) on levels of glutathione in the extrapyramidal system of the mouse. Neuropharmacology 1986; 25: 1071-4.

[260] Czlonkowska A, Kohutnicka M, Kurkowska-Jastrzebska I, Czlonkowski A. Microglial reaction in MPTP (1-methyl-4-phenyl-1,2,3,6-tetrahydropyridine) induced Parkinson's disease mice model. Neurodegeneration 1996; 5: 137-43.

[261] Burns RS, Markey SP, Phillips JM, *et al.* The neurotoxicity of 1-methyl-4-phenyl-1,2,3,6-tetrahydropyridine in the monkey and man. Can J Neurol Sci 1984; 11: 166-8.

[262] Jenner P, Rupniak NM, Rose S, *et al.* 1-Methyl-4-phenyl-1,2,3,6-tetrahydropyridine-induced parkinsonism in the common marmoset. Neurosci Lett 1984; 50: 85-90.

[263] Bedard PJ, Di PT, Falardeau P, *et al.* Chronic treatment with L-DOPA, but not bromocriptine induces dyskinesia in MPTP-parkinsonian monkeys. Correlation with [3H]spiperone binding. Brain Res 1986; 379: 294-9.

[264] Clarke CE, Sambrook MA, Mitchell IJ, *et al.* Levodopa induced dyskinesia and response fluctuations in primates rendered parkinsonian with 1-methyl-4-phenyl-1,2,3,6- tetrahydropyridine (MPTP). J Neurol Sci 1987; 78: 273-80.

[265] Schneider JS. Levodopa-induced dyskinesias in parkinsonian monkeys: relationship to extent of nigrostriatal damage. Pharmacol Biochem Behav 1989; 34: 193-6.

[266] Betarbet R, Sherer TB, MacKenzie G, *et al.* Chronic systemic pesticide exposure reproduces features of Parkinson's disease. Nat Neurosci 2000; 3: 1301-6.

[267] McCormack AL, Thiruchelvam M, Manning-Bog AB, *et al.* Environmental risk factors and Parkinson's disease: selective degeneration of nigral dopaminergic neurons caused by the herbicide paraquat. Neurobiol Dis 2002; 10: 119-27.

[268] Pascual A, Hidalgo-Figueroa M, Piruat JI, Pintado CO, Gomez-Diaz R, Lopez-Barneo J. Absolute requirement of GDNF for adult catecholaminergic neuron survival. Nat Neurosci 2008; 11(7): 755-61.

[269] Krieglstein K, Suter-Crazzolara C, Fischer WH, Unsicker K. TGF-beta superfamily members promote survival of midbrain dopaminergic neurons and protect them against MPP+ toxicity. EMBO J 1995; 14: 736-42.

[270] Hou JG, Lin LF, Mytilineou C. Glial cell line-derived neurotrophic factor exerts neurotrophic effects on dopaminergic neurons *in vitro* and promotes their survival and regrowth after damage by 1-methyl-4-phenylpyridinium. J Neurochem 1996; 66: 74-82.

[271] Eggert K, Schlegel J, Oertel W, Wurz C, Krieg JC, Vedder H. Glial cell linederived neurotrophic factor protects dopaminergic neurons from 6-hydroxydopamine toxicity *in vitro*. Neurosci Lett 1999; 269: 178-82.

[272]   Hoffer BJ, Hoffman A, Bowenkamp K, *et al.* Glial cell line-derived neurotrophic factor reverses toxin-induced injury to midbrain dopaminergic neurons *in vivo.* Neurosci Lett 1994; 182: 107-11.

[273]   Beck KD, Valverde J, Alexi T, *et al.* Mesencephalic dopaminergic neurons protected by GDNF from axotomy-induced degeneration in the adult brain. Nature 1995; 373: 339-341.

[274]   Bowenkamp KE, Hoffman AF, Gerhardt GA, *et al.* Glial cell line-derived neurotrophic factor supports survival of injured midbrain dopaminergic neurons. J Comp Neurol 1995; 355: 479-89.

[275]   Kearns CM, Gash DM. GDNF protects nigral dopamine neurons against 6-hydroxydopamine *in vivo.* Brain Res 1995; 672: 104-11.

[276]   Opacka-Juffry J, Ashworth S, Hume SP, Martin D, Brooks DJ, Blunt SB. GDNF protects against 6-OHDA nigrostriatal lesion: *in vivo* study with microdialysis and PET. Neuroreport 1995; 7: 348-52.

[277]   Sauer H, Rosenblad C, Björklund A. Glial cell line-derived neurotrophic factor but not transforming growth factor beta 3 prevents delayed degeneration of nigral dopaminergic neurons following striatal 6-hydroxydopamine lesion. Proc Natl Acad Sci USA 1995; 92: 8935-9.

[278]   Tomac A, Lindqvist E, Lin LF, *et al.* Protection and repair of the nigrostriatal dopaminergic system by GDNF *in vivo.* Nature 1995; 373: 335-9.

[279]   Bowenkamp KE, David D, Lapchak PL, *et al.* 6-hydroxydopamine induces the loss of the dopaminergic phenotype in substantia nigra neurons of the rat. A possible mechanism for restoration of the nigrostriatal circuit mediated by glial cell line-derived neurotrophic factor. Exp Brain Res 1996; 111: 1-7.

[280]   Martin D, Miller G, Cullen T, Fischer N, Dix D, Russell D. Intranigral or intrastriatal injections of GDNF: effects on monoamine levels and behavior in rats. Eur J Pharmacol 1996; 317: 247-56.

[281]   Martin D, Miller G, Fischer N, Diz D, Cullen T, Russell D. Glial cell line-derived neurotrophic factor: the lateral cerebral ventricle as a site of administration for stimulation of the substantia nigra dopamine system in rats. Eur J Neurosci 1996; 8: 1249-55.

[282]   Shults CW, Kimber T, Martin D. Intrastriatal injection of GDNF attenuates the effects of 6-hydroxydopamine. Neuroreport 1996; 7: 627-31.

[283]   Winkler C, Sauer H, Lee CS, Björklund A. Short-term GDNF treatment provides long-term rescue of lesioned nigral dopaminergic neurons in a rat model of Parkinson's disease. J Neurosci 1996; 16: 7206-15.

[284]   Bowenkamp KE, Lapchak PA, Hoffer BJ, Miller PJ, Bickford PC. Intracerebroventricular glial cell line-derived neurotrophic factor improves motor function and supports nigrostriatal dopamine neurons in bilaterally 6-hydroxydopamine lesioned rats. Exp Neurol 1997; 145: 104-17.

[285]   Kearns CM, Cass WA, Smoot K, Kryscio R, Gash DM. GDNF protection against 6-OHDA: time dependence and requirement for protein synthesis. J Neurosci 1997; 17: 7111-8.

[286]   Gash DM, Gerhardt GA, Hoffer BJ, Effects of glial cell line-derived neurotrophic factor on the nigrostriatal dopamine system in rodents and nonhuman primates. Adv Pharmacol 1998; 42: 911-5.

[287]   Rosenblad C, Martinez-Serrano A, Björklund A. Intrastriatal glial cell linederived neurotrophic factor promotes sprouting of spared nigrostriatal dopaminergic afferents and induces recovery of function in a rat model of Parkinson's disease. Neuroscience 1998; 82: 129-37.

[288]   Sullivan AM, Opacka-Juffry J, Blunt SB. Long-term protection of the rat nigrostriatal dopaminergic system by glial cell linederived neurotrophic factor against 6-hydroxydopamine *in vivo.* Eur J Neurosci 1998; 10: 57-63.

[289]   Rosenblad C, Kirik D, Devaux B, Moffat B, Phillips HS, Björklund A. Protection and regeneration of nigral dopaminergic neurons by neurturin or GDNF in a partial lesion model of Parkinson's disease after administration into the striatum or the lateral ventricle. Eur J Neurosci 1999; 11: 1554-66.

[290]   Aoi M, Date I, Tomita S, Ohmoto T. GDNF induces recovery of the nigrostriatal dopaminergic system in the rat brain following intracerebroventricular or intraparenchymal administration. Acta Neurochir 2000; 142: 805-10.

[291]   Aoi M, Date I, Tomita S, Ohmoto T. Single or continuous injection of glial cell line-derived neurotrophic factor in the striatum induces recovery of the nigrostriatal dopaminergic system. Neurol Res 2000; 22: 832-6.

[292]   Aoi M, Date I, Tomita S, Ohmoto T. The effect of intrastriatal single injection of GDNF on the nigrostriatal dopaminergic system in hemiparkinsonian rats: behavioral and histological studies using two different dosages. Neurosci Res 2000; 36: 319-25.

[293]   Kirik D, Rosenblad C, Björklund A. Preservation of a functional nigrostriatal dopamine pathway by GDNF in the intrastriatal 6-OHDA lesion model depends on the site of administration of the trophic factor. Eur J Neurosci 2000; 12: 3871-82.

[294]   Rosenblad C, Kirik D, Björklund A. Sequential administration of GDNF into the substantia nigra and striatum promotes dopamine neuron survival and axonal sprouting but not striatal reinnervation or functional recovery in the partial 6-OHDA lesion model. Exp Neurol 2000; 161: 503-16.

[295]   Fox CM, Gash DM, Smoot MK, Cass WA. Neuroprotective effects of GDNF against 6-OHDA in young and aged rats. Brain Res 2001; 896: 56-63.

[296]   Cohen AD, Zigmond MJ, Smith AD. Effects of intrastriatal GDNF on the response of dopamine neurons to 6-hydroxydopamine: Time course of protection and neurorestoration. Brain Res 2011; 1370: 80-8.

[297]   Gash DM, Zhang Z, Cass WA, *et al.* Morphological and functional effects of intranigrally administered GDNF in normal rhesus monkeys. J Comp Neurol 1995; 363: 345-58.

[298]   Gash DM, Zhang Z, Ovadia A, *et al.* Functional recovery in parkinsonian monkeys treated with GDNF. Nature 1996; 380: 252-5.

[299]   Miyoshi Y, Zhang Z, Ovadia A, *et al.* Glial cell line-derived neurotrophic factor–levodopa interactions and reduction of side effects in parkinsonian monkeys. Ann Neurol 1997; 42: 208-14.

[300]   Zhang Z, Miyoshi Y, Lapchak PA, *et al.* Dose response to intraventricular glial cell line-derived neurotrophic factor administration in parkinsonian monkeys. J Pharmacol Exp Ther 1997; 282: 1396-401.

[301]   Gerhardt GA, Cass WA, Huettl P, Brock S, Zhang Z, Gash DM. GDNF improves dopamine function in the substantia nigra but not the putamen of unilateral MPTP-lesioned rhesus monkeys. Brain Res 1999; 817: 163-71.

[302]   Costa S, Iravani MM, Pearce RK, Jenner P. Glial cell line-derived neurotrophic factor concentration dependently improves disability and motor activity in MPTP-treated common marmosets. Eur J Pharmacol 2001; 412: 45-50.

[303]   Iravani MM, Costa S, Jackson MJ, *et al.* GDNF reverses priming for dyskinesia in MPTP-treated, L-DOPA-primed common marmosets. Eur J Neurosci 2001; 13: 597-608.

[304]   Grondin R, Zhang Z, Yi A, *et al.* Chronic, controlled GDNF infusion promotes structural and functional recovery in advanced parkinsonian monkeys. Brain 2002; 125(10): 2191-201.

[305]   Maswood N, Grondin R, Zhang Z, *et al.* Effects of chronic intraputamenal infusion of glial cell line-derived neurotrophic factor (GDNF) in aged Rhesus monkeys. Neurobiol Aging 2002; 23: 881-9.

[306]   Ai Y, Markesbery W, Zhang Z, *et al.* Intraputamenal infusion of GDNF in aged rhesus monkeys: distribution and dopaminergic effects. J Comp Neurol 2003; 461: 250-61.

[307]   Grondin R, Cass WA, Zhang Z, Stanford JA, Gash DM, Gerhardt GA. Glial cell line-derived neurotrophic factor increases stimulus-evoked dopamine release and motor speed in aged rhesus monkeys. J Neurosci 2003; 23: 1974-80.

[308]   Nutt JG, Burchiel KJ, Comella CL, *et al.* Randomized, double-blind trial of glial cell line-derived neurotrophic factor (GDNF) in PD. Neurology 2003; 60: 69-73.

[309]   Kordower JH, Palfi S, Chen EY, *et al.* Clinicopathological findings following intraventricular glial-derived neurotrophic factor treatment in a patient with Parkinson's disease. Ann Neurol 1999; 46: 419-24.

[310]   Gill SS, Patel NK, Hotton GR, *et al.* Direct brain infusion of glial cell line-derived neurotrophic factor in Parkinson disease. Nat Med 2003; 9: 589-95.

[311]   Love S, Plaha P, Patel NK, Hotton GR, Brooks DJ, Gill SS. Glial cell line-derived neurotrophic factor induces neuronal sprouting in human brain. Nat Med 2005; 11: 703-4.

[312]   Patel NK, Bunnage M, Plaha P, Svendsen CN, Heywood P, Gill SS. Intraputamenal infusion of glial cell line-derived neurotrophic factor in PD: a two-year outcome study. Ann Neurol 2005; 57: 298-302.

[313]   Slevin JT, Gerhardt GA, Smith CD, Gash DM, Kryscio R, Young B. Improvement of bilateral motor functions in patients with Parkinson disease through the unilateral intraputaminal infusion of glial cell line-derived neurotrophic factor. J Neurosurg 2005; 102: 216-22.

[314] Lang AE, Gill S, Patel NK, *et al.* Randomized controlled trial of intraputamenal glial cell line-derived neurotrophic factor infusion in Parkinson disease. Ann Neurol 2006; 59: 459-66.

[315] Slevin JT, Gash DM, Smith CD, *et al.* Unilateral intraputamenal glial cell line-derived neurotrophic factor in patients with Parkinson disease: response to 1 year of treatment and 1 year of withdrawal. J Neurosurg 2007; 106: 614-20.

[316] Tatarewicz SM, Wei X, Gupta S, Masterman D, Swanson SJ, Moxness MS. Development of a maturing T-cell-mediated immune response in patients with idiopathic Parkinson's disease receiving r-metHuGDNF *via* continuous intraputaminal infusion. J Clin Immunol 2007; 27: 620-7.

[317] Hovland Jr DN, Boyd RB, Butt MT, *et al.* Six-month continuous intraputamenal infusion toxicity study of recombinant methionyl human glial cell line-derived neurotrophic factor (rmetHuGDNF) in rhesus monkeys. Toxicol Pathol 2007; 35: 676-92.

[318] Barker RA. Continuing trials of GDNF in Parkinson's disease. Lancet Neurol 2006; 5: 285-6.

[319] Lang AE, Langston JW, Stoessl AJ, *et al.* GDNF in treatment of Parkinson's disease: response to editorial. Lancet Neurol 2006; 5: 200-2.

[320] Penn RD, Dalvi A, Slevin J, *et al.* GDNF in treatment of Parkinson's disease: response to editorial. Lancet Neurol 2006; 5: 202-3.

[321] Salvatore MF, Ai Y, Fischer B, *et al.* Point source concentration of GDNF may explain failure of phase II clinical trial. Exp Neurol 2006; 202: 497-505.

[322] Sherer TB, Fiske BK, Svendsen CN, Lang AE, Langston JW. Crossroads in GDNF therapy for Parkinson's disease. Mov Disord 2006; 21: 136-41.

[323] Morrison PF, Lonser RR, Oldfield EH. Convective delivery of glial cell line-derived neurotrophic factor in the human putamen. J Neurosurg 2007; 107: 74-83.

[324] Evans JR, Barker RA. Neurotrophic factors as a therapeutic target for Parkinson's disease. Expert Opin Ther Targets 2008; 12: 437-47.

[325] Barker RA. Parkinson's disease and growth factors – are they the answer? Parkinsonism Relat Disord 2009; 15(3): S181-4.

[326] Richardson RM, Kells AP, Rosenbluth KH, *et al.* Interventional MRI-guided putaminal delivery of AAV2-GDNF for a planned clinical trial in Parkinson's disease. Mol Ther 2011; 19: 1048-57.

[327] Bilang-Bleuel A, Revah F, Colin P, *et al.* Intrastriatal injection of an adenoviral vector expressing glial-cell-line-derived neurotrophic factor prevents dopaminergic neuron degeneration and behavioral impairment in a ratmodel of Parkinson disease. Proc Natl Acad Sci USA 1997; 94: 8818-23.

[328] Choi-Lundberg DL, Lin Q, Chang YN, *et al.* Dopaminergic neurons protected from degeneration by GDNF gene therapy. Science 1997; 275(5301): 838-41.

[329] Kojima H, Abiru Y, Sakajiri K, *et al.* Adenovirus-mediated transduction with human glial cell line-derived neurotrophic factor gene prevents 1-methyl-4-phenyl-1,2,3,6-tetrahydropyridine-induced dopamine depletion in striatum of mouse brain. Biochem Biophys Res Commun 1997; 238: 569-73.

[330] Lapchak PA, Araujo DM, Hilt DC, Sheng J, Jiao S. Adenoviral vector-mediated GDNF gene therapy in a rodent lesion model of late stage Parkinson's disease. Brain Res 1997; 777: 153-60.

[331] Choi-Lundberg DL, Lin Q, Schallert T, *et al.* Behavioral and cellular protection of rat dopaminergic neurons by an adenoviral vector encoding glial cell line-derived neurotrophic factor. Exp Neurol 1998; 154: 261-75.

[332] Bohn MC, Choi-Lundberg DL, Davidson BL, *et al.* Adenovirus-mediated transgene expression in nonhuman primate brain. Hum Gene Ther 1999; 10: 1175-84.

[333] Kozlowski DA, Connor B, Tillerson JL, Schallert T, Bohn MC. Delivery of a GDNF gene into the substantia nigra after a progressive 6-OHDA lesion maintains functional nigrostriatal connections. Exp Neurol 2000; 166: 1-15.

[334] Connor B. Adenoviral vector-mediated delivery of glial cell line-derived neurotrophic factor provides neuroprotection in the aged parkinsonian rat. Clin Exp Pharmacol Physiol 2001; 28: 896-900.

[335] Gerin C. Behavioral improvement and dopamine release in a Parkinsonian rat model. Neurosci Lett 2002; 330: 5-8.

[336] Chen X, Liu W, Guoyuan Y, *et al.* Protective effects of intracerebral adenoviral-mediated GDNF gene transfer in a rat model of Parkinson's disease. Parkinsonism Relat Disord 2003; 10: 1-7.

[337] Do Thi NA, Saillour P, Ferrero L, Dedieu JF, Mallet J, Paunio T. Delivery of GDNF by an E1,E3/E4 deleted adenoviral vector and driven by a GFAP promoter prevents dopaminergic neuron degeneration in a rat model of Parkinson's disease. Gene Ther 2004; 11: 746-56.

[338]   Smith AD, Kozlowski DA, Bohn MC, Zigmond MJ. Effect of AdGDNF on dopaminergic neurotransmission in the striatum of 6-OHDA-treated rats. Exp Neurol 2005; 193: 420-6.

[339]   Zheng JS, Tang LL, Zheng SS, *et al.* Delayed gene therapy of glial cell line-derived neurotrophic factor is efficacious in a rat model of Parkinson's disease. Brain Res Mol Brain Res 2005; 134: 155-61.

[340]   Mandel RJ, Spratt SK, Snyder RO, Leff SE. Midbrain injection of recombinant adeno-associated virus encoding rat glial cell line-derived neurotrophic factor protects nigral neurons in a progressive 6-hydroxydopamine- induced degeneration model of Parkinson's disease in rats. Proc Natl Acad Sci USA 1997; 94: 14083-8.

[341]   Mandel RJ, Snyder RO, Leff SE. Recombinant adeno-associated viral vector-mediated glial cell line-derived neurotrophic factor gene transfer protects nigral dopamine neurons after onset of progressive degeneration in a rat model of Parkinson's disease. Exp Neurol 1999; 160: 205-14.

[342]   Kirik D, Rosenblad C, Björklund A, Mandel RJ. Long-term rAAVmediated gene transfer of GDNF in the rat Parkinson's model: intrastriatal but not intranigral transduction promotes functional regeneration in the lesioned nigrostriatal system. J Neurosci 2000b; 20: 4686-700.

[343]   McGrath J, Lintz E, Hoffer BJ, Gerhardt GA, Quintero EM, Granholm AC. Adeno-associated viral delivery of GDNF promotes recovery of dopaminergic phenotype following a unilateral 6-hydroxydopamine lesion. Cell Transplant 2002; 11: 215-27.

[344]   Wang L, Muramatsu S, Lu Y, *et al.* Delayed delivery of AAV-GDNF prevents nigral neurodegeneration and promotes functional recovery in a rat model of Parkinson's disease. Gene Ther 2002; 9: 381-9.

[345]   Eslamboli A, Cummings RM, Ridley RM, *et al.* Recombinant adenoassociated viral vector (rAAV) delivery of GDNF provides protection against 6-OHDA lesion in the common marmoset monkey (Callithrix jacchus). Exp Neurol 2003; 184: 536-48.

[346]   Eslamboli A, Georgievska B, Ridley RM, *et al.* Continuous low-level glial cell line-derived neurotrophic factor delivery using recombinant adeno-associated viral vectors provides neuroprotection and induces behavioral recovery in a primate model of Parkinson's disease. J Neurosci 2005; 25: 769-77.

[347]   Chen YH, Harvey BK, Hoffman AF, Wang Y, Chiang YH, Lupica CR. MPTP-induced deficits in striatal synaptic plasticity are prevented by glial cell line-derived neurotrophic factor expressed *via* an adeno-associated viral vector. FASEB J 2008; 22: 261-75.

[348]   Elsworth JD, Redmond Jr DE, Leranth C, *et al.* AAV2-mediated gene transfer of GDNF to the striatum of MPTP monkeys enhances the survival and outgrowth of co-implanted fetal dopamine neurons. Exp Neurol 2008; 211: 252-8.

[349]   Eberling JL, Kells AP, Pivirotto P, *et al.* Functional Effects of AAV2-GDNF on the dopaminergic nigrostriatal pathway in Parkinsonian rhesus monkeys. Hum Gene Ther 2009; 20: 511-8.

[350]   Johnston LC, Eberling J, Pivirotto P, Hadaczek P, Federoff HJ, Forsayeth J. Clinically relevant effects of AAV2-GDNF on the dopaminergic nigrostriatal pathway in aged Rhesus monkeys. Hum Gene Ther 2009; 20: 497-510.

[351]   Kells AP, Eberling J, Su X, *et al.* Regeneration of the MPTP-lesioned dopaminergic system after convection-enhanced delivery of AAV2-GDNF. J Neurosci 2010; 30: 9567-77.

[352]   Decressac M, Ulusoy A, Mattsson B, *et al.* 2011. GDNF fails to exert neuroprotection in a rat fag-synuclein model of Parkinson's disease. Brain 134: 2302-11.

[353]   Bensadoun JC, Deglon N, Tseng JL, Ridet JL, Zurn AD, Aebischer P. Lentiviral vectors as a gene delivery system in the mouse midbrain: cellular and behavioral improvements in a 6-OHDA model of Parkinson's disease using GDNF. Exp Neurol 2000; 164: 15-24.

[354]   Deglon N, Tseng JL, Bensadoun JC, *et al.* Self-inactivating lentiviral vectors with enhanced transgene expression as potential gene transfer system in Parkinson's disease. Hum Gene Ther 2000; 11: 179-90.

[355]   Kordower JH, Emborg ME, Bloch J, *et al.* Neurodegeneration prevented by lentiviral vector delivery of GDNF in primate models of Parkinson's disease. Science 2000; 290: 767-73.

[356]   Georgievska B, Kirik D, Rosenblad C, Lundberg C, Björklund A. Neuroprotection in the rat Parkinson model by intrastriatal GDNF gene transfer using a lentiviral vector. Neuroreport 2002; 13: 75-82.

[357]   Georgievska B, Kirik D, Björklund A. Aberrant sprouting and downregulation of tyrosine hydroxylase in lesioned nigrostriatal dopamine neurons induced by long-lasting overexpression of glial cell line derived neurotrophic factor in the striatum by lentiviral gene transfer. Exp Neurol 2002; 177: 461-74.

[358]   Palfi S, Leventhal L, Chu Y, *et al.* Lentivirally delivered glial cell line-derived neurotrophic factor increases the number of striatal dopaminergic neurons in primate models of nigrostriatal degeneration. J Neurosci 2002; 22: 4942-54.

[359]   Rosenblad C, Georgievska B, Kirik D. Long-term striatal overexpression of GDNF selectively downregulates tyrosine hydroxylase in the intact nigrostriatal dopamine system. Eur J Neurosci 2003; 17: 260-70.

[360]   Azzouz M, Ralph S, Wong LF, *et al.* Neuroprotection in a rat Parkinson model by GDNF gene therapy using EIAV vector. Neuroreport 2004; 15: 985-90.

[361]   Georgievska B, Jakobsson J, Persson E, Ericson C, Kirik D, Lundberg C. Regulated delivery of glial cell line-derived neurotrophic factor into rat striatum, using a tetracycline-dependent lentiviral vector. Hum Gene Ther 2004; 15: 934-44.

[362]   Georgievska B, Kirik D, Björklund A. Overexpression of glial cell line-derived neurotrophic factor using a lentiviral vector induces time- and dose-dependent downregulation of tyrosine hydroxylase in the intact nigrostriatal dopamine system. J Neurosci 2004; 24: 6437-45.

[363]   Lo Bianco C, Deglon N, Pralong W, Aebischer P. Lentiviral nigral delivery of GDNF does not prevent neurodegeneration in a genetic rat model of Parkinson's disease. Neurobiol Dis 2004; 17: 283-9.

[364]   Dowd E, Monville C, Torres EM, *et al.* Lentivector-mediated delivery of GDNF protects complex motor functions relevant to human Parkinsonism in a rat lesion model. Eur J Neurosci 2005; 22: 2587-95.

[365]   Sajadi A, Bauer M, Thony B, Aebischer P. Long-term glial cell line-derived neurotrophic factor overexpression in the intact nigrostriatal system in rats leads to a decrease of dopamine and increase of tetrahydrobiopterin production. J Neurochem 2005; 93: 1482-6.

[366]   Brizard M, Carcenac C, Bemelmans AP, Feuerstein C, Mallet J, Savasta M. Functional reinnervation from remainingDAterminals induced by GDNF lentivirus in a rat model of early Parkinson's disease. Neurobiol Dis 2006; 21: 90-101.

[367]   Emborg ME, Moirano J, Raschke J, *et al.* Response of aged parkinsonian monkeys to *in vivo* gene transfer of GDNF. Neurobiol Dis 2009; 36: 303-11.

[368]   Natsume A, Mata M, Goss J, *et al.* Bcl-2 and GDNF delivered by HSV-mediated gene transfer act additively to protect dopaminergic neurons from 6-OHDA-induced degeneration. Exp Neurol 2001; 169: 231-8.

[369]   Fink DJ, Glorioso J, Mata M. Therapeutic gene transfer with herpesbased vectors: studies in Parkinson's disease and motor nerve regeneration. Exp Neurol 2003; 184(1): S19-S24.

[370]   Monville C, Torres E, Thomas E, *et al.* HSV vector-delivery of GDNF in a rat model of PD: partial efficacy obscured by vector toxicity. Brain Res 2004; 1024: 1-15.

[371]   Sun M, Kong L, Wang X, Lu XG, Gao Q, Geller AI. Comparison of the capability of GDNF, BDNF, or both, to protect nigrostriatal neurons in a rat model of Parkinson's disease. Brain Res 2005; 1052: 119-29.

[372]   Drinkut A, Tereshchenko Y, Schulz JB, Bähr M, Kügler S. Efficient gene therapy for Parkinson's disease using astrocytes as hosts for localized neurotrophic factor delivery. Mol Ther 2012; 20(3): 534-43.

[373]   Gonzalez-Barrios JA, Lindahl M, Bannon MJ, *et al.* Neurotensin polyplex as an efficient carrier for delivering the human GDNF gene into nigral dopamine neurons of hemiparkinsonian rats. Mol Ther 2006; 14(6): 857-65.

[374]   Xia CF, Boado RJ, Zhang Y, Chu C, Pardridge WM. Intravenous glial-derived neurotrophic factor gene therapy of experimental Parkinson's disease with Trojan horse liposomes and a tyrosine hydroxylase promoter. J Gene Med 2008; 10: 306-15.

[375]   Huang R, Han L, Li J, *et al.* Neuroprotection in a 6-hydroxydopaminelesioned Parkinson model using lactoferrin-modified nanoparticles. J Gene Med 2009; 11: 754-63.

[376]   Huang R, Ke W, Liu Y, *et al.* Gene therapy using lactoferrin-modified nanoparticles in a rotenone-induced chronic Parkinson model. J Neurol Sci 2010; 290: 123-30.

[377]   Arango-Rodriguez ML, Navarro-Quiroga I, Gonzalez-Barrios JA, *et al.* Biophysical characteristics of neurotensin polyplex for *in vitro* and *in vivo* gene transfection. Biochim Biophys Acta 2006;1760: 1009-20.

[378]   Pardridge WM. Gene targeting *in vivo* with pegylated immunoliposomes. Methods Enzymol 2003; 373: 507-28.

[379]   Boado RJ, Zhang YF, Zhang Y, Pardridge WM. Humanization of anti-human insulin receptor antibody for drug targeting across the human blood-brain barrier. Biotechnol Bioeng 2007; 96: 381-91.

[380]   Huang R, Ke W, Liu Y, Jiang C, Pei Y. The use of lactoferrin as a ligand for targeting the polyamidoamine-based gene delivery system to the brain. Biomaterials 2008; 29: 238-46.

[381]   Huang R, Haojun Ma, Yubo Guo, *et al.* Angiopep-Conjugated Nanoparticles for Targeted Long-Term Gene Therapy of Parkinson's Disease. Pharm Res 2013; DOI 10.1007/s11095-013-1005-8.

[382]   Lindner MD, Winn SR, Baetge EE, *et al.* Implantation of encapsulated catecholamine and GDNF-producing cells in rats with unilateral dopamine depletions and parkinsonian symptoms. Exp Neurol 1995; 132: 62-76.

[383]   Tseng JL, Baetge EE, Zurn AD, Aebischer P. GDNF reduces drug-induced rotational behavior after medial forebrain bundle transection by a mechanismnot involving striatal dopamine. J Neurosci 1997; 17: 325-33.

[384]   Akerud P, Canals JM, Snyder EY, Arenas E. Neuroprotection through delivery of glial cell line-derived neurotrophic factor by neural stem cells in a mouse model of Parkinson's disease. J Neurosci 2001; 21: 8108-18.

[385]   Date I, Shingo T, Yoshida H, *et al.* Grafting of encapsulated genetically modified cells secreting GDNF into the striatum of parkinsonian model rats. Cell Transplant 2001; 10: 397-401.

[386]   Park KW, Eglitis MA, Mouradian MM. Protection of nigral neurons by GDNF-engineered marrow cell transplantation. Neurosci Res 2001; 40: 315-23.

[387]   Cunningham LA, Su C. Astrocyte delivery of glial cell line-derived neurotrophic factor in a mouse model of Parkinson's disease. Exp Neurol 2002; 174: 230-42.

[388]   Shingo T, Date I, Yoshida H, Ohmoto T. Neuroprotective and restorative effects of intrastriatal grafting of encapsulated GDNF-producing cells in a rat model of Parkinson's disease. J Neurosci Res 2002; 69: 946-54.

[389]   Kishima H, Poyot T, Bloch J, *et al.* Encapsulated GDNF-producing C2C12 cells for Parkinson's disease: a pre-clinical study in chronic MPTP-treated baboons. Neurobiol Dis 2004; 16: 428-39.

[390]   Duan D, Yang H, Zhang J, Zhang J, Xu Q. Long-term restoration of nigrostriatal system function by implanting GDNF genetically modified fibroblasts in a rat model of Parkinson's disease. Exp Brain Res 2005; 161: 316-24.

[391]   Ericson C, Georgievska B, Lundberg C. *Ex vivo* gene delivery of GDNF using primary astrocytes transduced with a lentiviral vector provides neuroprotection in a rat model of Parkinson's disease. Eur J Neurosci 2005; 22: 2755-64.

[392]   Yasuhara T, Shingo T, Muraoka K, *et al.* Early transplantation of an encapsulated glial cell line-derived neurotrophic factor-producing cell demonstrating strong neuroprotective effects in a rat model of Parkinson disease. J Neurosurg 2005; 102: 80-9.

[393]   Behrstock S, Ebert A, McHugh J, *et al.* Human neural progenitors deliver glial cell line-derived neurotrophic factor to parkinsonian rodents and aged primates. Gene Ther 2006; 13: 379-88.

[394]   Sajadi A, Bensadoun JC, Schneider BL, Lo Bianco C, Aebischer P. Transient striatal delivery of GDNF *via* encapsulated cells leads to sustained behavioral improvement in a bilateral model of Parkinson disease. Neurobiol Dis 2006; 22: 119-29.

[395]   Ebert AD, Beres AJ, Barber AE, Svendsen CN. Human neural progenitor cells over-expressing IGF-1 protect dopamine neurons and restore function in a rat model of Parkinson's disease. Exp Neurol 2008; 209: 213-23.

[396]   Biju K, Zhou Q, Li G, *et al.* Macrophage-mediated GDNF delivery protects against dopaminergic neurodegeneration: a therapeutic strategy for Parkinson's disease. Mol Ther 2010; 18: 1536-44.

[397] Piccini P, Brooks DJ, Bjorklund A, *et al.* Dopamine release from nigral transplants visualized *in vivo* in a Parkinson's patient. Nat Neurosci 1999; 2: 1137-40.

[398] Freed CR, Greene PE, Breeze RE, *et al.* Transplantation of embryonic dopamine neurons for severe Parkinson's disease. N Engl J Med 2001; 344: 710-9.

[399] Lindvall O, Hagell P. Cell therapy and transplantation in Parkinson's disease. Clin Chem Lab Med 2001; 39: 356-61.

[400] Redmond DE Jr. Cellular replacement therapy for Parkinson's disease—where we are today? Neuroscientist 2002; 8: 457-88.

[401] Kordower JH, Chu Y, Hauser RA, Freeman TB, Olanow CW. Lewy body-like pathology in longterm embryonic nigral transplants in Parkinson's disease. Nat Med 2008; 14: 504-6.

[402] Li JY, Englund E, Holton JL, *et al.* Lewy bodies in grafted neurons in subjects with Parkinson's disease suggest host-to-graft disease propagation. Nat Med 2008; 14: 501-3.

[403] Mendez I, Vinuela A, Astradsson A, *et al.* Dopamine neurons implanted into people with Parkinson's disease survive without pathology for 14 years. Nat Med 2008; 14: 507-9.

[404] Sautter J, Tseng JL, Braguglia D, *et al.* Implants of polymer-encapsulated genetically modified cells releasing glial cell line-derived neurotrophic factor improve survival, growth and function of fetal dopaminergic grafts. Exp Neurol 1998; 149: 230-6.

[405] Bauer M, Meyer M, Grimm L, *et al.* Nonviral glial cell-derived neurotrophic factor gene transfer enhances survival of cultured dopaminergic neurons and improves their function after transplantation in a rat model of Parkinson's disease. Hum Gene Ther 2000; 11: 1529-41.

[406] Ahn YH, Bensadoun JC, Aebischer P, *et al.* Increased fiber outgrowth from xeno-transplanted human embryonic dopaminergic neurons with co-implants of polymer-encapsulated genetically modified cells releasing glial cell line-derived neurotrophic factor. Brain Res Bull 2005; 66: 135-42.

[407] Ostenfeld T, Tai YT, Martin P, Deglon N, Aebischer P, Svendsen CN. Neurospheres modified to produce glial cell line-derived neurotrophic factor increase the survival of transplanted dopamine neurons. J Neurosci Res 2002; 69: 955-65.

[408] Georgievska B, Carlsson T, Lacar B, Winkler C, Kirik D. Dissociation between short-term increased graft survival and long-term functional improvements in Parkinsonian rats overexpressing glial cell line-derived neurotrophic factor. Eur J Neurosci 2004; 20: 3121-30.

[409] Yurek DM, Fletcher AM, Kowalczyk TH, Padegimas L, Cooper MJ. Compacted DNA nanoparticle gene transfer of GDNF to the rat striatum enhances the survival of grafted fetal dopamine neurons. Cell Transplant 2009; 18: 1183-96.

[410] Kotzbauer PT, Lampe PA, Heuckeroth RO, *et al.* Neurturin, a relative of glial-cell-line-derived neurotrophic factor. Nature 1996; 384: 467-70.

[411] Horger BA, Nishimura MC, Armanini MP, *et al.* Neurturin exerts potent actions on survival and function of midbrain dopaminergic neurons. J Neurosci 1998; 18: 4929-37.

[412] Creedon DJ, Tansey MG, Baloh RH, *et al.* Neurturin shares receptors and signal transduction pathways with glial cell line-derived neurotrophic factor in sympathetic neurons. Proc Natl Acad Sci USA 1997; 94: 7018-23.

[413] Tseng JL, Bruhn SL, Zurn AD, Aebischer P. Neurturin protects dopaminergic neurons following medial forebrain bundle axotomy. Neuroreport 1998; 9: 1817-22.

[414] Oiwa Y, Yoshimura R, Nakai K, Itakura T. Dopaminergic neuroprotection and regeneration by neurturin assessed by using behavioral, biochemical and histochemical measurements in a model of progressive Parkinson's disease. Brain Res 2002; 947: 271-83.

[415] Hamilton JF, Morrison PF, Chen MY, *et al.*, Heparin coinfusion during convection-enhanced delivery (CED) increases the distribution of the glial derived neurotrophic factor (GDNF) ligand family in rat striatum and enhances the pharmacological activity of neurturin. Exp Neurol 2001; 168: 155-61.

[416] Liu WG, Lu GQ, Li B, Chen SD. Dopaminergic neuroprotection by neurturin-expressing c17.2 neural stem cells in a rat model of Parkinson's disease. Parkinsonism Relat Disord 2007; 13: 77-88.

[417] Ye M, Wang XJ, Zhang YH, *et al.* Transplantation of bone marrow stromal cells containing the neurturin gene in rat model of Parkinson's disease. Brain Res 2007; 1142: 206-16.

[418] Fjord-Larsen L, Johansen JL, Kusk P, *et al.* Efficient *in vivo* protection of nigral dopaminergic neurons by lentiviral gene transfer of a modified Neurturin construct. Exp Neurol 2005; 195: 49-60.

[419] Gasmi M, Brandon EP, Herzog CD, *et al.* AAV2-mediated delivery of human neurturin to the rat nigrostriatal system: Long-term efficacy and tolerability of CERE-120 for Parkinson's disease. Neurobiol Dis 2007; 27: 67–76.

[420] Gasmi M, Herzog CD, Brandon EP, *et al.* Striatal delivery of neurturin by CERE-120, an AAV2 vector for the treatment of dopaminergic neuron degeneration in Parkinson's disease. Mol Ther 2007; 15: 62–8.

[421] Herzog CD, Dass B, Holden JE, *et al.* Striatal delivery of CERE-120, an AAV2 vector encoding human neurturin, enhances activity of the dopaminergic nigrostriatal system in aged monkeys. Mov Disord 2007; 22(8): 1124-32.

[422] Herzog CD, Dass B, Gasmi M, *et al.* Transgene expression, bioactivity and safety of CERE-120 (AAV2-Neurturin) following delivery to the monkey striatum. Mol Ther 2008; 16(10): 1737-44.

[423] Herzog CD, Brown L, Gammon D, *et al.* Expression, bioactivity and safety 1 year after adeno-associated viral vector type 2-mediated delivery of neurturin to the monkey nigrostriatal system support CERE-120 for Parkinson's disease. Neurosurgery 2009; 64(4): 602-12.

[424] Herzog CD, Bishop K, Brown L, Wilson A, Kordower JH, Bartus RT. Gene transfer provides a practical means for safe, longterm, targeted delivery of biologically active neurotrophic factor proteins for neurodegenerative diseases. Drug Deliv Transl Res 2011; 1: 361-82.

[425] Kordower JH, Herzog CD, Dass B, *et al.* Delivery of neurturin by AAV2 (CERE-120)-mediated gene transfer provides structural and functional neuroprotection and neurorestoration in MPTP-treated monkeys. Ann Neurol 2006; 60: 706-15.

[426] Marks WJ, Ostrem JL, Verhagen L, *et al.* Safety and tolerability of intraputaminal delivery of CERE-120 (adeno-associated virus serotype 2-neurturin) to patients with idiopathic Parkinson's disease: an open-label, phase I trial. Lancet Neurol 2008; 7: 400-8.

[427] Marks Jr WJ, Bartus RT, Siffert J, *et al.* Gene delivery of AAV2-neurturin for Parkinson's disease: a double-blind, randomised, controlled trial. Lancet Neurol 2010; 9: 1164-72.

[428] Bartus RT, Herzog CD, Chu Y, *et al.* Bioactivity of AAV2-neurturin gene therapy (CERE-120): differences between Parkinson's disease and nonhuman primate brains. Mov Disord 2011; 26: 27-36.

[429] Bartus RT, Brown L, Wilson A, *et al.* Properly scaled and targeted AAV2-NRTN (neurturin) to the substantia nigra is safe, effective and causes no weight loss: support for nigral targeting in Parkinson's disease. Neurobiol Dis 2011; 44: 38-52.

[430] Bartus RT. Translating the therapeutic potential of neurotrophic factors to clinical 'proof of concept': A personal saga achieving a career-long quest. Neurobiol Dis 2012; 48: 153-78.

[431] Bartus RT, Baumann TL, Brown L, Kruegel BR, Ostrove JM, Herzog CD. Advancing neurotrophic factors as treatments for age-related neurodegenerative diseases: developing and demonstrating "clinical proof-of-concept" for AAV-neurturin (CERE-120) in Parkinson's Disease. Neurobiol Aging 2013; 34: 35–61.

[432] Lewis TB, Standaert DG. Parkinson's Disease, Primates and Gene Therapy: Vive la Différence? Mov Dis 2011; 26(1): 2-3.

[433] Herzog CD, Brown L, Chu Y, Baumann T, Kordower JH, Bartus RT. Robust, stable, targeted, long-term neurturin expression and enhanced tyrosine hydroxylase labeling in Parkinson's disease brain 4 years following delivery of CERE-120 (AAV2-neurturin) to the human putamen. Neuroscience 2012; 2012 Oct 13-17; New Orleans, LA, USA. Washington, D.C.: Society for Neuroscience. Program No. 755.01. Online.

[434] Baumann T, Lang AE, Lozano AM, Olanow CW, Bartus RT. AAV-2-neurturin (CERE-120) and Parkinson's disease: the safety and feasibility of combined substantia nigral and putaminal stereotactic targeting *via* a Phase 1/2b clinical trial in advanced Parkinson's disease. 16th Intl. Congress of Movement Disorders Society; 2012 June 17-21; Dublin, Ireland. Abstract No. 9.

[435] Bartus RT, Baumann TL, Siffert J, *et al.* Safety/feasibility of targeting the substantia nigra with AAV2-neurturin in Parkinson patients. Neurology 2013; 80: 1698-701.

[436] Ceregene reports data from Parkinson's disease Phase 2b study. http://www.ceregene.com/press. Accessed 19 April 2013.

[437] Baloh RH, Tansey MG, Lampe PA, *et al.* Artemin, a novel member of the GDNF ligand family, supports peripheral and central neurons and signals through the GFRalpha3-RET receptor complex. Neuron 1998; 21: 1291-302.

[438] Milbrandt J, de Sauvage FJ, Fahrner TJ, *et al.* Persephin, a novel neurotrophic factor related to GDNF and neurturin. Neuron 1998; 20: 245-53.

[439] Rosenblad C, Gronborg M, Hansen C, *et al. In vivo* protection of nigral dopamine neurons by lentiviral gene transfer of the novel GDNF-family member neublastin/artemin3. Mol Cell Neurosci 2000; 15, 199-214.

[440] Åkerud P, Holm PC, Castelo-Branco G, Sousa K, Rodriguez FJ, Arenas E. Persephin-overexpressing neural stem cells regulate the function of nigral dopaminergic neurons and prevent their degeneration in a model of Parkinson's Disease. Mol Cell Neurosci 2002; 21: 205-22.

[441] Hyman C, Hofer M, Barde YA, *et al.* BDNF is a trophic factor for dopaminergic neurons of the substantia nigra. Nature 1991; 350: 230-3.

[442] Knusel B, Winslow JW, Rosenthal A, *et al.* Promotion of central cholinergic and dopaminergic neurons differentiation by brain-derived neurotrophic factor but not neurotrophin-3. Proc Natl Acad Sci USA 1991; 88: 961-5.

[443] Beck KD, Knusel B, Winslow JW, *et al.* Pretreatment of dopaminergic neurons in culture with BDNF attenuates toxicity of 1-methyl-4-phenylpyridinium. Neurodegeneration 1992; 1: 27-36.

[444] Skaper SD, Negro A, Facci L, Toso RD. Brain-derived neurotrophic factor selectively rescues mesencephalic dopaminergic neurons from 2,4,5-trihydroxyphenylalanine-induced injury. J Neurosci Res 1993; 34: 478-87.

[445] Spina M, Squinto S, Miller J, Lindsay RM, Hyman C. Brain-derived neurotrophic factor protects dopamine neurons against 6-hydroxydopamine and N-methyl-4-phenylpyridium ion toxicity: involvement of the gluthathione system. J Neurochem 1992; 59: 99-106.

[446] Altar CA, Boylan CB, Jackson C, *et al.* Brain-derived neurotrophic factor augments rotational behavior and nigrastriatal dopamine turnover *in vivo.* Proc Natl Acad Sci USA 1992; 89: 11347-51.

[447] Martin-Iverson MT, Todd KG, Altar CA. Brain-derived neurotrophic factor and neurotrophin-3 activate striatal dopamine and serotonin metabolism and related behaviors: interactions with amphetamine. J Neurosci 1994; 14: 1262-70.

[448] Martin-Iverson MT, Altar CA. Spontaneous behaviors of rats are differentially affected by substantia nigra infusions of brain-derived neurotrophic factor and neurotrophin-3. Eur J Neurosci 1996; 8: 1696-706.

[449] Wiegand SJ, Anderson K, Alexander C, *et al.* Receptor binding and axonal transport of 125I-labeled neurotrophins in the basal ganglia and related brain regions. Mov Disord 1992; 7(1): 63.

[450] Shen RY, Altar CA, Chiodo LA. Brain-derived neurotrophic factor increases the electrical activity of pars compacta dopamine neurons *in vivo.* Proc Natl Acad Sci USA 1994; 91: 8920-4.

[451] Altar CA, Boylan CB, Fritsche M, *et al.* Efficacy of brain-derived neurotrophic factor and neurotrophin-3 on neurochemical and behavioral deficits associated with partial nigrostriatal dopamine lesions. J Neurochem 1994; 63: 1021-32.

[452] Hagg T. Neurotrophins prevent death and differentially affect tyrosine hydroxylase of adult rat nigrostriatal neurons *in vivo.* Exp Neurol 1998; 149: 183-92.

[453] Yan Q, Rosenfeld RD, Matheson CR, *et al.* Expression of brain-derived neurotrophic factor protein in the adult rat central nervous system. Neuroscience 1997; 78(2): 431-48.

[454] Yan Q, Radeke MJ, Matheson CR, Talvenheimo J, Welcher AA, Feinstein SC. Immunocytochemical localization of TrkB in the central nervous system of the adult rat. J Comp Neurol 1997; 378(1): 13557.

[455] Mogi M, Togari A, Kondo T, *et al.* Brain-derived growth factor and nerve growth factor concentrations are decreased in the substantia nigra in Parkinson's disease. Neurosci Lett 1999; 270: 45-8.

[456] Parain K, Murer MG, Yan Q, *et al.* Reduced expression of brain-derived neurotrophic factor protein in Parkinson's disease substantia nigra. Neuroreport 1999; 10: 557-61.

[457] Guillin O, Diaz J, Carroll P, Griffon N, Schwartz JC, Sokoloff P. BDNF controls dopamine D3 receptor expression and triggers behavioural sensitization. Nature 2001; 411: 86-9.

[458]   Zhang X, Andren PE, Svenningsson P. Repeated l-DOPA treatment increases c-fos and BDNF mRNAs in the subthalamic nucleus in the 6-OHDA rat model of Parkinson's disease. Brain Res 2006; 1095(1): 207-10.

[459]   Gyarfas T, Knuuttila J, Lindholm P, Rantamaki T, Castren E. Regulation of brainderived neurotrophic factor (BDNF) and cerebral dopamine neurotrophic factor (CDNF) by anti-parkinsonian drug therapy *in vivo*. Cell Mol Neurobiol 2010; 30: 361-8.

[460]   Bousquet M, Gibrat C, Saint-Pierre M, Julien C, Calon F, Cicchetti F. Modulation of brain-derived neurotrophic factor as a potential neuroprotective mechanism of action of omega-3 fatty acids in a parkinsonian animal model. Prog Neuropsychopharmacol Biol Psychiatry 2009; 13: 1401-8.

[461]   Frim DM, Uhler TA, Galpern WR, Beal MF, Breakefield XO, Isacson O. Implanted fibroblasts genetically engineered to produce brain-derived neurotrophic factor prevent l-methyl-4-phenylpyridinium toxicity to dopaminergic neurons in the rat. Proc Natl Acad Sci USA 1994; 91: 5104-8.

[462]   Galpern WR, Frim DM, Tatter SB, Altar CA, Beal MF, Isacson O. Cell-mediated delivery of brain-derived neurotrophic factor enhances dopamine levels in an MPP+ rat model of substantia nigra degeneration. Cell Transplant 1996; 5: 225-32.

[463]   Levivier M, Przedborski S, Bencsics C, Kang UJ. Intrastriatal implantation of fibroblasts genetically engineered to produce brain-derived neurotrophic factor prevents degeneration of dopaminergic neurons in a rat model of Parkinson's disease. J Neurosci 1995; 15(12): 7810-20.

[464]   Yoshimoto Y, Lin Q, Collier TJ, Frim DM, Breakefield XO, Bohn MC. Astrocytes retrovirally transduced with BDNF elicit behavioral improvement in a rat model of Parkinson's disease. Brain Res 1995; 691: 25-36.

[465]   Klein RL, Lewis MH, Muzyczka N, Meyer EM. Prevention of 6-hydroxydopamine-induced rotational behavior by BDNF somatic gene transfer. Brain Res 1999; 847(2): 314-20.

[466]   Stahl K, Mylonakou MN, Skare Ø, Amiry-Moghaddam M, Torp R. Cytoprotective effects of growth factors: BDNF more potent than GDNF in an organotypic culture model of Parkinson's disease. Brain Res 2011; 1378: 105-18.

[467]   Fu A, Zhang M, Gao F, Xu X, Chen Z. A Novel Peptide Delivers Plasmids across Blood-Brain Barrier into Neuronal Cells as a Single-Component Transfer Vector. PLoS ONE 2013; 8: e59642.

[468]   Fu A, Wang Y, Zhan L, Zhou R. Targeted Delivery of Proteins into the Central Nervous System Mediated by Rabies Virus Glycoprotein-Derived Peptide. Pharm Res 2012; 29: 1562-9.

[469]   Krieglstein K, Suter-Crazzolara C, Hötten G, Pohl J, Unsicker K. Trophic and protective effects of growth/differentiation factor 5, a member of the transforming growth factor-b superfamily, on midbrain dopaminergic neurones. J Neurosci Res 1995b; 42: 724-32.

[470]   Lingor P, Unsicker K, Krieglstein K. Mid brain dopaminergic neurons are protected from radical induced damage by GDF-5 application. Short communication. J Neural Transm 1999; 106: 139-44.

[471]   O'Keeffe GW, Dockery P, Sullivan AM. Effects of growth/differentiation factor-5 on the survival and morphology of embryonic rat midbrain dopaminergic neurones *in vitro*. J Neurocytol 2004a; 33: 479-88.

[472]   Wood TK, McDermott KW, Sullivan AM. Differential effects of growth/differentiation factor 5 and glial cell line-derived neurotrophic factor on dopaminergic neurons and astroglia in cultures of embryonic rat midbrain. J Neurosci Res 2005; 80: 759-66.

[473]   Clayton KB, Sullivan AM. Differential effects of GDF5 on the medial and lateral rat ventral mesencephalon. Neurosci Lett 2007; 427: 132-7.

[474]   Sullivan AM, Opacka-Juffry J, Hötten G, Pohl J, Blunt SB. Growth/differentiation factor 5 protects nigrostriatal dopaminergic neurones in a rat model of Parkinson's disease. Neurosci Lett 1997; 233: 73-6.

[475]   Sullivan AM, Opacka-Juffry J, Pohl J, Blunt SB. Neuroprotective protective effects of growth/differentiation factor 5 depend on the site of administration. Brain Res 1999; 818: 176-9.

[476]   Hurley FM, Costello DJ, Sullivan AM. Neuroprotective effects of delayed administration of growth/differentiation factor-5 in the partial lesion model of Parkinson's disease. Exp Neurol 2004; 185: 281-9.

[477]   O'Sullivan DB, Harrison PT, Sullivan AM. Effects of GDF5 overexpression on embryonic rat dopaminergic neurones *in vitro* and *in vivo*. J Neural Transm 2010; 117: 559-72.

[478] Lindholm P, Saarma M. Novel CDNF/MANF family of neurotrophic factors. Dev Neurobiol 2010; 70: 360-71.
[479] Lindholm P, Voutilainen MH, Lauren J, *et al.* Novel neurotrophic factor CDNF protects and rescues midbrain dopamine neurons *in vivo*. Nature 2007; 448: 73-7.
[480] Voutilainen MH, Back S, Porsti E, *et al.* Mesencephalic astrocyte-derived neurotrophic factor is neurorestorative in rat model of Parkinson's disease. J Neurosci 2009; 29: 9651-9.
[481] Voutilainen MH, Back S, Peränen J, *et al.* Chronic infusion of CDNF prevents 6-OHDA-induced deficits in a rat model of Parkinson's disease. Exp Neurol 2011; 228: 99-108.
[482] Airavaara M, Harvey BK, Voutilainen MH, *et al.* CDNF protects the nigrosriatal dopamine system and promotes recovery after MPTP treatment in mice. Cell Transplant 2012; 21: 1213-23.
[483] Bäck S, Peränen J, Galli E, *et al.* Gene therapy with AAV2-CDNF provides functional benefits in a rat model of Parkinson's disease. Brain Behav 2013; 3: 75-88.
[484] Ren X, Zhang T, Gong X, Hu G, Ding W, Wang X. AAV2-mediated striatum delivery of human CDNF prevents the deterioration of midbrain dopamine neurons in a 6-hydroxydopamine induced parkinsonian rat model. Exp Neurol 2013; 248C: 148-56.
[485] Xue YQ, B. F. BF, Zhao LR, *et al.* AAV9-mediated erythropoietin gene delivery into the brain protects nigral dopaminergic neurons in a rat model of Parkinson's disease. Gene Ther 2010; 17: 83-94.
[486] Shen Y, Muramatsu SI, Ikeguchi K, *et al.* Triple transduction with adeno-associated virus vectors expressing tyrosine hydroxylase, aromatic-L-amino-acid decarboxylase and GTP cyclohydrolase I for gene therapy of Parkinson's disease. Hum Gene Ther 2000; 11: 1509-19.
[487] Muramatsu S-I, Fujimoto K-I, Ikeguchi K, *et al.* Behavioral recovery in a primate model of Parkinson's disease by triple transduction of striatal cells with adeno associated viral vectors expressing dopamine-synthesizing enzymes. Hum Gene Ther 2002; 13: 345-54.
[488] Azzouz M, Martin-Rendon E, Barber RD, *et al.* Multicistronic lentiviral vector-mediated striatal gene transfer of aromatic L-amino acid decarboxylase, tyrosine hydroxylase and GTP cyclohydrolase I induces sustained transgene expression, dopamine production and functional improvement in a rat model of Parkinson's disease. J Neurosci 2002; 22: 10302-12.
[489] Moffat M, Harmon S, Haycock J, O'Malley KL. L-Dopa and dopamine-producing gene cassettes for gene therapy approaches to Parkinson's disease. Exp Neurol 1997; 144: 69-73.
[490] Jarraya B, Boulet S, Ralph GS, *et al.* Dopamine gene therapy for Parkinson's disease in a nonhuman primate without associated dyskinesia. Sci Transl Med 2009; 1: 2ra4.
[491] Oxford BioMedica Announces Successful Completion of Prosavin® Phase I/II Study in Parkinson's Disease. http://www.oxfordbiomedica.co.uk/press-releases/. Accessed 16 April 2012.
[492] Sun M, Kong L,Wang X, Lu XG, Gao Q, Geller AI. Comparison of the capability of GDNF, BDNF, or both, to protect nigrostriatal neurons in a rat model of Parkinson's disease. Brain Res 2005. 1052; 119-29.
[493] Lloyd K, Hornykiewicz O. Parkinson's disease: activity of L-dopa decarboxylase in discrete brain regions. Science 1970; 170: 1212-3.
[494] Ichinose H, Ohye T, Fujita K, *et al.* Quantification of mRNA of tyrosine hydroxylase and aromatic L-amino acid decarboxylase in the substantia nigra in Parkinson's disease and schizophrenia. J Neural Transm Park Dis Dement Sect 1994; 8: 149-58.
[495] Nagatsu T, Sawada M. Biochemistry of postmortem brains in Parkinson's disease: historical overview and future prospects. J Neural Transm Suppl 2007; 72: 113-20.
[496] Leff SE, Spratt SK, Snyder RO, Mandel RJ. Long-term restoration of striatal L-aromatic amino acid decarboxylase activity using recombinant adeno-associated viral vector gene transfer in a rodent model of Parkinson's disease. Neuroscience 1999; 92: 185-96.
[497] Sanchez-Pernaute R, Harvey-White J, Cunningham J, Bankiewicz KS. Functional effect of adeno-associated virus mediated gene transfer of aromatic L-amino acid decarboxylase into the striatum of 6-OHDA-lesioned rats. Mol Ther 2001; 4: 324-30.
[498] Bankiewicz KS, Eberling JL, Kohutnicka M, *et al.* Convection-enhanced delivery of AAV vector in parkinsonian monkeys; *in vivo* detection of gene expression and restoration of dopaminergic function using pro-drug approach. Exp Neurol 2000; 164: 2-14.
[499] Bankiewicz KS, Forsayeth J, Eberling JL, *et al.* Long-term clinical improvement in MPTP-lesioned primates after gene therapy with AAV-hAADC. Mol Ther 2006; 14: 564-70.

[500]   Eberling JL, Jagust WJ, Christine CW, *et al.* Results from a phase I safety trial of hAADC gene therapy for Parkinson disease. Neurology 2008; 70: 1980-3.

[501]   Christine CW, Starr PA, Larson PS, *et al.* Safety and tolerability of putaminal AADC gene therapy for Parkinson disease. Neurology 2009; 73: 1662-9.

[502]   Muramatsu S, Fujimoto K, Kato S, *et al.* A phase I study of aromatic L-amino acid decarboxylase gene therapy for Parkinson's disease. Mol Ther 2010; 18: 1731-5.

[503]   Kish SJ, Tong J, Hornykiewicz O, *et al.* Preferential loss of serotonin markers in caudate *versus* putamen in Parkinson's disease. Brain 2008; 131: 120-31.

[504]   Melamed E, Hefti F, Wurtman RJ. Nonaminergic striatal neurons convert exogenous L-dopa to dopamine in parkinsonism. Ann Neurol 1980; 8: 558-63.

[505]   Mura A, Jackson D, Manley MS, Young SJ, Groves PM. Aromatic L-amino acid decarboxylase immunoreactive cells in the rat striatum: a possible site for the conversion of exogenous L-DOPA to dopamine. Brain Res 1995; 704: 51-60.

[506]   Bertler A, Falck B, Owman C, Rosengrenn E. The localization of monoaminergic blood-brain barrier mechanisms. Pharmacol Rev 1966; 18: 369-85.

[507]   Juorio AV, Walz W, Sloley BD. Absence of decarboxylation of some aromatic-Lamino acids by cultured astrocytes. Brain Res 1987; 426: 183-6.

[508]   Juorio AV, Li XM, Walz W, Paterson IA. Decarboxylation of L-dopa by cultured mouse astrocytes. Brain Res 1993; 626: 306-9.

[509]   Mandel RJ, Rendahl KG, Spratt SK, Snyder RO, Cohen LK, Leff SE. Characterization of intrastriatal recombinant adenoassociated virus-mediated gene transfer of human tyrosine hydroxylase and human GTP-cyclohydrolase I in a rat model of Parkinson's disease. J Neurosci 1998; 18: 4271-84.

[510]   Kirik D, Georgievska B, Burger C, *et al.* Reversal of motor impairments in parkinsonian rats by continuous intrastriatal delivery of L-dopa using rAAV-mediated gene transfer. Proc Natl Acad Sci USA 2002; 99: 4708-13.

[511]   Carlsson T, Winkler C, Burger C, *et al.* Reversal of dyskinesias in an animal model of Parkinson's disease by continuous LDOPA delivery using rAAV vectors. Brain 2005; 128: 559-69.

[512]   Björklund T, Carlsson T, Cederfjäll EA, Carta M, Kirik D. Optimized adeno-associated viral vector-mediated striatal DOPA delivery restores sensorimotor function and prevents dyskinesias in a model of advanced Parkinson's disease. Brain 2010; 133: 496-511.

[513]   Leriche L, Björklund T, Breysse N, *et al.* Positron emission tomography imaging demonstrates correlation between behavioral recovery and correction of dopamine neurotransmission after gene therapy. J Neurosci 2009; 29: 1544-53.

[514]   Cederfjäll E, Sahin G, Kirik D, Björklund T. Design of a Single AAV Vector for Coexpression of TH and GCH1 to Establish Continuous DOPA Synthesis in a Rat Model of Parkinson's Disease. Mol Ther 2012; 20: 1315-26.

[515]   Cederfjäll E, Nilsson N, Sahin G, *et al.* Continuous DOPA synthesis from a single AAV: dosing and efficacy in models of Parkinson's disease. Sci Rep 2013; 3: 2157.

[516]   Gill SS, Heywood P. Bilateral dorsolateral subthalamotomy for advanced Parkinson's disease. Lancet 1997; 350: 1224.

[517]   Alvarez L, Macias R, Guridi J, *et al.* Dorsal subthalamotomy for Parkinson's disease. Mov Disord 2001; 16: 72-8.

[518]   Limousin P, Pollak P, Benazzouz A, *et al.* Effect of parkinsonian signs and symptoms of bilateral subthalamic nucleus stimulation. Lancet 1995; 345: 91-5.

[519]   Deep Brain Stimulation for Parkinson's Disease Study Group. Deep-brain stimulation of the subthalamic nucleus or the pars interna of the globus pallidus in Parkinson's disease. N Engl J Med 2001; 345: 956-63.

[520]   Benabid AL, Chabardes S, Seigneuret E, *et al.* Functional neurosurgery: past, present and future. Clin Neurosurg 2005; 52: 265-70.

[521]   Levy R, Lang AE, Dostrovsky JO, *et al.* Lidocaine and muscimol microinjections in subthalamic nucleus reverse Parkinsonian symptoms. Brain 2001; 124: 2105-18.

[522]   Benabid AL, Chabardes S, Mitrofanis J, Pollak P. Deep brain stimulation of the subthalamic nucleus for the treatment of Parkinson's disease. Lancet Neurol 2009; 8: 67-81.

[523]   During MJ, Kaplitt MG, Stern MB, Eidelberg D. Subthalamic GAD gene transfer in Parkinson disease patients who are candidates for deep brain stimulation. Hum Gene Ther 2001; 12: 1589-91.

[524]   Luo J, Kaplitt MG, Fitzsimons HL, *et al.* Subthalamic GAD gene therapy in a Parkinson's disease rat model. Science 2002; 298: 425-9.

[525]   Lee B, Lee H, Nam YR, Oh JH, Cho YH, Chang JW. Enhanced expression of glutamate decarboxylase 65 improves symptoms of rat parkinsonian models. Gene Ther 2005; 12: 1215-22.

[526]   Emborg ME, Carbon M, Holden JE, *et al.* Subthalamic glutamic acid decarboxylase gene therapy: Changes in motor function and cortical metabolism. J Cereb Blood Flow Metab 2007; 27: 501-9.

[527]   Kaplitt MG, Feigin A, Tang C, *et al.* Safety and tolerability of gene therapy with an adeno-associated virus (AAV) borne GAD gene for Parkinson's disease: An open label, phase I trial. Lancet 2007; 369: 2097-105.

[528]   Feigin A, Kaplitt MG, Tang C, *et al.* Modulation of metabolic brain networks after subthalamic gene therapy for Parkinson's disease. Proc Natl Acad Sci USA 2007; 104: 19559-64.

[529]   LeWitt PA, Rezai AR, Leehey MA, *et al.* AAV2-GAD gene therapy for advanced Parkinson's disease: A double-blind, sham-surgery controlled, randomised trial. Lancet Neurol 2011; 10: 309-19.

[530]   Follett KA, Weaver FM, Stern M, *et al.* Pallidal *versus* subthalamic deep-brain stimulation for Parkinson's disease. N Engl J Med 2010; 362: 2077-91.

[531]   Zheng D, Jiang X, Zhao J, Duan D, Zhao H, Xu Q. Subthalamic hGAD65 Gene Therapy and Striatum TH Gene Transfer in a Parkinson's Disease Rat Model. Neural Plast 2013; doi: 10.1155/2013/263287.

[532]   Mouradian MM. MicroRNAs in Parkinson's disease. Neurobiol Dis 2012; 46: 279-84.

[533]   Eriksen JL, Wszolek Z, Petrucelli L. Molecular pathogenesis of Parkinson disease. Arch Neurol 2005; 62: 353-7.

[534]   Spillantini MG, Schmidt ML, Lee VM, Trojanowski JQ, Jakes R, Goedert M. Alpha-synuclein in Lewy bodies. Nature 1997; 388; 839-40.

[535]   Mouradian MM. Recent advances in the genetics and pathogenesis of Parkinson disease. Neurology 2002; 58: 179-85.

[536]   Eriksen JL, Dawson TM, Dickson DW, Petrucelli L. Caught in the act: Alpha-synuclein is the culprit in Parkinson's disease. Neuron 2003; 40: 453-6.

[537]   Lim KL, Dawson VL, Dawson TM. The cast of molecular characters in Parkinson's disease: Felons, conspirators and suspects. Ann NY Acad Sci 2003; 991: 80-92.

[538]   Singleton AB, Farrer M, Johnson J, *et al.* Alpha-synuclein locus triplication causes Parkinson's disease. Science 2003; 302: 841.

[539]   Farrer M, Kachergus J, Forno L, *et al.* Comparison of kindreds with parkinsonism and alpha-synuclein genomic multiplications. Ann Neurol 2004; 55: 174-9.

[540]   Nishioka K, Hayashi S, Farrer MJ, *et al.* Clinical heterogeneity of a-synuclein gene duplication in Parkinson's disease. Ann Neurol 2006; 59: 298-309.

[541]   Maraganore DM, de Andrade M, Elbaz A, *et al.* Collaborative analysis of a-synuclein gene promoter variability and Parkinson disease. JAMA 2006; 296: 661-70.

[542]   Chiba-Falek O, Lopez GJ, Nussbaum RL. Levels of alpha-synuclein mRNA in sporadic Parkinson disease patients. Mov Disord 2006; 21: 1703-8.

[543]   Zhou W, Hurlbert MS, Schaack J, Prasad KN, Freed CR. Overexpression of human alpha-synuclein causes dopamine neuron death in rat primary culture and immortalized mesencephalon-derived cells. Brain Res 2000; 866: 33-43.

[544]   Junn E, Mouradian MM. Human alpha-synuclein over-expression increases intracellular reactive oxygen species levels and susceptibility to dopamine. Neurosci Lett 2002; 320: 146-50.

[545]   Zhou W, Schaack J, Zawada WM, Freed CR. Overexpression of human alpha-synuclein causes dopamine neuron death in primary human mesencephalic culture. Brain Res 2002; 926: 42-50.

[546]   Orth M, Tabrizi SJ, Schapira AH, Cooper JM. Alpha-synuclein expression in HEK293 cells enhances the mitochondrial sensitivity to rotenone. Neurosci Lett 2003; 351: 29–32.

[547]   Moussa CE, Wersinger C, Tomita Y, Sidhu A. Differential cytotoxicity of human wild type and mutant alpha-synuclein in human neuroblastoma SH-SY5Y cells in the presence of dopamine. Biochemistry 2004; 43: 5539-50.

[548]   Kirik D, Rosenblad C, Burger C, *et al.* Parkinson-like neurodegeneration induced by targeted overexpression of alpha-synuclein in the nigrostriatal system. J Neurosci 2002; 22: 2780-91.

[549]   Yamada M, Iwatsubo T, Mizuno Y, Mochizuki H. Overexpression of alpha-synuclein in rat substantia nigra results in loss of dopaminergic neurons, phosphorylation of alpha-synuclein and activation of

caspase-9: resemblance to pathogenetic changes in Parkinson's disease. J Neurochem 2004; 91: 451-61.

[550]   Sapru MK, Yates JW, Hogan S, Jiang L, Halter J, Bohn MC. Silencing of human alpha-synuclein *in vitro* and in rat brain using lentiviral-mediated RNAi. Exp Neurol 2006; 198: 382-90.

[551]   Junn E, Lee KW, Jeong BS, Chan TW, Im JY, Mouradian MM. Repression of a-synuclein expression and toxicity by microRNA-7. Proc Natl Acad Sci USA 2009; 106: 13052-7.

[552]   Fountaine TM, Wade-Martins R. RNA interference-mediated knockdown of alpha-synuclein protects human dopaminergic neuroblastoma cells from MPP(?) toxicity and reduces dopamine transport. J Neurosci Res 2007; 85: 351-63.

[553]   McCormack AL, Mak SK, Henderson JM, Bumcrot D, Farrer MJ, Di Monte DA. a-synuclein suppression by targeted small interfering RNA in the primate substantia nigra. PloS One 2010; 5: e12122.

[554]   Hayashita-Kinoh H, Yamada M, Yokota T, Mizuno Y, Mochizuki H. Down-regulation of a-synuclein expression can rescue dopaminergic cells from cell death in the substantia nigra of Parkinson's disease rat model. Biochem Biophys Res Commun 2006. 341: 1088-95.

[555]   Abeliovich A, Schmitz Y, Farinas I, *et al.* Mice lacking alpha-synuclein display functional deficits in the nigrostriatal dopamine system. Neuron 2000; 25: 239-52.

[556]   Gorbatyuk OS, Li S, Nash K, *et al. In vivo* RNAi-mediated a-synuclein silencing induces nigrostriatal degeneration. Mol Ther 2010; 18: 1450-7.

[557]   Zhou C, Emadi S, Sierks MR, Messer A. A human single-chain Fv intrabody blocks aberrant cellular effects of overexpressed alpha-synuclein. Mol Ther 2004; 10: 1023-31.

[558]   Emadi S, Barkhordarian H, Wang MS, Schulz P, Sierks MR. Isolation of a human single chain antibody fragment against oligomeric α-synuclein that inhibits aggregation and prevents α- synuclein induced toxicity. J Mol Biol 2007; 368: 1132-44.

[559]   Danzer KM, Haasen D, Karow AR, *et al.* Different species of alpha-synuclein oligomers induce calcium influx and seeding. J Neurosci 2007; 27: 9220-32.

[560]   Lynch SM, Zhou C, Messer A. An scFv intrabody against the nonamyloid component of α-synuclein reduces intracellular aggregation and toxicity. J Mol Biol 2008; 377: 136-47.

[561]   Giasson BI, Murray IV, Trojanowski JQ, Lee VM. A hydrophobic stretch of 12 amino acid residues in the middle of alpha-synuclein is essential for filament assembly. J Biol Chem 2001; 276: 2380-6.

[562]   Periquet M, Fulga T, Myllykangas L, Schlossmacher MG, Feany MB. Aggregated alphasynuclein mediates dopaminergic neurotoxicity *in vivo.* J Neurosci 2007; 27: 3338-46.

[563]   Joshi SN, Butler DC, Messer A. Fusion to a highly charged proteasomal retargeting sequence increases soluble cytoplasmic expression and efficacy of diverse anti-synuclein intrabodies. MAbs 2012; 4: 686-93.

[564]   Shimura H, Hattori N, Kubo Si, *et al.* Familial Parkinson disease gene product, parkin, is a ubiquitin-protein ligase. Nat Genet 2000; 25: 302-5.

[565]   Chung KK, Thomas B, Li X, *et al.* S-nitrosylation of parkin regulates ubiquitination and compromises parkin's protective function. Science 2004; 304: 1328-31.

[566]   Yao D, Gu Z, Nakamura T, *et al.* Nitrosative stress linked to sporadic Parkinson's disease: *S*-nitrosylation of parkin regulates its E3 ubiquitin ligase activity. Proc Natl Acad Sci USA 2004; 101: 10810-4.

[567]   LaVoie MJ, Ostaszewski BL, Weihofen A, Schlossmacher MG, Selkoe DJ. Dopamine covalently modifies and functionally inactivates parkin. Nat Med 2005; 11: 1214-21.

[568]   Hyun DH, Lee M, Hattori N, *et al.* Effect of wild-type or mutant Parkin on oxidative damage, nitric oxide, antioxidant defenses and the proteasome. J Biol Chem 2002; 277: 28572-7.

[569]   Narendra D, Tanaka A, Suen DF, Youle RJ. Parkin is recruited selectively to impaired mitochondria and promotes their autophagy. J Cell Biol 2008. 183: 795-803.

[570]   Yang Y, Nishimura I, Imai Y, Takahashi R, Lu B. Parkin suppresses dopaminergic neuron-selective neurotoxicity induced by Pael-R in Drosophila. Neuron 2003; 37: 911-24.

[571]   Park J, Lee SB, Lee S, *et al.* Mitochondrial dysfunction in *Drosophila PINK1* mutants is complemented by parkin. Nature 2006; 441:1157-61.

[572]   Clark IE, Dodson MW, Jiang C, *et al. Drosophila PINK1* is required for mitochondrial function and interacts genetically with parkin. Nature 2006; 441: 1162-6.

[573]   Yang Y, Gehrke S, Imai Y, *et al*. Mitochondrial pathology and muscle and dopaminergic neuron degeneration caused by inactivation of *Drosophila Pink1* is rescued by parkin. Proc Natl Acad Sci USA 2006; 103: 10793-8.

[574]   Lo Bianco C, Schneider BL, Bauer M, *et al*. Lentiviral vector delivery of parkin prevents dopaminergic degeneration in an a-synuclein rat model of Parkinson's disease. Proc Natl Acad Sci USA 2004; 101: 17510-5.

[575]   Yamada M, Mizuno Y, Mochizuki H. Parkin gene therapy for a-synucleinopathy: A rat model of Parkinson's disease. Hum Gene Ther 2005; 16: 262-70.

[576]   Kirik D, Bjorklund A. Parkinson's disease: Viral vector delivery of parkin generates model results in rats. Gene Ther 2005; 12: 727-9.

[577]   Vercammen L, Van der Perren A, Vaudano E, *et al*. Parkin protects against neurotoxicity in the 6-hydroxydopamine rat model for Parkinson's disease. Mol Ther 2006; 14: 716-23.

[578]   Manfredsson FP, Burger C, Sullivan LF, Muzyczka N, Lewin AS, Mandel RJ. rAAV-mediated nigral human parkin over-expression partially ameliorates motor deficits *via* enhanced dopamine neurotransmission in a rat model of Parkinson's disease. Exp Neurol 2007; 207: 289-301.

[579]   Paterna JC, Leng A, Weber E, Feldon J, Büeler H. DJ-1 and Parkin modulate dopamine-dependent behavior and inhibit MPTP-induced nigral dopamine neuron loss in mice. Mol Ther 2007; 15: 698-704.

[580]   Yasuda T, Hayakawa H, Nihira T, *et al*. Parkin-mediated protection of dopaminergic neurons in a chronic MPTP-minipump mouse model of Parkinson disease. J Neuropathol Exp Neurol 2011; 70: 686-97.

[581]   Yasuda T, Miyachi S, Kitagawa R, *et al*. Neuronal specificity of a-synuclein toxicity and effect of Parkin co-expression in primates. Neuroscience 2007; 144: 743-53.

[582]   Mochizuki H. Parkin gene therapy. Parkinsonism Relat Disord 2009; 15: S43-S45.

[583]   Ulusoy A, Kirik D. Can overexpression of parkin provide a novel strategy for neuroprotection in Parkinson's disease? Exp Neurol 2008; 212: 258-60.

[584]   Ridet JL, Bensadoun JC, Déglon N, Aebischer P, Zurn AD. Lentivirus-mediated expression of glutathione peroxidase: neuroprotection in murine models of Parkinson's disease. Neurobiol Dis 2006; 21: 29 -34.

[585]   Hirsch EC. Mechanism and consequences of nerve cell death in Parkinson's disease. J Neural Transm 1999; Suppl. 56: 127-37.

[586]   Sulzer D. alpha-Synuclein and cytosolic dopamine: stabilizing a bad situation. Nat Med 2001; 7: 1280-2.

[587]   Lotharius J, Brundin P. Impaired dopamine storage resulting from alpha-synuclein mutations may contribute to the pathogenesis of Parkinson's disease. Hum Mol Genet 2002; 11: 2395-407.

[588]   Lotharius J, Falsig J, Van Beek J. *et al*. Progressive degeneration of human mesencephalic neuron-derived cells triggered by dopamine-dependent oxidative stress is dependent on the mixed-lineage kinase pathway. J Neurosci 2005; 25: 6329-42.

[589]   Mosharov EV, Staal RG, Bove J, *et al*. Alpha-synuclein overexpression increases cytosolic catecholamine concentration. J Neurosci 2006; 26: 9304-11.

[590]   Caudle WM, Richardson JR, Wang MZ, *et al*. Reduced vesicular storage of dopamine causes progressive nigrostriatal neurodegeneration. J Neurosci 2007; 27: 8138-48.

[591]   Schuldiner S. A molecular glimpse of vesicular monoamine transporters. J Neurochem 1994; 62: 2067-78.

[592]   Vergo S, Johansen JL, Leist M, Lotharius J. Vesicular monoamine transporter 2 regulates the sensitivity of rat dopaminergic neurons to disturbed cytosolic dopamine levels. Brain Res 2007; 1185: 18-32.

[593]   Abou-Sleiman PM, Muqit MM, Wood NW. Expanding insights of mitochondrial dysfunction in Parkinson's disease. Nat Rev Neurosci 2006; 7: 207-19.

[594]   Bueler H. Impaired mitochondrial dynamics and function in the pathogenesis of Parkinson's disease. Exp Neurol 2009; 218: 235-46.

[595]   Parker WD Jr, Boyson SJ, Parks JK. Abnormalities of the electron transport chain in idiopathic Parkinson's disease. Ann Neurol 1989; 26: 719-23.

[596] Betarbet R, Canet-Aviles RM, Sherer TB, *et al.* Intersecting pathways to neurodegeneration in Parkinson's disease: Effects of the pesticide rotenone on DJ-1, a-synuclein and the ubiquitin-proteasome system. Neurobiol Dis 2006; 22: 404-20.

[597] Sherer TB, Betarbet R, Testa CM, *et al.* Mechanism of toxicity in rotenone models of Parkinson's disease. J Neurosci 2003; 23: 10756-64.

[598] Barber-Singh J, Seo BB, Nakamaru-Ogiso E, Lau YS, Matsuno-Yagi A, Yagi T. Neuroprotective effect of long-termNDI1 gene expression in a chronic mouse model of Parkinson disorder. Rejuvenation Res 2009; 12: 259-67.

[599] Seo BB, Nakamaru-Ogiso E, Cruz P, Flotte TR, Yagi T, Matsuno-Yagi A. Functional expression of the single subunit NADH dehydrogenase in mitochondria *in vivo*: A potential therapy for complex I deficiencies. Hum Gene Ther 2004; 15: 887-95.

[600] Seo BB, Nakamaru-Ogiso E, Flotte TR, Matsuno-Yagi A, Yagi T. *In vivo* complementation of complex I by the yeast Ndi1 enzyme: possible application for treatment of Parkinson disease. J Biol Chem 2006; 281: 14250-5.

CHAPTER 2

Cellular Cysteine Network (CYSTEINET): Pharmacological Intervention in Brain Aging and Neurodegenerative Diseases

Marcos Arturo Martínez Banaclocha*

Pathology Service at the Lluis Alcanyis Hospital, Játiva, Valencia, Spain

Abstract: Reactive species have been regarded as by-products of cellular metabolism that cause oxidative damage contributing to aging, cancer and neurodegenerative diseases. However, accumulated evidence support the notion that reactive species mediate intracellular signals that regulate physiological functions including posttranslational protein modifications with important functional implications. Cysteine thiol groups of proteins are particularly susceptible to oxidative modifications by oxygen, nitrogen and sulfur species and they can be oxidized to several different products, including disulfide, sulfenic acid, sulfinic acid, sulfonic acid, S-nitrosothiols and S-glutathione, which have critical roles in cellular redox homeostasis. Since there are many cysteine-bearing proteins and cysteine-dependent enzymes susceptible to oxidative modifications that may contribute to cellular function and dysfunction, this chapter reviews the role of oxidative-changed proteins at cysteine residues in aging and some frequent neurodegenerative diseases. The concept of a cellular cysteine network (CYSTEINET) is advanced as a functional and structural matrix of interconnected proteins that in conjunction with reactive species and glutathione can regulate the cellular bioenergetic metabolism, the redox homeostasis, and the cellular survival. This network may represent an ancestral down-top system composed by a complex matrix of proteins with very different cellular functions, but bearing the same regulatory thiol radical. In this context and based on scientific evidences, current therapeutic and potential mechanism of action of some particular thiol bearing substances are revised.

Keywords: Acetylcysteine, aging, Alzheimer, cysteine, cysteinet, free radical, glutathione, hydrogen sulfide, mitochondria, neurodegeneration, nitric oxide, oxygen, Parkinson, reactive species, redox homeostasis, thiol.

INTRODUCTION

Normal cell metabolism generates oxygen, nitrogen and sulfur reactive species (ROS, RNS, and RSS respectively) that are finely controlled under physiological conditions. The disturbance of this cellular exquisite control results in oxidative

***Corresponding author Marcos Arturo Martínez Banaclocha:** Pathology Service at the Lluis Alcanyis Hospital, Játiva, Valencia, Spain; Tel: (+34)962289532; Fax: (+34)962289272; E-mail: martinez_marben@gva.es

stress that causes damage on all types of cellular macromolecules. If this oxidative stress is prolonged over time and antioxidant mechanisms become insufficient, the cellular damage may result in aging and a range of human diseases including cancer. At physiological conditions, reactive species can function not only as intracellular second messengers with regulatory roles in many cellular metabolic processes, but also as an ancestral biochemical network that control cellular survival, regeneration and death. An example of this class of bio-chemical matrix is reviewed and proposed in this chapter, which is named the cellular cysteine network (CYSTEINET). This cellular network is based on the finding that thiol groups of the amino acid cysteine are distributed through the majority of the proteins that have key roles in metabolic as well as structural functions in cells. In this chapter I elaborate the hypothesis that brain aging and age-related neurodegenerative diseases can be linked by oxidative modification of cysteine residues in a wide range of proteins with key cellular functions, which form a widespread cysteine-based cellular network (CYSTEINET). This network may represent an ancestral down-top system composed by a complex matrix of proteins with very different cellular functions, but bearing the same regulatory thiol radical. These proteins include kinases, proteases, antioxidant enzymes, phosphatases, and other structural proteins that use cysteine residues as cellular sensors that can entangle, at very short time scale (nanoseconds), a lot of very different metabolic, respiratory, transport and mechanical functions in the cell.

Why Cysteine?

Since the emergence of life, redox reactions have been necessary to fuel metabolism and growth in both prokaryotic and eukaryotic living organisms. The eukaryotic cells use energy-converting enzymes evolved from ancestral enzymes that worked in very low oxygen concentration or in anaerobic conditions. These ancestral proteins did not have to deal with the side reactions related with high oxygen concentrations. Then, different redox adaptations have evolved to avoid the deleterious side-reactions of oxygen. The 1-electron reduction of oxygen by reducing components of different cellular electron-transfer pathways can produce reactive oxygen species (ROS), which may have toxic effects on cellular macromolecules but also can regulate the cellular metabolism directly as well as playing a role in the cellular second messenger systems.

The control of the redox microenvironment occurs in virtue of the interaction of reactive oxygen, nitrogen and sulfur species as well as the modifications of cysteine-containing proteins and peptides. Then, cysteine is one of the major regulators of redox homeostasis in biological systems [1-3], including cysteine dependent enzymatic and non-enzymatic compounds that form an intricate system working to maintain redox homeostasis [4]. The "redox hypothesis" postulates that the disruption of the function and homeostasis of thiol systems is the key central feature of oxidative stress that contributes to aging and age-related disease [4].

Cysteine is a special amino acid because it contains a reactive sulphydryl or thiol group (SH). Free cysteine possesses low reactivity to undergo redox transitions [5] but the thiol group of cysteine is more reactive than the thioether of methionine from which cysteine is synthesized (Fig. **1**). Cysteine participates in a variety of reactions because of its ability to exist in different oxidation states, including thiol, disulfide, sulfenate, sulfinate, sulfonate and the thiyl radical (Fig. **2**). This allows cysteine to have a diversity of roles such as structural, contractile, metal-binding and catalytic activities.

It has been shown that cysteine occurrence in proteins appears to correlate positively with the complexity of the organism, ranging between 2.26% in mammals to 0.5% in some members of the Archeabacteria order. Likewise, the comparison of cysteine residues in ribosomal proteins suggests that evolution takes advantage by increasing the use of this amino acid in proteins. In metal-binding proteins and oxidoreductases studied from the majority of organisms, there are two cysteine residues separated by two amino acids. This finding suggests that cysteine appeared in ancient metal-binding proteins first and it was introduced into other proteins later during the evolution [6].

Cysteine is derived from the essential amino acid methionine and thus, it is not considered as essential amino acid. [7]. Cysteine and methionine amount in the diet must be sufficiently high to meet the needs of protein synthesis and the production of other essential molecules that include glutathione, coenzyme A, taurine and inorganic sulfur (Fig. **1**). On the other hand, cysteine levels must also be below the threshold of cytotoxicity. Elevated tissue cysteine levels may lead to autooxidation to form cystine and ROS, oxidation of protein thiol groups, neurotoxicity mediated by NMDA- type glutamate receptors, membrane cystine/glutamate antiporter activity or excessive production of hydrogen sulfide (H_2S). The toxicity of cysteine has been demonstrated in several experimental models and chronically high levels of cysteine have been associated with human diseases including Parkinson and Alzheimer's diseases [8-12]. Cysteine dioxygenase maintain the levels of cysteine catalyzing the addition of molecular oxygen to the thiol group of cysteine to generate cysteine sulfinic acid. This steep represents an irreversible loss of cysteine, which is shuttled into several pathways (Fig. **1**). The liver has the highest amount of cysteine dioxygenase protein expression and activity and regulates the excess of cysteine obtained through the diet generating cysteine sulfinate, the biosynthetic precursor of the essential metabolites sulfate, hypotaurine and taurine [13-15].

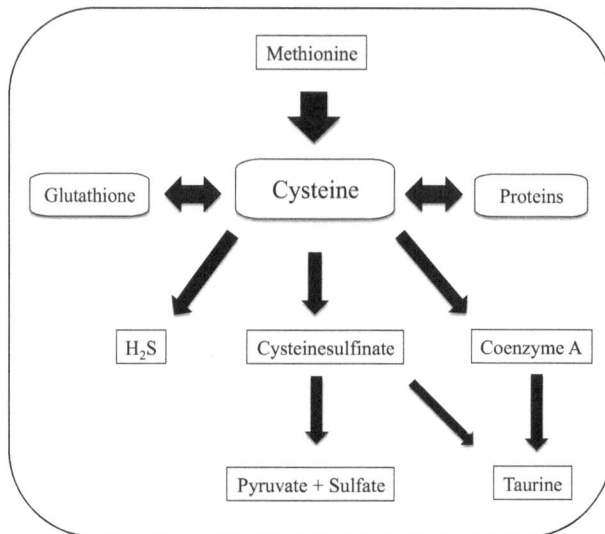

Fig. (1). Principal cysteine pathways in cellular metabolism. The liver removes dietary cysteine by converting it to glutathione, which is released into the circulation. Glutathione synthesis occurs in all cells but it is decreased when cysteine residues are needed for protein synthesis. Glutamate-cysteine ligase, also known as γ-glutamylcysteine synthase, catalyzes the first step in glutathione synthesis, playing an important role in the regulation of the flux of cysteine to glutathione. (Hydrogen sulfide: H2S).

The reactivity of a thiol depends on the accessibility and pKa of the thiol group. The pKa of a thiol is defined as the pH at which 50% of that thiol is in the deprotonated state. Thus, the pKa is important in determining the specificity of cysteine modifications by reactive species [16]. Most cysteines are not reactive to oxidants because their microenvironment makes them less nucleophilic or they are buried within the tertiary or quaternary protein structure. At physiological conditions, the typical pKa of a cysteine residue thiol is approximately 8.5, which is too high to be reactive [17]. In contrast, some redox proteins possess a reactive cysteine that is stabilized by neighbouring basic residues (lysine or arginine) [18]. The sulfur atom of these reactive cysteine residues are very versatile since can undergo redox transitions into any oxidation state between +6 and -2 [19].

On the other hand, cysteine is considered the major extracellular antioxidant found principally as disulfide cystine because the extracellular environment is relatively oxidizing [20]. This pool is integrated with the major intracellular antioxidant glutathione although the two are not in equilibrium [21]. The principal roles of cysteine in cells are:

a)   Reduction of disulfide bonds, which are implicated in the reversible formation and destruction of structural disulfides in proteins and peptides, the regulation of enzyme activities and the maintenance of the redox homeostasis. Besides, the thioredoxin system (NAPDH/thioredoxin/thioredoxin-reductase) and the glutathione system (NADPH/glutathione/glutathione-reductase) are the major thiol dependent redox pathways present in cells.

b)   Formation of disulfide bonds (Fig. **2**), which stabilizes the conformation of proteins. In eukaryotic cells, protein disulfide bond formation takes place within the lumen of the endoplasmic reticulum and is secreted to the extracellular space. The redox state inside endoplasmic reticulum is more oxidizing than that of the cytosol allowing the formation of disulfide bridges. The major redox system in the cytosol as well as in the lumen of the endoplasmic reticulum and mitochondria is the glutathione system.

c)   Protein S-glutathionylation, which affects enzyme activities and has important roles in different metabolic pathways (Fig. **2**).

d)   Antioxidant cellular protection: redox homeostasis is maintained and also maintains a wide range of antioxidant systems that work together to reach an adequate balance for respiration, ATP synthesis and cell signaling.

e)   Redox signaling mediated by hydrogen peroxide and other reactive species like nitric oxide and hydrogen sulfide, which have mild oxidative actions and work as intracellular messengers (Fig. **2**).

Cysteine Transport in Mammalian Cells

Cystine-Glutamate Antiporter

There are various routes for the transport of cystine and cysteine into the cell [22-24]. The cystine-glutamate antiporter system mediates the exchange of one molecule of extracellular cystine for one molecule of intracellular glutamate. Since the levels of cysteine in plasma and extracellular fluids are low and the

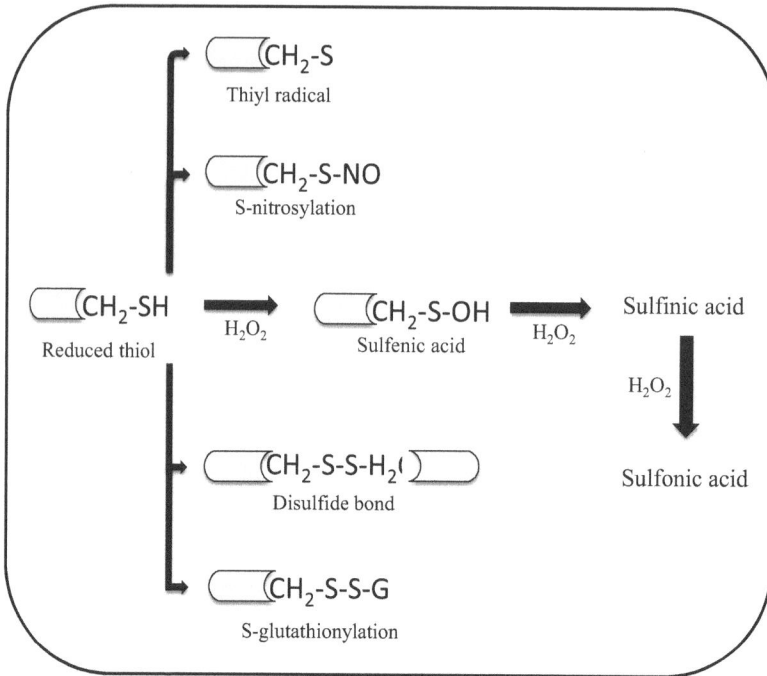

Fig. (2). Principal oxidative modification of cysteine residues in cells. Oxidation may covalently modify specific reactive cysteine residues within many proteins. In fact, hydrogen peroxide generates sulfenic, sulfinic, sulfonic acids as well as disulfide bonds. Nitric oxide oxidizes susceptible cysteine resulting in S-nitrosylation. S-glutathionylation is caused by reaction with oxidized glutathione. Free radicals that interact with cysteine can generate thiyl radical. With the exception of sulfonic acid all cysteine modifications can be reversible. At physiological conditions, there is an overlap and interrelationship among the cellular antioxidant systems that allows them to reverse different forms of cysteine oxidative modifications.

capacity to synthesize L-cysteine from methionine through trans-sulfuration is limited, the transport of L-cystine represents the principal source for cellular intake of sulfur-containing amino acids [24]. At physiological pH, cystine and glutamate were transported in anionic forms [25]. Besides, cystine and cysteine can be transported by the excitatory amino acid transporters [26-30]. Cystine-glutamate antiporter system plays a significant role in the brain (Fig. **3**), where the demand for glutathione is high [31, 32]. These types of transporters are localized in such a manner that cystine is transported into the astrocytes and cysteine is preferentially transported into neurons [33-36]. This "cystine/cysteine cycle" begins with the cystine-glutamate antiporter system that mediates the uptake of cystine into astrocytes, a process that is driven by the high intracellular glutamate concentration present into these cells (Fig. **3**). In astrocytes, cystine is rapidly reduced to cysteine and serves as a rate-limiting precursor in the synthesis of glutathione [37]. The efflux of glutathione and cysteine from astrocytes provides an extracellular source of cysteine that is transported into neurons to support glutathione generation in these cells [38-40]. Moreover, cysteine is capable of contributing to oxidative protection compensating low levels of extracellular glutathione [40]. Then, changes in glutathione concentrations may modulate the activity of cystine-glutamate antiporter system, which is fundamental for redox homeostasis in neurons.

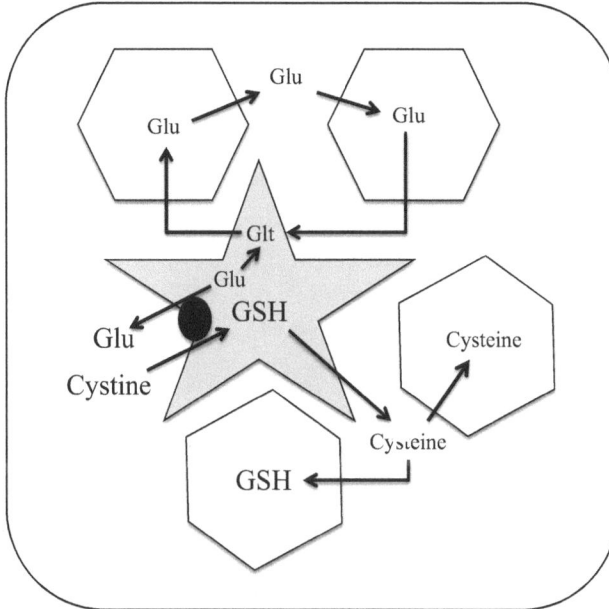

Fig. (3). Cystine-glutamate antiporter and glutathione systems in the brain. Hexagons represent synaptic terminals. Central star represent astrocytes. Cystine is transported into astrocytes and is reduced to cysteine that can be used for glutathione synthesis or can be released to the extracellular space. Glutathione can be released by astrocytes *via* multi-drug resistance-associated protein 1 transporters, where it is degraded by γ-glutamyl-transpeptidase and aminopeptidase N, providing a source of cysteine that is transported into neurons by the neuronal glutamate transporter EAAT3 to support glutathione synthesis in neurons. In addition, cysteine may also oxidize in the oxygen-rich extracellular space, thereby supplying cystine to support back cystine-glutamate antiporter system. Glu: glutamic acid; Glt: glutamine; GSH: glutathione.

Glutathione can modulate the activity of many proteins that have reactive cysteine residues increasing or decreasing the protein activity [41]. Examples of this regulatory effect include, among others, the excitatory amino acid transporters and NMDA receptors within the excitatory glutamate system [42, 43]; GABA and glycine receptors within inhibitory systems [44-46]; transient receptor potential channels [47]; acid-sensing ion channels and potassium channels [48, 49]. Cystine-glutamate antiporter system may contribute to the antioxidant capacity of cells maintaining glutathione levels and glutamate homeostasis, and counteracting the potential toxic effects of glutamate release [50]. Thus, up-regulation of cystine-glutamate antiporter system can be a protective mechanism against oxidative stress although, under certain circumstances, it can contribute to excitotoxicity.

CYSTEINE REACTIVITY IN PROTEINS

Posttranslational modification of proteins is a much-regulated process that increases the complexity and variability of thousand of proteins that are differentially expressed in specific tissues. These modifications of proteins can include principally phosphorylation, acetylation, allosteric and redox regulation. Amino acids that can be reversibly modified include methionine, tryptophan, histidine, tyrosine and cysteine, but only the later has a thiol group that is deprotonated at physiological pH, which enhance its reactivity. Among the redox posttranslational modifications of proteins, the regulations of thiol groups of cysteine residues have many functional implications. These include reversible modifications of the catalytic sites in enzymes including those of the mitochondrial respiratory chain, structural and contractile proteins that support the cytoskeleton and motion of cells, the folding and transport of proteins through membranes, the modulation of intracellular and extracellular signals, the regulation of DNA synthesis and expression and the storage and consumption of metabolic fuels. The redox modification of cysteine thiols results in a range of sulfur-containing products that can be reversible or irreversible (Fig. **2**). The reversibility of these oxidative modifications of cysteine thiols allows their ability to function as switches in many regulatory pathways as well as at many structural sites into proteins. The sensitivity of thiol groups of cysteine residues is highly variable depending on the microenvironmental conditions. The pKa for low-molecular weight thiols as cysteine and glutathione are 8.3 and 8.8, respectively. Low pKa protein thiols, particularly those ionized at physiological pH, are considered "reactive cysteines" [51]. As previously mentioned, micro-environmental conditions that facilitate thiol ionization in cysteine residues of proteins are the proximity to basic amino acids, the hydrogen bonding, and the location of cysteine at the N-terminal end of a α-helix.

Moreover, there are oxidants, such as hydrogen peroxide that react exclusively with the thiolate anion [52-57]. Indeed, it has been demonstrated that reaction rate constants of hydrogen peroxide with two reactive cysteine bearing proteins with very similar pKa like peroxiredoxin 2 (pKa $\approx$ 5–6) and protein tyrosine phosphatase1B (pKa $\approx$ 5.4) are so different as 2×10^7 M^{-1} s^{-1} and 20 M^{-1} s^{-1} respectively, due to microenvironmental conditions [55]. This means that there is an oxidant-dependent sensitivity of cysteine that is associated with the place and other microenvironmental conditions in each particular protein. Then, reactive cysteines (low pKa protein thiols) and oxidant-sensitive cysteines are not synonymous from a redox point of view [58].

CELLULAR GENERATION OF REACTIVE SPECIES AND CYSTEINE MODIFICATIONS

Reactive species form part of the redox system contributing to cell metabolism, morphology and signaling pathways [59]. The difference between physiological function and pathological damage of reactive species is fundamentally due to the disturbance of the exquisite cellular redox balance. In general, reactive species that have low reactivity are more likely to participate in cell signaling than in cellular damage [60, 61]. This concept is applicable to the reactivity of reactive oxygen, nitrogen and sulfur species, which have been reviewed elsewhere [62, 63].

Reactive Oxygen Species (ROS)

There are numerous sites of ROS generation within the cell that can be divided in mitochondrial and non-mitochondrial sources.

Non-Mitochondrial Generation of ROS

ROS can be generated during the catalytic action of multiple enzymes, such as peroxidases, xanthine oxidase, cyclooxygenases and aldehyde oxidases [64]. Cytochrome P450 reductase can also generate peroxide radical during the metabolism of hormones and drugs in the endoplasmic reticulum [65].

Hydrogen peroxide is produced by many enzymes including xanthine, monoamine and D-amino acid oxidases as well as by the peroxisomal pathway for β-oxidation of fatty acids [66]. Multiple extracellular signals that include growth factors, cytokines, G-protein-coupled receptor agonists and mechanical distortion of cells can generate ROS through the activation of NADPH oxidase complex (NOX complex) [67]. This family of membrane-bound enzymes is widely expressed and evolutionarily conserved [68, 69]. The function of these NADPH-dependent oxidases is the regulation of ROS generation, required for the coordination of many metabolic pathways and the maintenance of the stem cell populations [56, 67]. ROS production by NOX complex is mediated by one of seven enzymatic systems (NOX1-5, Duox1, and Duox2) that have differential cellular and tissue specific expressions.

Moreover, NOX complex activation requires the association with FAD cofactor, distinct membrane and cytoplasmic co-activator proteins and the binding of calcium to some intracellular domains allowing the pathway and isoforms specificities [70, 71]. Activation of NOX complex results the transport of electrons, from cytoplasmic NADPH through FAD and heme cofactors, across plasma and intracellular membranes to produce superoxide anion radical on the extracellular surface, which is dismutated to hydrogen peroxide and molecular oxygen *via* extracellular superoxide dismutase [58, 72]. The translocation of electrons from the cytoplasm across biological membranes with the concomitant release of protons from NADPH result in local acidification and in a positive charge accumulation on the ROS-producing face that may promote electron transfer through NOX, which can regulate the amplitude and duration of NOX signals themselves.

Many cell types express multiple NOX isoforms with specific roles in a given signaling pathway [73]. NOX isoforms are expressed and localized into distinct subcellular compartments contributing to restrict hydrogen peroxide specificity as a cellular signal mediator [74].

Mitochondrial Generation of ROS

ROS are generated principally as by-products of mitochondrial respiration and metabolism. Nutrients (glucose, fatty acids, amino acids) are converted into metabolic intermediates that are metabolized and decarboxylated by eight different enzymes in the tricarboxylic acid cycle (TCAc) into the mitochondrial matrix. In the mitochondrial inner membrane there is an enzymatic machinery composed by the electron transport chain that transport electrons from reduced metabolic compounds (NADH and FADH2) through four protein complexes (I-IV) in which the molecular oxygen serves as the terminal electron acceptor resulting in the generation of water (Fig. **4**).

The energy released during the electron transfer in the mitochondrial respiratory chain is used to establish a proton gradient across the mitochondrial inner membrane. This proton gradient drives the production of ATP by the mitochondrial complex V (ATP synthase). In this process some electrons (1-2%) escape from mitochondrial complexes I and III, which result in the production of superoxide anion radical (O_2^-), either in the matrix or in the inter-membrane space of the mitochondria (Fig. **4**). Superoxide radical is quickly dismutated to hydrogen peroxide that is a mild oxidant and has the longest cellular half-life (1ms) allowing its diffusion through membranes [62, 75, 76]. There are other mitochondrial sources of ROS but it seems that hydrogen peroxide emitted from mitochondria modulates many intracellular and intercellular pathways through the reversible oxidation of protein cysteine thiols [77, 78]. The control of mitochondrial superoxide anion radical and hydrogen peroxide is carried out through the regulation of the flux of the metabolic pathways that regulate the flow of electrons into the electron transport chain [79, 80]. Cellular levels and half-lives for these ROS can vary considerably depending on the cell type, nutritional and environmental conditions [76].

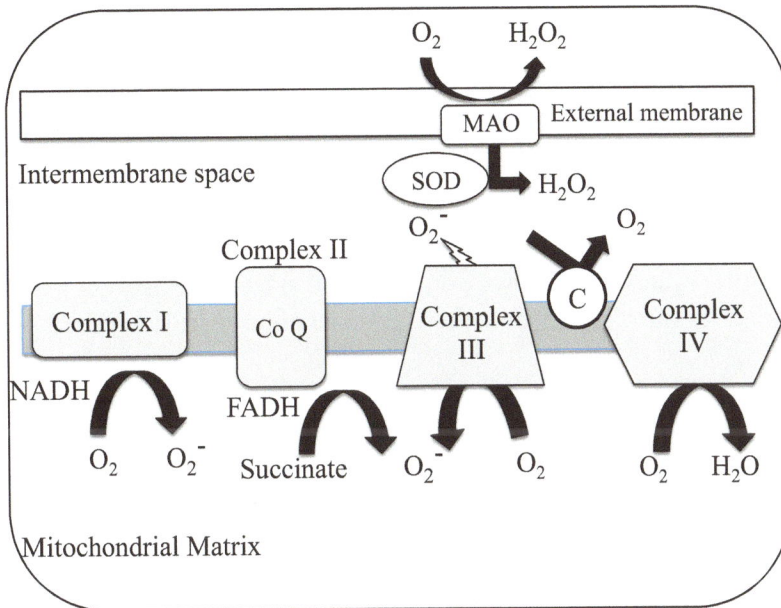

Fig. (4). Schematic representation of superoxide anion generation in the mitochondrial respiratory chain. Various respiratory complexes leak electrons to oxygen producing superoxide anion radical (O_2^-). Superoxide radical may reduce cytochrome c, in the intermembrane space, or may be dismutated to hydrogen peroxide (H_2O_2) and oxygen in the matrix and the intermembrane space. Increased steady state concentrations of superoxide radical may reduce transition metals, which in turn react with hydrogen peroxide to produce hydroxyl radicals. Besides, superoxide radical can react with nitric oxide to form peroxynitrite. Both, hydroxyl radical and peroxynitrite are strong oxidants that indiscriminately react with proteins, lipids and nucleic acids. (MAO: monoamine oxidase; SOD: superoxide dismutase; C: Cytochrome c).

Accumulation of hydrogen peroxide and superoxide anion radical into the cell can lead to the generation of hydroxyl radical that has a high reactivity with cellular macromolecules. Likewise, high concentration of hydrogen peroxide can oxidize thiol groups to sulfinic and sulfonic acids (Fig. **2**), which may inactivate proteins irreversibly [81]. Similarly, superoxide radical disassembles Fe-S clusters in various TCA cycle enzymes and in respiratory complexes and can also combine with nitric oxide to form peroxynitrite by-products [82]. Then, although mitochondria use ROS for intra and intercellular signals, ROS overproduction is associated with a number of pathologies including aging and neurodegenerative disorders. The balance of ATP and ROS production into the mitochondria has a central role in cellular homeostasis (Fig. **5**). In this process, redox modifications including S-oxidation (sulfenylation and sulfinylation), S-glutathionylation and S-nitrosylation converge on the mitochondria to regulate a number of processes ranging from the oxidative phosphorylation to morphology and cell death [83, 84]. This is accomplished because mitochondria have a microenvironment that allows redox signaling *via* thiol cysteine oxidation. As mentioned above, although the pKa of the thiol group on free cysteine is around 8, in some tyrosine phosphatases or similar enzymes, the microenvironmental conditions of the cysteine residues can modify the pKa around 4. These reactive cysteine residues may be easily oxidized to a sulfenic form that is unstable and is further oxidized to sulfinic or sulfonic species, which inactivate the enzyme. Other possibilities for post-translational cysteine modifications include nitrosylation, glutathionylation, or the formation of an intermolecular or intramolecular disulfide bond.

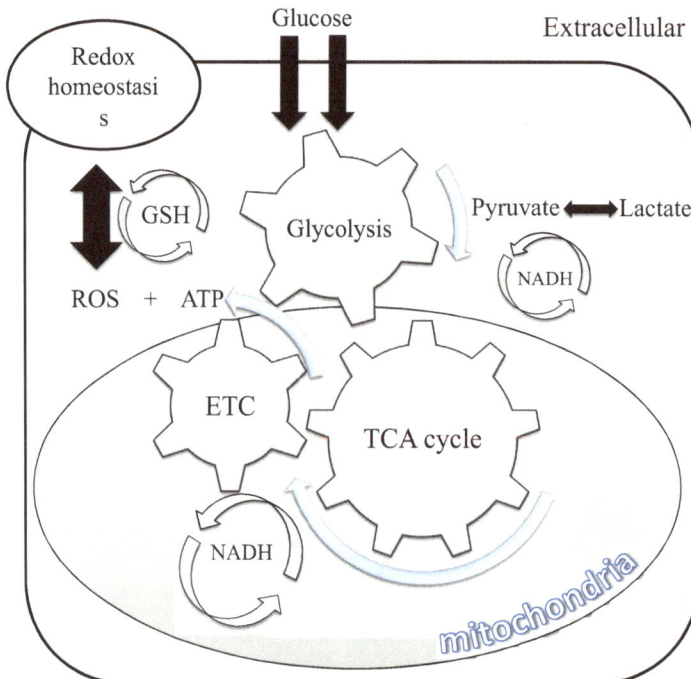

Fig. (5). Redox homeostasis and cellular metabolism interdependence. Cellular energy is principally obtained from oxidation of glucose in the mitochondrial tricarboxylic acid (TCA) cycle that is coupled to the electron transport chain (ETC) to generate ATP and ROS for maintaining cellular work and redox homeostasis that feedback regulate cellular metabolism.

Computational analysis through large-scale proteomic approaches has suggested that reactive and regulatory cysteine residues might exist in over 500 proteins, extending this form of redox regulation to a wide range of enzymatic activities [85, 86]. Likewise recently, it has been developed an algorithm that use three parameters for prediction of the thiol oxidation behaviour. These parameters are the distance to the nearest cysteine sulfur atom, the solvent accessibility and pKa. The algorithm was optimized to correctly classify cysteines in 80.1% of the oxidation-susceptible cysteine thiols. The classifier developed from these parameters, named the Cysteine Oxidation Prediction Algorithm (COPA) predicts the oxidation-susceptible sites of protein cysteines susceptible to redox-mediated regulation and it identifies possible enzyme catalytic sites with reactive cysteine thiols [87].

Modifications of Cysteine-Bearing Proteins by ROS

Cysteine residues of proteins are very sensitive to hydrogen peroxide and other ROS oxidations, constituting a widespread cellular sensor of redox homeostasis and signaling. Depending on the microenvironmental conditions, the generation of sulfenic acid associated with cysteine-bearing proteins can be stabilized in virtue of several factors that include the absence of thiols proximal to the site of formation and the accessibility to low-molecular-weight thiols such as glutathione [88].

Reaction of sulfenic acid with a protein thiol or glutathione produces a disulfide bridge or a protein-S-GSH disulfide, respectively. In proteins without neighbouring cysteine residues, the nitrogen atom of an amide can react with sulfenic acid forming a cyclic sulfenamide by S-N bonds [89]. The formation of disulfide and sulfenamide bridges can protect the protein against new oxidations that can be irreversible. Disulfide and S−N bonds can be reduced through the activity of thioredoxin/thioredoxin reductase or glutaredoxin/glutaredoxin reductase systems contributing to the restoration of protein thiol groups [58]. In the presence of high hydrogen peroxide levels, sulfenic acid can be further oxidized to sulfinic acid [58]. The oxidation of cysteine thiols can also occur by one-electron redox pathways to give thiyl radicals, which undergo distinct sets of reactions (Fig. **2**). All these types of thiol cysteine modifications can regulate a diversity of cellular proteins with metabolic, structural and receptor-signaling functions.

<u>*Cytosolic Enzymes Regulated by Thiol Cysteine*</u>

Changes in redox homeostasis can have both specific and widespread effects, which can or not be simultaneous. It is possible that when systemic changes in the cellular redox microenvironment occur, cysteine modification would result in modifications of many thiol-containing proteins in the cytosol. The sensitivity and specificity of the amino acid cysteine in multiple proteins is dependent on the fact that oxidative molecules can modify only one or two cysteine residues within the 3-D structure of the protein. Likewise, not all oxidative changes in protein cysteines will contribute to any change in its activity or cellular function. However, it is possible that a domino effect occurs in the cytoplasm, through oxidative modifications of thiols, which can entangle a wide range of different metabolic pathways. This domino effect may have metabolic and homeostatic regulatory functions because of the existence of the proposed cysteinet.

Protein tyrosine kinases (PTK) are a family of enzymes that contain a conserved active-site cysteine (His-Cys-X-X-Gly-X-X-Arg-Ser/Thr), which because of its microenvironment has a low pKa at physiological pH. Protein tyrosine phosphatase 1b is a classical example of such PTK redox regulation through the oxidation of cysteine residues in the catalytic centre. Protein tyrosine phosphatase 1b is reversibly oxidized by hydrogen peroxide to form sulfenic acid as well as by S-glutathionylation [90, 91]. The net result of the oxidative change is an increase in tyrosine phosphorylation at physiological conditions during insulin signaling.

Creatine kinase is another enzyme regulated by cysteine oxidative modification with important implications in the energy transfer because it catalyzes the transfer of high-energy phosphate from ATP to creatine. There are mitochondrial and cytosolic isoforms of this protein that contains up to four cysteines, but only one appears to be necessary and sufficient for full creatine kinase activity [92]. Creatine kinase can also be modified by reactive nitrogen species, disulfide formation and S-glutathionylation [92].

Calpains are cysteine-dependent proteases that are involved in several physiological pathways [93]. They have a catalytic triad, related to papain, consisting of cysteine, histidine and asparagine residues. Calpains exist in multiple tissue-specific isoforms, but they are primarily found in different muscle types. Similar to other thiol-dependent enzymes, the active-site cysteine generates an anionic sulfur group during catalysis that is susceptible to oxidative modification. Various members of this family of enzymes can be inhibited by a variety of oxidants, such as hydrogen peroxide and nitric oxide, resulting in reversible enzyme inactivation [92].

Mitochondrial Enzymes Regulated by Thiol Cysteine

Mitochondrial enzymes are regulated by cysteine oxidation modulating its structure, activity and metabolism. Enzymes in the TCA cycle can be either sulfenylated or S-glutathionylated. Aconitase and pyruvate dehydrogenase can be reversibly inactivated by hydrogen peroxide through thiol group oxidation [76]. Succinate dehydrogenase can also be modified by S-glutathionylation [76]. L-carnitine/acyl-carnitine carrier has also been found to be S-glutathionylated on Cys136 and Cys155, respectively [76]. The oxidative modification of these enzymes by S-glutathionylation is reversible and occurs under physiological conditions, but it is increased under conditions of oxidative stress when GSH/GSSG ratio into de mitochondria is low. For instance, aconitase can be reversibly deactivated by S-glutathionylation [94].

Oxidative phosphorylation is also modulated by oxidative modification of mitochondrial enzymatic complexes at cysteine residues. ATP synthase (Complex V) can be S-glutathionylated on Cys294 of the α-subunit located in the F1 hydrophilic part of the Complex. Cys294 may also react with a neighbouring Cys103 residue to form a disulfide bridge [76]. S-glutathionylation blocks nucleotide binding to the complex, which results in a decrease in the production of ATP and the bioenergetic capacity of the mitochondria. Moreover, S-glutathionylation of Complex I limits NADH production and it decreases electron flow through the respiratory chain diminishing ROS generation, which protects this enzymatic complex from irreversible oxidation by ROS. Thus, reversible S-glutathionylation allows the control of mitochondrial ROS generation and signaling. Complex I of the mitochondrial respiratory chain can also undergo sulfenylation decreasing its activity [95]. Sulfenylation may suffer further oxidation that irreversibly deactivates Complex I. However, it seems that cysteine residues that are oxidized in Complex I can be protected from further oxidation by S-glutathionylation [76]. Thus, as the result of oxidative modifications of specific cysteine residues in mitochondrial enzymes, redox signals can control the mitochondrial ROS generation and the cellular bioenergetic.

Mitochondrial Structural Proteins Regulated by Thiol Cysteine

Mitochondrial morphology is closely related with its function that is dependent on mitochondrial biogenesis [95]. The proteins involved in mitochondrial fusion are the integral membrane GTPases Mitofusin 1, Mitofusin 2 and autosomal dominant Optical Atrophy 1 (Opa1) [96]. Moreover, the dynamin family of GTPases Drp-1 mediates mitochondrial fission. S-glutathionylation of mitofusin proteins is required to induce mitochondrial hyperfusion, which is regulated by the redox micro-environmental signals [97-99]. Through these mechanisms mitochondrial morphology and fusion can prevent oxidative damage and the induction of intrinsic apoptotic cascades *via* preservation of mitochondrial redox homeostasis.

Mitochondrial Permeability and Thiol Cysteine

The mitochondrial permeability transition pore (MPTP) is a non-selective pore that allows the free diffusion of molecules of 1.5 KDa in size [100]. ROS and chemical toxins that induce ROS overproduction can modify mitochondrial permeability transition, being hydrogen peroxide the ROS that most induces MPTP [101, 102]. Then, sulfenylation and S-glutathionylation have the ability to induce programmed cell death in the mitochondria through the modulation of MPTP opening [103]. Pore opening was shown to require the disulfide bond formation of specific cysteine in the structural component of the MPTP, adenine nucleotide translocator (ANT) [104]. The opening of this translocator exports the ATP from the mitochondrial matrix and imports the ADP into the matrix [105]. ANT can be S-glutathionylated maintaining a S-glutathionylated state in mitochondria from primary astrocytes and rat cortex [106]. Moreover, it is possible that other components of MPTP may be required to sense surrounding changes in redox signals and they may participate in the pore formation following cysteine modifications.

Neuronal Regeneration by Thiol Cysteine

Nucleoredoxin is an oxidoreductase enzyme localized in the nucleus and in the cytoplasm of cells that contains a pair of reactive cysteine in its catalytic centre that plays an essential role in repulsive axon guidance caused by semaphorin [107]. Stimulation of neurons with semaphorin resulted in the generation of hydrogen peroxide in the neuron growth cones and the oxidation of the collapsin response mediator protein-2 at Cys504 in the carboxy-terminal region of the protein. This produced the homo-oligomerization of the protein, which is essential for the semaphorin response [108, 109].

Regulation of Apoptosis by Thiol Cysteine

Hydrogen peroxide is a key mediator of apoptosis. B-cell lymphoma-2 (Bcl-2) is an oxidative stress protein that regulates apoptosis. A recent study has shown that hydrogen peroxide induce apoptosis in epithelial cells concomitant with cysteine oxidation and down-regulation of Bcl-2. In addition, inhibition of Bcl-2 oxidation by antioxidants or by site-directed mutagenesis at Cys158 and Cys229 inhibited these effects on Bcl-2 and apoptosis. Likewise, immunoprecipitation and confocal microscopic studies have shown that Bcl-2 interacts with mitogen-activated protein kinase to suppress apoptosis *via* the modulation of cysteine oxidation of Bcl-2. These results demonstrate a critical role of Bcl-2 cysteine oxidation in the regulation of apoptosis [110].

Regulation of Growth Factor Receptors by Thiol Cysteine

Epidermal growth factor receptor (EGFR) exemplifies the family of receptor tyrosine kinases that mediate numerous cellular processes including growth, proliferation and differentiation. In addition to ligand-dependent activation and concomitant tyrosine phosphorylation, the activation of EGFR results in the generation of hydrogen peroxide by NADPH-dependent oxidases that mediate intracellular signals through the modification of specific thiol Cys797 within the EGFR active site [111]. Other consequence of growth factor-mediated generation of intracellular hydrogen peroxide may be the control of actin microfilament function, which may result in changes in the morphology and motile activity of cells. Hydrogen peroxide may act directly on the polymerization of non-muscle ß /γ-actin. Oxidation of ß /γ-actin can cause a complete loss of polymerization that can be reversed by the thioredoxin system. Further, oxidation of the actin impedes its interaction with profilin and causes depolymerization of filamentous actin. The anti-parallel homo-dimeric structure of oxidized ß/γ-actin is connected by an intermolecular disulfide bond involving Cys374, which is the cysteine residue most reactive with hydrogen peroxide, highlighting the specificity of this oxidation in cellular functions [112].

Regulation of Transcription by Thiol Cysteine

It has been shown that disulfide bridges control the formation of the homotrimer of the heat shock transcription factor 1 (HSF1) in mammals, which is translocated into the nucleus where it activates the transcription of both Hsp70 and Hsp90 proteins [113]. *In vitro* studies have shown that heat-induced bonding between the Cys36 and Cys103 residues of HSF1 forms an intermolecular disulfide bond that causes HSF1 to trimerize and bond to DNA. On the other hand, disulfide intramolecular bond (in which participate the Cys153, Cys373, and Cys378 residues) formation inhibits this trimerization, thereby preventing DNA binding. Thus, HSF1 activation is regulated positively by intermolecular disulfide bond formation and negatively by intramolecular disulfide bond formation. Moreover, these two disulfide bonds have different sensitivities to redox conditions. Under oxidizing conditions the intramolecular disulfide bond is cleaved but not the intermolecular bond, improving DNA binding of HSF1. However, under reducing conditions both disulfide bonds are cleaved inhibiting HSF1 activation.

Nuclear factor (erythroid-derived 2)-like 2 (Nrf2) is a transcription factor that in humans regulates the expression of antioxidant proteins protecting against oxidative damage. Nrf2 is kept in the cytoplasm by Kelch like-ECH- associated protein 1 (Keap1) and Cullin-3, which degrade Nrf2 by ubiquitination [114]. As a principal factor in the upregulation of antioxidant enzymes, the Nrf2-Keap1 interaction responds to changes in the redox state. Oxidative stress disrupts critical cysteine residues in Keap1, disrupting the Keap1- Cullin-3 ubiquitination system. Then, Nrf2 is not ubiquitinated and translocated into the nucleus. Into the nucleus, it combines with a small Maf protein and binds to the antioxidant response element in the upstream promoter region of many antioxidant genes initiating their transcription [115].

Reactive Nitrogen Species (RNS)

Nitric oxide is the principal reactive nitrogen species in cells with an estimated concentration and half-life of 0.1-5 nM and 0.1-2 s, respectively [116]. Nitric oxide was the first gas (NO) known as a second messenger in mammals where it regulates vasodilation, proliferation, apoptosis, angiogenesis and host immune responses [117]. Proteins and other small nitrogen molecules may be donors of nitric oxide, which is metabolized by autooxidation to nitrite and nitrate. The oxidation of nitric oxide generates products that play important roles in physiological and pathological processes. Indeed, nitric oxide reacts rapidly with superoxide anion radical to generate peroxynitrite that is highly reactive with cellular macromolecules [118]. However, S-nitrosylation is a reversible posttranslational modification of proteins where a NO group is covalently linked to a cysteine thiol group to form an S-nitroso derivative. S-nitrosylation can affect protein activity mediating nitric oxide signaling [86]. Proteins also can be denitrosylated by thioredoxin and S-nitroso-glutathione reductase systems, protein disulfide isomerase (PDI) and other enzymes [58].

Synthesis of Nitric Oxide

Enzymatic generation of nitric oxide is predominantly mediated by nitric oxide synthase (NOS), which catalyze the formation of nitric oxide from NADPH, molecular oxygen, and L-arginine [119]. There are three known NOS isoforms: the inducible NOS (iNOS) expressed in a wide range of cell types and tissues, the endothelial NOS (eNOS) expressed primarily in endothelial cells where nitric oxide functions to regulate vasodilation, and the neuronal NOS (nNOS) that is expressed primarily in the brain where nitric oxide is involved in neurotransmission [120]. NOS isoforms can be classified as those that are activated in a $Ca2^{+}$-dependent (eNOS, nNOS) and independent (iNOS) manner. All NOS isoforms have a C-terminal tail that appears to regulate its activity [120]. NOS activity is regulated also by phosphorylation and protein-protein interactions [121].

Metabolism of Nitric Oxide

Nitric oxide auto-oxidizes to NO_2^- but this reaction can be catalyzed by ceruloplasmin in plasma [122]. In contrast, there is increasing evidence for the enzymatic reduction of NO_2^- to regenerate nitric oxide by xanthine oxidase or through reaction with deoxyhemoglobin in the vasculature [123]. NO_2^- reduction could also facilitate nitric oxide release at sites distant from NOS action.

Modification of Cysteine-Bearing Proteins by RNS

Nitric oxide reacts in cells with other radicals as superoxide anion radical and metals. Nitric oxide binds to soluble guanylyl cyclase in the vascular smooth muscle cells to promote vasodilation. Guanylyl cyclase activation can also stimulates mitochondrial biogenesis in brown adipose tissue and it can regulate other mitochondrial proteins including Complex IV of the electron transport chain, where nitric oxide inhibits cellular respiration and ROS overproduction under hypoxic conditions [124]. In addition to the metal (iron-sulfur clusters) associated regulation of proteins, nitric oxide can modify protein cysteine residues by S-nitrosylation, depending on the cysteine reactivity, the local microenvironmental conditions and the proximity to the oxidant source [58]. S-nitrosothiol generation involves complex chemical reactions without direct reaction of nitric oxide with other thiols. The S-nitrosothiol group is not ionisable and can undergo hydrolysis to give sulfenic acid or react with a thiol [58]. The chemical reactions of S-nitrosothiol with a protein thiol or glutathione do not always generate a mixed disulphide, but can also undergo transnitrosylation [125]. S-nitrosothiol can be reduced by glutathione to give the free thiol and S-nitrosoglutathione, which is newly reduced to regenerate GSH and release HNO by S- nitrosoglutathione reductases. Protein S-nitrosothiols can also be reduced by the thioredoxin/thioredoxin reductase system [58]. S-nitrosylation has been implicated in the regulation of proteins involved in many cellular functions including neuronal transmission [58]. S-nitrosylation is implicated in synaptic plasticity because nNOS are recruited to the membrane through its interaction with postsynaptic density-95 (PSD-95), which physically links NOS to NMDAR [126]. PSD-95 is localized to the membrane through a reversible cycling of S-palmitoylation of two N-terminal cysteine residues. Likewise, nNOS activation mediates S-nitrosylation of the same cysteine residues in PSD-95 preventing S-palmitoylation and reducing PSD-95 resulting in neuronal activation. Thus, S-nitrosylation of PSD-95 regulates the duration of NMDA signaling. Moreover, nNOS-derived nitric oxide can also regulate neural cells by gene transcription mechanisms [58]. Then, dysregulation of S-nitrosylation has been implicated in the pathogenesis of some neurodegenerative disorders.

On the other hand, cysteine residues play a key role in the regulation of the biological activity of S100 proteins. Ten of them, including S100A1 and S100B, possess a conserved cysteine residue near their C-termini. Nitric oxide mediates in a number of physiological and pathological events in which S100 proteins play important roles, such as neuronal development and synaptic transmission [127-129]. S100B has been identified as the extracellular neurite extension factor in vertebrate brains and contains an additional cysteine residue, which may be susceptible to thiolic redox modification [130]. Experiments using reducing agents in neuronal protein preparations have demonstrated to inhibit the neurite extension and neuronal survival. S-nitrosylation is a predominant modification of the conserved C-terminal cysteine in S100B that takes place in a metal binding-dependent way and it provides a mechanism for redox homeostasis and metal signal transduction interaction [131]. Then, the observed coincidence of S100 proteins and nitric oxide synthesis may be physiological in neurons, in which cysteine S-nitrosylation of S-100 proteins provides a simple, but efficient mechanism for redox signaling [132], integrating this fundamental group of proteins in the cellular "cysteinet".

Reactive Sulfur Species (RSS)

Protein and low molecular weight thiols oxidation generate a wide range of sulfur-containing products including disulfides, thiosulfinates, sulfenic acids and S-nitrosothiols, each of which are able to propagate redox modifications allowing the oxidation of other thiols in a similar way to ROS and RNS. These chemically reactive forms of cysteine can be classified as reactive sulfur species [133, 134]. In addition to reactive cysteine species in proteins, peptides and glutathione, there are inorganic sulfur-containing molecules that are classified as RSS. These inorganic species are hydrogen sulfide (H_2S) and related species HS^- and S^{2-} [133-135].

Hydrosulfide anion (HS^-) is the predominant sulfur-containing species in extracellular fluids and plasma at physiological pH, whereas into the cell, hydrogen sulfide and hydrosulfide anion are nearly equal [136-139]. Hydrosulfide anions are powerful reductants and strong nucleophiles as evidenced by their reaction with S-nitrosothiols to release nitric oxide [140]. Hydrogen sulfide can scavenge ROS and RNS modulating redox signals and homeostasis [141, 142]. However, in the presence of molecular oxygen, hydrogen sulfide generates free radicals by autooxidation [143]. Hydrogen sulfide has been implicated in a number of physiological processes like vasodilation, synaptic modulation, neuroprotection, regulation of inflammation, cell proliferation and cell survival [144-146]. In addition to carbon monoxide and nitric oxide, hydrogen sulfide forms the so-called gaseous signaling molecules or gasotransmitters whose generation and metabolism are regulated by enzymes [147-149]. Other inorganic RSS in cells are thiocyanate, thiocyanogen, trithiocyanate and hypothiocyanite [133, 150].

Synthesis of Hydrogen Sulfide

Hydrogen sulfide is synthesized in mammals from cysteine by the enzymes cystathionine β-synthase, cystathionine γ-lyase and 3-mercaptopyruvate sulfurtransferase (3-MST) (Fig. **1**). 3-MST is mainly localized in mitochondria and the other two enzymes exist in the cytosol [151-153]. Besides, hydrogen sulfide can be produced non-enzymatically from naturally occurring and therapeutic compounds [144, 154]. Recently Kimura and co-workers demonstrated that 3-MST depends on a biological dithiol-thioredoxin (Trx) or dihydrolipoic acid for the production of hydrogen sulfide from 3-mercaptopyruvate [155]. In the brain, hydrogen sulfide is produced mainly in astrocytes, which contain larger amounts of glutathione than neurons. Mitochondrial glutathione has been shown to act as a "sulfide buffer" when hydrogen sulfide starts to build up in the cell. Indeed, ethylmalonic encephalopathy responds well to treatment with high doses of N-acetylcysteine [156]. In the brain, hydrogen sulfide may function as neuromodulator and it has been implicated in the memory process [157, 158].

The hydrogen sulfide-cysteine-glutathione cycle has been well documented in the biomedical literature [159-164]. Hydrogen sulfide can increase intracellular glutathione by the enhancement of cellular glutamate uptake, increasing γ-glutamylcysteine synthase levels and cystine transporter activity. Moreover, hydrogen sulfide can mediate the reduction of cystine into cysteine in the extracellular space and the transport of cysteine into cells by the cysteine transporter. Finally, hydrogen sulfide can increase the levels of nuclear transcription factor Nrf2, which in turn upregulates glutathione synthesis and transport, and the decrease in the activity of glutathione-catabolizing enzymes [165-168]. Hydrogen sulfide-cysteine-glutathione cycle is dependent on the fact that hydrogen sulfide and L-serine act as co-substrates of cystathionine to yield L-cysteine [136]. Through this connection, a hydrogen sulfide pro-drug may function as a precursor of L-cysteine and glutathione.

Metabolism of Hydrogen Sulfide

Hydrogen sulfide accumulation is toxic and it is metabolized in the mitochondria by an enzymatic oxidation process to generate persulfide, sulfite and sulfate. The electrons from hydrogen sulfide oxidation funnel directly into the complex III of the electron transport chain through the sulfide: quinone oxidoreductase [137, 169-172].

Modification of Cysteine-Bearing Proteins by RSS

Hydrogen sulfide regulates biological processes through S-sulfhydration/persulfide modification of cysteine residues in proteins like those of many enzymes, including sulfurtransferases and cysteine desulfurases. These enzymes release sulfide for the production of sulfur-containing vitamins and cofactors, including iron-sulfur clusters. These mechanisms maintain free hydrogen sulfide at physiological concentrations [173-175]. Besides, hydrogen sulfide can regulate biological processes through neutralization of reactive electrophiles and intercepting reactive electrophiles [176, 177]. Hydrogen sulfide can modulate protein activity through S-sulfhydration [178]. S-sulfhydryls contain two electrophilic centers that can undergo reaction with a second protein thiol resulting in a disulfide bond or it can facilitate trans-sulfhydration in specific proteins depending on the pKa of the reactive cysteine and the micro-environmental conditions as previously mentioned. Moreover, hydrogen sulfide regulates the activity of the mitochondrial complex IV (cytochrome c oxidase) and some NOS isoforms by metal chelation [179-182]. The inhibition of cytochrome c oxidase activity by hydrogen sulfide decreases the cellular metabolic rate and transport through membrane channels, including the cystine /glutamate antiporter system stimulating cystine uptake and glutathione synthesis to modulate cellular redox homeostasis [141, 183, 184]. Thus, hydrogen sulfide is a fundamental mediator of the proposed "cysteinet" (Fig. **6**).

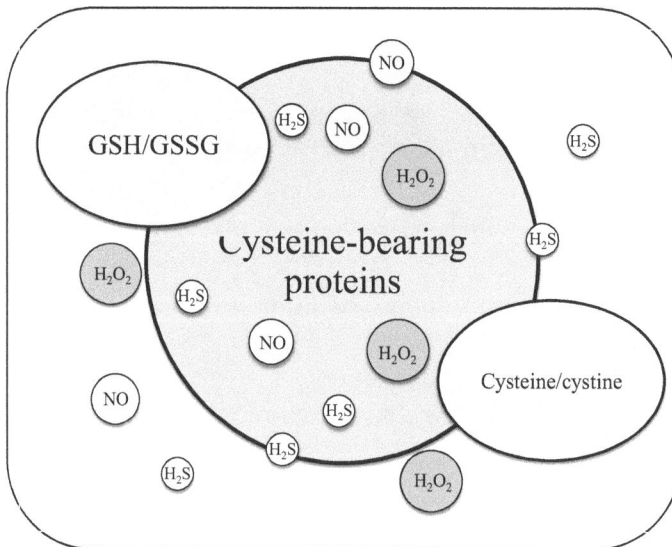

Fig. (6). Major components of the cellular cysteine network "CYSTEINET": a) cysteine-bearing proteins and peptides, including all cysteine-enzymatic systems; b) GSH/GSSG and cysteine/cystine systems; c) cellular reactive species (hydrogen peroxide: H_2O_2; nitric oxide: NO; and hydrogen sulfide: H_2S).

CELLULAR DEFENSE AGAINST REACTIVE SPECIES

The regulation of redox homeostasis is dependent on cofactors (FADH/H2, NADH and NADPH) as well as the control of reactive species generation and metabolism. To reach this exquisite control, the cell has an arsenal of defensive mechanisms that maintain reactive species within physiological non-toxic concentrations. These mechanisms include enzymatic systems as well as non-enzymatic pathways such as cysteine, glutathione and vitamins (vitamins C and E).

Antioxidant Enzymatic Systems

There are many enzymatic pathways that metabolize ROS. Superoxide dismutase (SOD) produces the dismutation of superoxide anion radical to hydrogen peroxide. There are three main forms of SOD in mammalian cells, MnSOD (SOD2) localized into the mitochondria, and two types of CuZnSOD that are found in intracellular (SOD1) and extracellular (SOD3) fluids (Fig. **7**). These enzymes are responsible for the conversion of superoxide anion radical to hydrogen peroxide preventing the formation of peroxynitrite as the result of the reaction between superoxide anion radical and nitric oxide [185]. Later, peroxiredoxin and glutathione peroxidase systems maintain hydrogen peroxide at physiological levels into the cells (Fig. **7**). These enzymatic systems metabolize hydrogen peroxide to water and molecular oxygen by the oxidation of reactive cysteine within these proteins [186, 187]. The reactive cysteines of these enzymes are reduced back by thioredoxin/thioredoxin reductase or glutathione/glutathione reductase systems using reducing equivalents from NADPH.

Catalase is another intracellular enzyme that can metabolize hydrogen peroxide to water (Fig. **7**). Catalase and peroxidases are enzymes that remove hydrogen peroxide to decrease the generation, through the Fenton reaction, of hydroxyl radicals that are highly reactive. It has been suggested that these antioxidant enzymatic systems exist in a hierarchical network to regulate the cysteine proteome [188].

Mammalian cells contain two thioredoxin systems (formed by thioredoxin/ thioredoxin reductase), one in the cytosol and the other in the mitochondria [189]. Thioredoxin is a protein that protects cells from oxidative stress by reducing intracellular oxidized proteins. Thioredoxin possesses a pair of reactive cysteines in its catalytic centre, which reduces its target proteins by transferring disulfide bonds to the cysteine pair forming a disulfide bond in it. Its reducing activity is maintained by NADPH and thioredoxin reductase [190]. Thioredoxin can also scavenge hydroxyl radicals, and also it participates in the regulation of apoptosis suppressing the signal regulating kinase-1 (Ask1) activity. However, under the effect of TNFα, thioredoxin becomes oxidized and it forms intra-molecular disulfide bonds, losing its ability to associate with Ask1. Then, free Ask1 can be activated and induce apoptosis [191-193]. Glutaredoxin system is composed by mono and dithiol glutaredoxin, glutathione, glutathione reductase and NADPH [194]. There are several glutaredoxin isoforms; glutaredoxin 1 is localized mainly in the cytosol and the intermembrane space of the mitochondria whereas glutaredoxin 2 is localized into the matrix of the mitochondria and into the nucleus of the cell [195]. Glutaredoxin 5 is localized in the mitochondria where it is required for the activity of iron-sulfur enzymes.

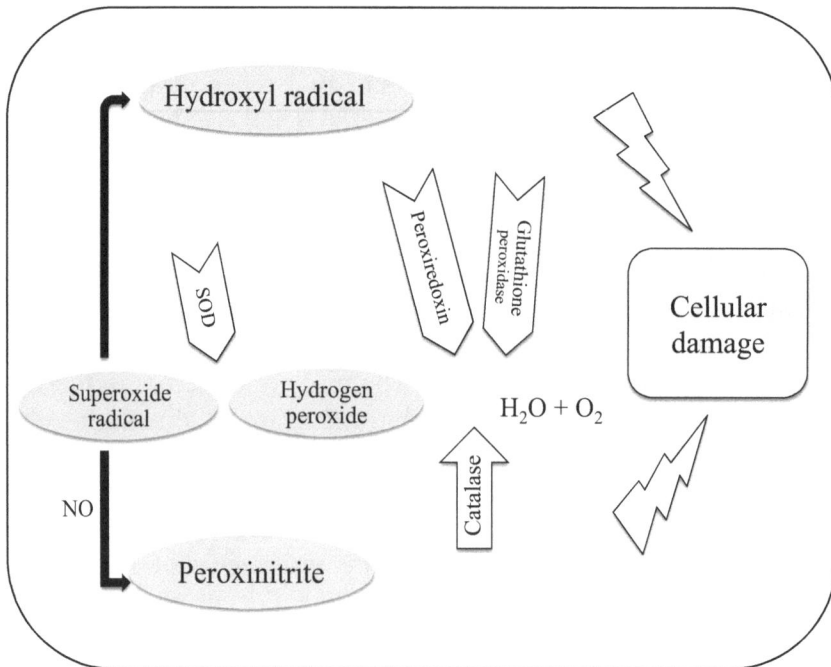

Fig. (7). Hydrogen peroxide is considered a key signal molecule in redox homeostasis. Superoxide dismutase (SOD) is responsible for the conversion of superoxide anion radical to hydrogen peroxide preventing the formation of peroxynitrite and hydroxyl radical. Peroxiredoxin and glutathione peroxidase systems maintain hydrogen peroxide at physiological levels, metabolizing hydrogen peroxide to water and molecular oxygen. Catalase is another intracellular enzyme that can metabolize hydrogen peroxide to water.

Glutathione peroxidases are seleno-cysteine-containing proteins primarily involved in the reduction of hydrogen peroxide and lipid peroxides with six isoforms found throughout the cell in the cytosol, nucleus and mitochondria [196].

Peroxiredoxins are ubiquitously expressed in mammalian cells for metabolize hydrogen peroxide, lipid hydroperoxides as well as peroxynitrite and they have the ability to protect proteins against oxidative damage in a thiol dependent manner.

In peroxiredoxins, the reactive cysteine is oxidized to sulfenic acid, which then reacts to form an intermolecular disulphide that is reduced by thioredoxin. Sulfenic acid may further be oxidized to sulfinic acid that can be reduced by sulfiredoxin [197]. Specifically, peroxiredoxin family contains a reactive cysteine thiol in the active site where sulfenic acid is generated [198-200]. In the presence of hydrogen peroxide, Cys47 gets oxidized to the sulfenic acid intermediate (Cys-SOH) and Cys170 is responsible to form a disulfide bond with Cys47 (Cys47-S-S-Cys170) in order to resolve the sulfenic acid [201]. The enzyme thioredoxin completes the catalytic cycle by reducing the disulfide bond between Cys47 and Cys170. If peroxiredoxin is hyperoxidized in the presence of high levels of hydrogen peroxide, the sulfinic acid formation inactivates its function. In many proteins, sulfinic acid formation is irreversible, but peroxiredoxin can be reduced by sulfiredoxin action [202]. Thus, peroxiredoxin plays a crucial role in redox signaling by regulating hydrogen peroxide levels, but in turn is regulated by hydrogen peroxide itself [203]. Ascorbic acid may be used by some peroxiredoxins to regenerate the reactive cysteine thiols [204].

Antioxidant Non-Enzymatic Systems

Glutathione is a hydrosoluble tripeptide (γ-glutamyl-cysteinyl-glycine) that exists in two states, an oxidized (GSSG) and a reduced (GSH) form, which is dependent on the redox transition of the cysteine thiol group playing a significant role in redox homeostasis inside the cells [4]. In addition, glutathione regenerate various antioxidant systems, it serves directly as an antioxidant, and it provides a source for the amino acid cysteine. On the other hand, cysteine bioavailability is the rate-limiting step for the synthesis of glutathione and it is transported into the cell by sodium dependent mechanisms [205]. The GSH/GSSG ratio is critical to cell survival playing a key role in the detoxification of a variety of electrophilic compounds and peroxides. Besides, glutathione participates in the glyoxalase system, reduction of ribonucleotides to deoxyribonucleotides and the regulation of protein and gene expression through redox modification of the cysteine thiol groups [206]. Glutathione peroxidase detoxifies peroxides with glutathione acting as an electron donor in the reduction reaction, producing GSSG as the end product. GSSG is newly reduced by glutathione reductase in a process that requires NADPH.

As previously mentioned, glutathione participates in the post-translational modification of many proteins by the control of the thiolic groups of reactive cysteine providing proteins its capacity to form disulfide bonds and to fold into different functional conformations. S-glutathionylation is based on either thiol-disulfide exchange through protein thiolate and glutathione disulfide or the reaction between an oxidized thiol to sulfenic acid with the reduced form of glutathione. This regulation ability named glutathionylation occurs on a large number of proteins and in may cases can alter their biological activity [207]. For instance, a relative increase in reduced glutathione induces glutathionylation of the cysteine located in the DNA binding site of c-jun and it produces a disulfide bond between cysteines proximal to the leucine zipper motif. The result is a direct and reversible control of transcriptional regulation through this stress kinase pathway [208]. Glutathione can exert negative regulation on cellular protein kinase c isoenzymes through cysteine residues placed at or near the active center of the enzyme but this function is removed when reduced glutathione levels are depleted by exposure to ROS [209]. Additionally, the function of p53 protein in the regulation of cell cycle is mediated by the modification of its cysteine residues [210]. Indeed, thiol antioxidants can induce apoptosis in transformed cell lines but not in normal cells, suggesting that direct redox control of p53 determines cell death under stress conditions [211]. Finally, glutathione and associated enzymes play important roles in cellular immunity. Indeed, glutathione concentrations in antigen-presenting cells will determine what type lymphocytes (Th1 or Th2) will predominate in the immune cellular response [209]. Since intracellular GSH/GSSG ratio is high at physiological conditions [212, 213], S-glutathionylation has been recognized as an important posttranslational modification of proteins playing major regulatory functions in cells [214, 215]. Protein thiols and glutathione can be activated by reactive species to further react with other thiols. Therefore, S-glutathionylation of proteins can transduce redox signals generated by ROS, RNS and RSS, depending on the reactivity of their cysteine thiol groups [216, 217]. As in other processes, two relevant factors in protein susceptibility for S-glutathionylation are thiol steric accessibility and thiol pKa that depend on the 3-D folding of the protein and the vicinity of basic amino acids in the side protein chain respectively. Ascorbic acid (Vitamin C) is an essential nutrient in humans that serves as a hydrosoluble antioxidant and cofactor for regeneration of some enzymatically important thiols. Vitamin C is important in the removal of superoxide, hydroxyl and peroxyl radicals. The monodehydroascorbate produced by the scavenging of these radicals is transformed to ascorbate by enzymatic activities that use NADH, NADPH or glutathione [218]. Vitamin E is a group of eight related compounds, among which α-tocopherol shows the greatest biological activity. It is a liposoluble antioxidant whose main role is the prevention of membrane lipid peroxidation. Oxidized vitamin E is regenerated by either ascorbic acid or ß-carotene.

The above-mentioned findings suggest a key role for thiol/disulfide balance in the regulation of cellular functions and metabolism through the synchronic control of multiple key proteins. ROS, RNS and RSS can modify GSH/GSSG ratio in many places of the cell influencing a variety of key metabolic proteins by oxidation and disulfide exchange reactions at specific cysteine residues. Then, it is conceivable that substances like N- acetylcysteine that can induce changes in the GSH/GSSG ratio and total glutathione (GST) levels, would have biological implications in the treatment of disorders with altered redox homeostasis, aging and neurodegenerative diseases.

AGING, NEURODEGENERATIVE DISORDERS AND CYSTEINE

Aging

The free-radical theory of aging postulates that oxidative damage to macromolecules produced by ROS, RNS and RSS, plays an important role in senescence [219]. This theory is supported by the large amount of experimental data that demonstrate a direct correlation between the resistance to oxidative damage and the lifespan in different organisms [220]. In general, the over-expression of several antioxidant enzymes makes animals more resistant to oxidative damage, but some animal models failed to increase their lifespan [221]. Besides, deterioration of the cellular antioxidant mechanisms has been implicated in the aging process. The protection of cysteine thiols of proteins may be central in aging and age-related diseases because of the key regulatory role of glutathione in cellular antioxidant systems and redox homeostasis [222, 223]. Thus, aging increases the rate of irreversible modifications of proteins, which result in the irreversible damage of key structural and metabolic proteins interfering normal cellular functions and homeostasis. From a cellular point of view, aging may be defined as the progressive deterioration of the cellular bioenergetic capacity associated with the inability to regulate metabolic homeostasis including redox balance and reparative mechanisms. Furthermore, mitochondrial genes that encode proteins involved in the oxidative phosphorylation and other key proteins in the mitochondrial function may also contribute to aging and the pathogenesis of multiple neurodegenerative disorders [224]. It has been postulated that mitochondrial DNA mutations accumulated in brain cells during the aging process is a fundamental factor in senescence [225].

On the other hand, clinical and epidemiological studies support the beneficial effects of dietary restriction in aging and age associated neurodegenerative disorders [226]. Dietary restriction can stabilize mitochondrial function and it reduces oxidative damage in neurons of rodents increasing the resistance of cells to different types of genetic and environmental factors [227]. Dietary restriction can also induce the expression of genes that encode proteins that promote neuronal survival like heat shock protein-70 (HSP-70) in cerebral neurons of rats and mice [228]. Increased HSP-70 levels can protect neurons against excitotoxic and oxidative damage contributing to the neuroprotective effect of dietary restriction [229]. A study in healthy humans aged 19-85 measured the ratios of cysteine/cystine and GSH/GSSG in plasma. A linear decrease in the cysteine/cystine ratio was observed throughout life while GSH/GSSG ratio was not altered until age 45, after which there was an enhanced oxidation at a nearly linear rate associated with an age-related decrease in the content of total glutathione [230]. Down-regulation of the γ-glutamyl cysteine synthase during aging may contribute to the decrease in the synthesis of glutathione as well as to a low GSH regeneration by glutathione reductase. In addition, high glutathione degradation may contribute to lower glutathione levels in senescence. There are many protective enzymes that use GSH to counteract the deleterious effects associated with aging.

Many important metabolic enzymes can be inactivated by oxidation [231]. Aconitase III participates in the citric acid cycle and it possesses an active site iron-sulfur cluster that is sensitive to inactivation by the superoxide anion radical [232]. Carbonic anhydrase III has two reactive thiols that are susceptible to the conversion of cysteine to sulfinic acid or cysteic acid in the presence of high hydrogen peroxide levels. These reactions may be inhibited by S-glutathionylation of the reactive cysteine residues [233]. Glutathionylation of specific proteins has been shown to have a role in many neurodegenerative disorders, including Alzheimer's disease, Friedreich's ataxia, amyotrophic lateral sclerosis, Huntington's disease and Parkinson's disease, which are characterized by a progressive loss of neurons [234]. These disorders are characterized by protein dysfunctions and mitochondrial damage that result in the overproduction of ROS, RNS and RSS. Consequently, the levels of GSH have been reported to be reduced in AD and PD [235]. Specifically, PD is characterized by early glutathione depletion in dopaminergic cells in the substantia nigra. In particular, glutathionylation of mitochondrial proteins has been shown to be involved in neurodegenerative disorders. Glutaredoxin is responsible for the deglutathionyla-tion of proteins and this process is decreased in post-mortem AD brains [236]. In a mouse model of PD (acute treatment with 1-methyl-4-phenyl-1,2,3,6-tetrahydropyridine, which targets mitochondrial Complex I) glutaredoxin was increased in brains of the animals [237]. Besides, glutathionylation of the tau protein causes its polymerization and the generation of neurofibrillary tangles found in AD [238].

Reactive species can induce cell membrane damage having a key role in senescence. Fenton reactions of metals may promote oxygen toxicity generating hydroxyl radicals, which can initiate lipid peroxidation in the cellular membranes. In humans the total body content of iron increases with age and it has been proposed that this increase may promote oxidative damage during aging and age-associated disorders [239]. In healthy conditions, the potential oxidative activity of iron is regulated through its binding to proteins such as ferritin and transferrin [240]. Aging has been associated with more oxidized forms of macromolecules, particularly in the mitochondria where the majority of ROS, RNS and RSS are produced within the cell [241]. The activities of various complexes of the mitochondrial respiratory chain decrease with age and the protection of thiol groups with N-acetylcysteine prevented such decrease [242]. Moreover, mitochondrial peroxiredoxin III was found to be present more oxidized and inactive in aged rats due to oxidative sulfonation of Cys109 in the catalytic site of the enzyme [243]. The age-related decline in GSH levels may result in reduced availability for S-glutathionylation within the mitochondria. S-glutathionylation has both inactivating effects on the function of proteins, but protective effects against irreversible damage on proteins in other conditions [244, 245]. Finally, S-nitrosylation of proteins linked to neurodegenerative diseases, cardiovascular diseases, obesity and diabetes has been related to aging [246]. Succination in age-related disease has not been studied in great detail, but mitochondria are important sites for hydrogen sulfide metabolism and they have high sensitivity to hydrogen sulfide signaling [247]. Then, nitrosylation, sulfonylation, glutathionylation and succination of cysteine-bearing proteins represent efficient posttranslational modifications that control important mitochondrial and cellular functions, but at the same time, they are the Achilles' heels for senescence and age-associated neurodegenerative disorders.

Age-Associated Neurodegenerative Diseases and Cysteine Dysregulation

Parkinson's Disease

Parkinson's Disease (PD) is a progressive neurodegenerative disorder that results in impaired motor and cognitive functions that affect nearly 1% of individuals over the age of 65 [248]. The cause of the disease is the degeneration of dopaminergic neurons in the substantia nigra, pars compacta region, of the midbrain [249]. The pathogenesis of PD seems to be multifactorial including environmental factors that act on genetically susceptible individuals as they age [250, 251]. Aging is the single most important risk factor for PD and it contributes to the progression of the disease through oxidative damage and impairment of the mitochondrial bioenergetic capacity and antioxidant ability in the brain (Fig. **8**) [252-258].

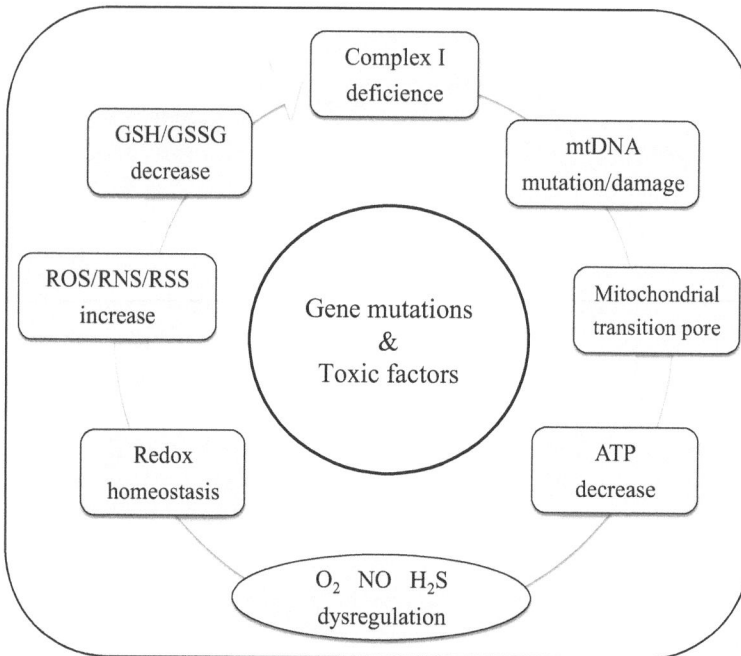

Fig. (8). Pathogenic factors in Parkinson's disease. A vicious circle develops in dopaminergic neurons that cause oxidative damage and bioenergetic deficiency, resulting in dopaminergic neurodegeneration.

More than 95% of PD cases are sporadic, some of which appear to be correlated with exposures to toxic substances that can induce ROS, RNS and RSS generation within neuronal cells raising the hypothesis that oxidative damage contributes to PD pathogenesis through the alteration of the function of PD-associated proteins [259].

Six genes identified as α-synuclein (SNCA), ubiquitin C-terminal hydrolase like 1 (UCH-L1), parkin (PARK2), LRRK 2, PINK 1 and DJ-1 have been linked to PD and/or to neurodegeneration of the parkinsonian type. However, mutations in these genes are responsible for only a small number of patients with PD. The LRRK 2 gene (PARK8) is the most common cause (5-7%) of familial PD to date [260].

The impact of environmental agents on the risk of PD has been thoroughly studied. In fact, PD patients show abnormalities of oxidative phosphorylation that impair their mitochondrial energy metabolism increasing ROS generation, which closely resembles that attributable to 1-methyl, 4-phenyl, 1,2,3,6-tetrahydropyridine (MPTP) [261-263], although this impairment is apparently constitutive in origin [264, 265]. Schapira and co-workers were the first to report that mitochondrial Complex I activity was selectively reduced in the substantia nigra of patients with PD [266]. The Complex I impairment is worse in more advanced cases and seems to affect also non-nigral brain areas, muscle and fibroblasts of PD subjects [265, 266]. Swerdlow and co-workers carried out an elegant experiment using cybrid cells to confirm that the mitochondria in PD were at least 20% less efficient in Complex I activity, produced higher levels of ROS and rendered their host cells more susceptible to MPTP-induced cell death [265]. They suggested that Complex I defect was in the mitochondrial DNA of PD patients caused by parental inheritance or by oxidative damage on the mitochondrial DNA. This constitutive defect in Complex I may explain why some individuals develop PD following toxin exposure, while others do not. MPTP is not toxic by itself but into the brain it is transformed to the toxic product MPP+ (1-methyl-4-phenylpyridinium) by MAO-B. MPP+ is selectively taken up by the dopaminergic neurons of the substantia nigra and selectively taken up by their mitochondria where it inhibits Complex I activity [263]. Then, a vicious circle develops, which causes oxidative damage and bioenergetic deficiency into neurons (Fig. **8**).

ROS such as hydrogen peroxide are generated during dopamine metabolism, which are removed by glutathione protecting dopaminergic neurons from oxidative damage. Indeed, the progression of PD is associated with a depletion of glutathione levels and a consequent increase in ROS within the substantia nigra [267, 268]. In a clinical study, improvements in patients with PD were observed following administration of GSH [269], however, prolonged treatments with this compound or other agents that can cross the blood brain barrier remains to be investigated [270] In fact, the earliest reported biochemical change identified in the substantia nigra of early PD patients is a significant depletion of reduced glutathione, which may promote morphological mitochondrial damage by ROS [271, 272]. The mitochondrial glutathione is dependent on the uptake from the cytosol since mitochondria lack the enzymes necessaries for glutathione synthesis [273].

As previously mentioned, glutathione plays an important role in scavenging ROS and RNS, and in recycling other antioxidants. Glutathione is regenerated to its thiol-reduced form (>98%) by glutathione reductase, which maintains optimal GSH/GSSG ratios into the cell [273]. The magnitude of glutathione depletion seems to parallel the severity of the disease and it is the earliest known indicator of the substantia nigra dysfunction, preceding the losses in both mitochondrial Complex I activity and striatal dopamine content (Fig. **7**) [274]. Since Complex I impairment results in the generation of ROS (Fig. **4**) in agreement with reports of elevated markers of oxidative damage to lipids, proteins and DNA in the substantia nigra of patients with PD [275, 276], and it has been demonstrated that exposure of mitochondrial membranes to nitric oxide resulted in selective and persistent inhibition of Complex I activity *via* S-nitrosylation of critical cysteine thiol groups in this enzymatic complex, the inhibition of Complex I activity may be reversible by restoring mitochondrial GSH levels [277-279]. Besides, it has been shown that Complex I inhibition following prolonged dopaminergic glutathione depletion *in vitro* was reversible with dithiothreitol, suggesting that Complex I decreased activity involve a reversible cysteine thiol modification. Then, it seems that glutathione depletion in the substantia nigra of PD patients results in increased ROS and RNS generation leading to Complex I inhibition with subsequent mitochondrial dysfunction that significantly affects glutathione synthesis closing the vicious circle (Fig. **7**) that ultimately leads to dopaminergic neurodegeneration [279].

In addition to S-nitrosylation of mitochondrial Complex I [280], other proteins and enzymes have been implicated in the development of PD. Likewise, protein disulphide isomerase (PDI), an endoplasmic reticulum enzyme that facilitates proper protein folding, is S-nitrosylated in brain samples of sporadic Parkinson and Alzheimer's disease patients [281]. S-nitrosylation of PDI inhibits its activity, resulting in activation of endoplasmic reticulum stress pathways contributing to neurodegenerative disorders. When dopaminergic cells are treated with rotenone, a pesticide implicated in the pathogenesis of PD that inhibit the mitochondrial oxidative phosphorylation, it was observed an increase in S-nitrosylation of PDI associated with a decrease in the PDI chaperone activity.

S-nitrosylation of dynamin-related protein 1 (Drp1) may be associated to the pathological fragmentation of mitochondria and consequent loss of synapses in neurodegenerative disorders [282]. Drp1 controls mitochondrial fission and is responsible for proper neuronal development. Mitochondrial fusion/fission is essential for maintaining mitochondrial integrity and disruption of these processes leads to mitochondrial dysfunction contributing to the development of PD and AD [283, 284]. Nitric oxide produced as the result of the β-amyloid associated inflammatory response may nitrosylates some cysteine on Drp1 resulting in mitochondrial fission, synaptic loss and neuronal damage [285].

S-nitrosylation of peroxiredoxin 2 (Prx2), the most abundant isoform in mammalian neurons, is increased in PD patients compared to control patients [286]. The reactive cysteine residues in Prx2 reduce peroxides to H_2O. Oxidized cysteines of Prx2 can form an intermolecular disulfide bond with another Prx2 molecule, they can undergo reduction/regeneration back to reduced thiol by thioredoxin, or they can further be oxidized to produce a sulfinic or sulfonic acid, which inactivate the protein. In this case, Prx2 does not react with hydrogen peroxide and the normal redox cycle to detoxify ROS is altered contributing to dopaminergic neurodegeneration.

Deletions or point mutations in the protein DJ-1 (PARK7) have been shown to be responsible for an early-onset, autosomal-recessive form of PD as well as sporadic form of PD [287]. Sequence analysis of PARK7 has revealed three potentially reactive cysteine residues (Cys106, Cys46 and Cys53) susceptible to S-nitrosylation suggesting that S-nitrosylation of this protein can disrupt their antioxidant action in dopaminergic neurons [288, 289]. In addition, oxidative stress may increase the accumulation of toxic forms of α-synuclein through oxidative ligation to dopamine playing a central role in PD. This suggests that increased oxidative stress due to early glutathione deficiency in the substantia nigra may lead to enhanced toxicity of α-synuclein in dopaminergic neurons [290].

Parkin (PARK2) is a RING domain-containing E3 ubiquitin ligase involved in proteasome-dependent degradation of proteins. Parkin is also important for mitochondrial degradation by lysosome-dependent autophagy or mitophagy [291, 292]. Parkin has multiple cysteine residues in its RING domain and elsewhere [293, 294], which can react with nitric oxide to form S-nitrosylated parkin. This S-nitrosylation interferes with the neuroprotective function of the protein. Indeed, S-nitrosylation of parkin has been found in brains of patients with Lewy body disease and PD [294, 295]. These findings support the concept that posttranslational changes of proteins through oxidative modification of cysteine thiols may contribute to the pathogenesis of sporadic PD.

It has been shown that glyceraldehyde-3-phosphate dehydrogenase (GAPDH) can be S-nitrosylated resulting in a loss of the enzymatic activity [296]. Moreover, in conjunction with E3 ubiquitin protein ligase 1, it forms a signal complex to promote cell death and neurodegeneration [296]. E3 ubiquitin ligase is involved in ubiquitination and proteasome-mediated degradation of specific proteins. The activity of this ligase is implicated in the regulation of the cellular response to hypoxia and apoptosis. This enzyme has also be implicated in the development of certain forms of PD and it is S-nitrosylated in PD brains inhibiting its activity and interfering with ubiquitination of proteins, which may contributing to the selective neurodegeneration in PD [295].

Inhibitors of apoptosis (IAPs) are a family of proteins that regulate cell survival through binding to caspases to repress their catalytic activity [297, 298]. The protein X-linked inhibitor of apoptosis (XIAP) is the most commonly expressed and the most potent endogenous caspase inhibitor among the IAPs, and it can be S-nitrosylated in several neurodegenerative disorders. In fact, recent studies have shown a significant increase of S-nitrosylated XIAP in brain samples from PD, AD and HD patients possibly promoting apoptotic signaling [299, 300].

Alzheimer's Disease

Alzheimer's disease (AD) is an age-related neurodegenerative disorder that accounts for 60% to 70% of cases of dementia. The cognitive impairments of patients with AD are associated with synaptic deficits and loss, which have been shown early events in the pathogenesis and progression of AD [301-303]. Most AD cases (>95%) are sporadic and occur in elderly people over 60 years old. Then, like in PD, the principal risk factor for sporadic AD is aging.

Several studies have shown that AD brains exhibit increased levels of ROS, RNS and RSS that affect proteins that are critical to neuronal survival contributing to the disease pathogenesis [304]. Besides, β-amyloid can increase nitric oxide production inducing mitochondrial fission, synaptic loss and neuronal damage *via* S-nitrosylation of some proteins. Energy demand in neurons depends on mitochondrial dynamics, consisting of fission and fusion mechanisms to generate new mitochondria. Recent studies have revealed that β-amyloid accumulates inside AD brain mitochondria including synaptic mitochondria. Likewise, β-amyloid levels in mitochondrial are associated with abnormalities of mitochondrial structure and function [305-310]. Indeed, aberrant S-nitrosylation of Drp1 hyper-activates Drp1 causing a dramatic increase in mitochondrial fission. This alteration of mitochondrial dynamic contributes to synaptic loss and subsequent neurodegeneration [304]. Like in PD, S-nitrosylated Drp1 levels are significantly increased in post-mortem AD patient brains compared to controls [304].

Cyclin dependent kinase 5 (Cdk5) in neurons has many function including cell survival, neuronal migration, dendritic spine density and synaptic plasticity [304]. It has been shown that nitric oxide may modify Cdk5 by S-nitrosylation in a ß-amyloid as well as NMDAR-dependent manner in neurons, contributing to dendritic spine loss, which correlate to clinical dementia in AD [311]. Moreover, S-nitrosylated Cdk5 levels are significantly increased in post-mortem AD brains compared to age matched controls [311].

Apolipoprotein E has a critical influence on aging and age-related neurodegenerative disorders [312]. There are three alleles of apolipoprotein E that differ in two amino acids. ApoE2 contains a cysteine in each position, ApoE3 contains a cysteine in one of the positions, but Apo E4 does not contain a cysteine in either position. Individuals with an E4 allele have a reduced lifespan [313] and they have an increased risk to develop AD [314]. The mechanism by which ApoE4 accelerates brain aging may involve decreased antioxidant and neuroprotective properties of this apolipoprotein isoform, since cysteine residues in ApoE2 and ApoE3 may bind to and detoxify 4-hydroxynonenal, a cytotoxic product of lipid peroxidation [315]. Moreover, the different isoforms of apolipoprotein E are potential sites for redox modification and regulation. Indeed, apolipoprotein E isoforms can bind to neuronal NOS and ApoE2 and ApoE3 can be S-nitrosylated in human hippocampus, possibly interfering with the lipid metabolism in AD [316].

As mentioned previously, brain produces endogenous hydrogen sulfide from cysteine and patients with AD have decreased levels of hydrogen sulfide in the brain compared with the brains of age-matched controls [157, 317, 318]. In fact, brains of patients with AD have reduced cystathionine β-synthase activity [319]. This defect is crucial in glutaminergic neurons since the production of hydrogen sulfide by cystathionine β-synthase enhances NMDAR-mediated currents, which are fundamental for neuronal function [157].

PDI is one of the enzymes that catalyze disulfide bond formation in the endoplasmic reticulum [320, 321]. Several lines of evidence have implicated S-nitrosylation of PDI in human AD brains compared to control brains [304]. S-nitrosylation of PDI facilitates further oxidation of cysteine residues to sulfenic, sulfinic and sulfonic acid derivatives. These redox modifications compromise PDI chaperone/protein folding leading to protein misfolding and accumulation in the endoplasmic reticulum [281].

The above-mentioned studies highlight some of the roles of oxidative modification of proteins and how they can alter diverse cellular functions, including mitochondrial dynamics, synapse loss, endoplasmic reticulum protein folding, signal transduction pathways and lipid metabolism, thereby contributing to the progression of AD. In the next section, I review the scientific evidence that support the pharmacological intervention against aging and age-associated neurodegenerative diseases, based on the "cysteinet" paradigm (Fig. **6**).

PHARMACOLOGICAL INTERVENTION IN BRAIN AGING AND NEURODEGENERATIVE DISEASES

Multiple treatments have been used in an attempt to counteract the deleterious effects of brain aging. Specifically, antioxidants and supplements such as creatine, which enhances cellular energy, may have beneficial effects against aging [322]. Besides, recent studies have evaluated the effects of caloric restriction to increase the resistance of neurons to age-related disease [228, 323, 324]. Strategies against apoptosis using agents that target key proteins in this process are other potentially effective anti-aging treatments [325, 326]. Statins have been proposed to protect the brain against aging and age-related neurodegenerative disease through actions on the vasculature or directly on neurons [327]. If we consider that aging and age-related disorders result from the progressive deterioration of the bioenergetic capacity of cells associated with an inability to regulate metabolic homeostasis including redox balance and reparative antioxidant mechanisms (Fig. **9**), none of the previously mentioned interventions have reached these objectives. In this sense, anti-aging strategies must correct the imbalance between decreased ATP generation and reactive species overproduction that evolves to the disruption of cellular homeostasis with time. Cellular aging is mediated by the inability of the mitochondria to maintain bio-energetic/redox homeostasis and then, brain aging and age-related neurodegenerative diseases are mediated by the dysregulation of the normal mitochondria biogenesis, structure and metabolism or a combination of these processes. The mitochondrial theory of aging proposed by Harman [328] and later complemented by others [329-331] is based on the findings that there is an age-related decline in ATP generation associated with a mitochondrial ROS overproduction. Besides, there is an age-related decrease in the activity of several mitochondrial ROS-scavenging enzymes and an accumulation of mutations in mitochondrial DNA that may affect even stem cells. It has been proposed that a vicious circle develops because somatic mtDNA mutations impair mitochondrial respiratory chain function and oxidative phosphorylation, which in turn results in a further increase in ROS generation and accumulation of oxidative damage to proteins, lipids and DNA (Fig. **9**) [331]. According to this theory, mitochondria play a key role in initiating and mediating the oxidative stress that drives the aging process.

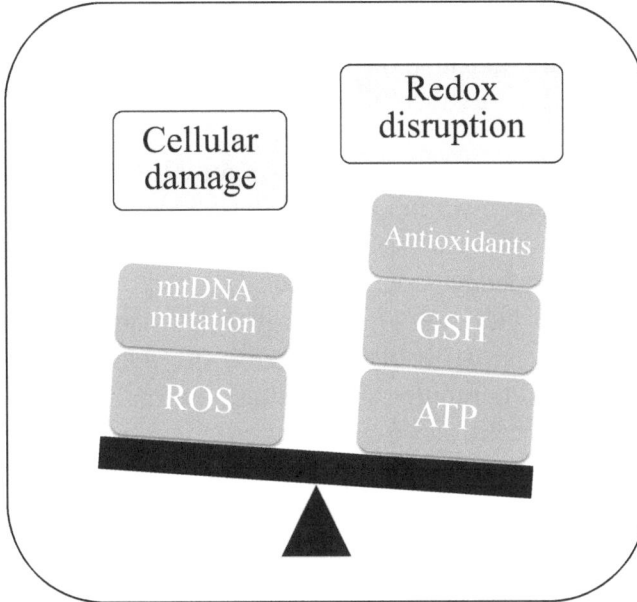

Fig. (9). Mitochondrial disequilibrium between ATP generation and ROS overproduction in aging that may affect postmitotic as well as stem cells.

N-Acetylcysteine and the Regulation of Cysteinet

Among antioxidant strategies, supplement of reduced glutathione is of limited value since it is oxidized at physiological conditions. The product of glutathione will provide cysteine that may be used back to gastrointestinal synthesis of glutathione. However, there is sufficient clinical evidence that thiol-containing compounds such as N-acetylcysteine (NAC) can rescue patients from acute exposure to oxidative stress (acetaminophen overdose). NAC is also effective in the chronic treatment of old patients with obstructive pulmonary disease with important chemo-preventive properties in lung cancer [332]. In large groups of patients with chronic obstructive lung disease NAC was a safe agent with minor side effects even when it was prescribed for a prolonged period of time. NAC has beneficial effects in many different diseases including cancer, cardiovascular diseases, ophthalmic diseases, HIV infections, acetaminophen toxicity, metal toxicity, cerebral ischemia and hemorrhage, traumatic brain injury and even psychiatric diseases [333-341]. Many clinical trials have used NAC as an adjunctive treatment in various psychiatric disorders with beneficial effects on the clinical outcomes in most cases [342, 343].

In this chapter, I propose that NAC is a singular compound that has beneficial effects in brain aging and neurodegenerative disorders through the regulation of "cysteinet", not only through classical antioxidant properties, but also contributing to the restoration and maintenance of the cellular redox homeostasis and cysteine network of intracellular and extracellular proteins.

In regard to the oral NAC administration, a minimal dose of 600 mg/day and a maximum of 1800 mg/day would be into the range of use, but it would be necessary to follow up closely the optimal dose of administration to get the minimal effective therapeutic concentration in each clinical condition and disorder. The possible lack of efficacy of NAC can be explained by its pharmacokinetic, which is highly dependent on the administration route. After oral administration, NAC is rapidly absorbed and an oral dose of 600 mg results in plasma concentration peaks of 4.6 µM (60 min) that quickly decrease to 2.5 µM at 90 min. Concentrations of 16 and 35 µM after oral administration of 600 mg and 1200 mg/day respectively have been reported. The plasma half-life is estimated to be about 2.5 hours and no NAC is detectable 10-12 h after oral administration [344]. The plasma pharmacokinetics of oral NAC administration and the relative bioavailability of two regimens were studied in 12 adult subjects. On two different occasions in a crossover and balanced fashion the subjects were administered orally either a single dose of 600 mg of NAC as effervescent tablets or 4 repeated dose of NAC as granules in sachets at the regimen of 200 mg. With the 600 mg dose maximal concentrations were greater than with a single 200 mg dose; after summing up the values of these parameters for the 200 mg dose no significant differences were observed in comparison to the single 600 mg dose in maximal concentrations [344]. It has been calculated that after oral administration, reduced NAC has a terminal half-life of 6.25 hours and it is rapidly metabolized and incorporated into proteins with only low levels of oxidized NAC detectable for several hours [345].

The pharmacokinetics after oral administration of 200, 600 or 1200 mg of NAC was studied in 10 healthy subjects. Normalized maximal plasma concentration was significantly higher after a 600 mg dose than after a 200 mg dose and the bioavailability of NAC significantly increased with the increase of the dose. In an extension of this study, 600 mg of NAC was given twice a day for 5 days and the plasma concentrations were followed after the morning dose on day 6. No differences in the pharmacokinetic parameters were observed in comparison with the single 600 mg dose suggesting that the beneficial effects observed after repeated dosing cannot be ascribed to an accumulation of NAC in plasma [346]. Therefore, it is conceivable that the principal therapeutic function of NAC could be mediated by the regulation of the thiol groups of cellular and extracellular proteins.

On the other hand, intravenous infusion of NAC rapidly forms disulfides in plasma, with half-life around 6 hours [347, 348]. The proportion of NAC bound to proteins is relatively low during steady-state infusion (30-40%) but increases rapidly after completion of infusion (60-70%) [349]. Although renal clearance of NAC has been reported to be around 0.190- 0.211 L/h per kg, up to 70% of the total body clearance of NAC is non-renal. NAC is generally safe and well tolerated, and the most frequently reported side effects are nausea, vomiting and diarrhoea.

Finally, as with any antioxidant nutrient, NAC may have potential pro-oxidant activity and therefore it has been not recommended in the absence of confirmed oxidative stress [350].

N-Acetylcysteine and Aging

Although the mechanisms of NAC actions are not completely understood, the direct antioxidant properties of NAC, reconstitution of glutathione pool as well as protein-associated thiols interactions have been suggested as the major mechanisms [273, 351]. NAC does not need the alanine-serine-cysteine system to enter the cells [352], since NAC is a membrane-permeable cysteine precursor that does not require active transport. Into the cell, NAC is rapidly hydrolysed to yield cysteine [338].

Modification of protein-associated cysteine residues might explain many effects of NAC including improvement of metabolic function, modulation of immune system and anti-aging properties. However, modifications of cysteine residues in proteins by NAC have not been systematically studied yet. NAC can alter posttranslational modifications of important plasma proteins like transthyretin (TTR) by interacting with the protein-associated cysteine residue *in vitro* and *in vivo* [353]. The interaction of NAC with TTR seems to be dose-dependent with a biphasic character of NAC-TTR interaction [353]. Moreover, it has been shown that the administration of NAC can ameliorate the onset and severity of premature aging in Bmal1-deficient mice. Bmal1 is a circadian clock protein implicated in tissue homeostasis by the direct regulation of ROS and it acts as transcription factor of key components of the circadian clock. Administration of NAC attenuated the development of the age-related phenotype of Bmal1-/- mice such as development of cataracts, and it extended the average and maximal lifespan of the animals [354]. Another possible role of NAC in aging may be related with the uptake of glutamate in the brain that is carried out, under physiological conditions, by astrocytes and neurons that express excitatory amino-acid carrier-1, which can also transport cysteine necessary for glutathione synthesis. In a mice deficient model for these transporters, NAC reversed glutathione reduction and oxidative damage associated with the deficiency. These results suggested that excitatory amino-acid carrier-1 might be essential for cysteine uptake and glutathione synthesis in neurons [355]. In humans, NAC is a scavenger of free radicals as well as a major contributor to restore cellular glutathione in muscle cells. Some studies have demonstrated possible roles for NAC to minimize fatigue as well as the prevention of apoptosis secondary to exhaustive exercise [356]. Likewise, it has been well documented that infusion of NAC can attenuate muscle fatigue and it can enhance the overall redox status inside the cell [357].

Therefore, future studies are needed considering the essential role of cysteine residues for the maintenance of protein structure and function and the importance of NAC in various clinical and experimental setting. The present chapter emphasizes the singularity of NAC as the unique substance that, at present, is widely used in clinical practice, can cross the blood brain barrier (BBB) and can directly interact with key cysteine-bearing proteins into the brain counteracting brain aging and age-related neurodegenerative diseases. NAC can be potentially used to restore the dysregulation of redox homeostasis through the replenishment of mitochondrial soluble and protein-linked thiols, which may restore the mitochondrial bioenergetic capacity and ROS physiological levels. Much remains to be investigated in terms of redox signaling and the modulation of mitochondrial function, but based on present knowledge, it is clear that redox signaling plays a central role in the modulation of mitochondrial biology and physiology.

NAC can maintain the availability of cysteine in the blood for glutathione replenishment and it is the most commonly used substance in order to maintain the biosynthesis of glutathione as a result of their proven safety and efficacy [358]. Then, many studies have explored if NAC can improve glutathione bioavailability and maintain the cellular redox state and bioenergetic ability. Since the first time it was reported the potential beneficial effects of NAC in the treatment of age-related mitochondrial neurodegenerative diseases, many studies have corroborated the concept that this thiol bearing substance can cross the BBB having important regulatory roles in brain, which go further from its known antioxidant role [351]. Our studies in synaptic mitochondria isolated from aged mice chronically treated with NAC demonstrated, fifteen years ago, the anti-aging properties of NAC, acting fundamentally by increasing ATP levels through the activation of the mitochondrial complexes of the respiratory chain and oxidative phosphorylation as well as restoring GSH levels and decreasing lipid and protein oxidation in presynaptic terminals [359-363].

The redox homeostasis is the principal mechanism through which reactive species (ROS, RNS and RSS) are integrated into the cellular regulation of metabolic pathways (Fig. **6**), which are mediated by cysteine-bearing proteins. Since NAC affects redox-sensitive signal transduction and gene expression pathways, its functions on cell signaling should be also considered. Interestingly, NAC can directly modulate the activity of common transcription factors both *in vitro* and *in vivo* [343]. NAC treatment suppressed NF-kB activation in oxidative stressed cultured cells and in clinical sepsis reducing subsequent cytokine production. NF-kB is naturally bound to its inhibitor (I-kB) that prevents its nuclear translocation. Dissociation of I- kB following its phosphorylation by specific kinase of NF- kB (IKK) allows NF- kB transport to the nucleus [343]. These effects of NAC are mediated through is ability to regulate and modulate the cysteine matrix of proteins "cysteinet", which I propose is the cellular network that allows many of the pluripotential effect of this thiolic substance. Thus, a probable beneficial action of NAC administration may be the prevention of age-related protein oxidation, misfolding and accumulation through the regulation of cysteine oxidative modification associated with aging. The 3-D structures of proteins are related to their function and conformational changes occur in many proteins when they accumulate in tissues [364]. The transition from α-helix to β-sheet is characteristic of amyloid as well as other type of proteinaceous deposits. Conformational changes are most likely to occur in proteins that have repetitive amino acid sequences, such as polyglutamine in Huntington's disease. Under physiological conditions, chaperones help proteins that have problems to achieve their native configuration, but during aging or increased mutations, the fine balance among the synthesis, folding and degradation of proteins will decrease resulting in the accumulation of misfolded proteins. Indeed, accumulation of misfolded proteins can contribute to the pathogenesis of age-related neurodegenerative disease such as Alzheimer's, Parkinson's and Huntington's diseases, which may be potentially restored by NAC administration.

N-Acetylcysteine and Parkinson's Disease

Having in account the accumulation of oxidative damage in PD patients, some clinical studies have been performed using antioxidants in the treatment of PD with controversial results [365-367]. The central implication of glutathione deficiency in PD has stimulated many investigations to find new potential approaches for maintain or restore glutathione levels in these patients. Moreover, the use of glutathione as a therapeutic agent is limited by its very short half-life in human plasma (<3 min) and difficulty to cross cell membranes, being necessary high doses to reach therapeutic levels [368].

Under physiological conditions, the cellular availability of cysteine is considered the rate-limiting factor in the synthesis of glutathione. However, cysteine is toxic at high concentrations as the result of free radicals generation during cysteine autooxidation [369]. As a consequence, compounds that can be metabolized to cysteine must be used as pro-drugs to increase neuronal glutathione concentrations.

NAC is the simplest cysteine pro-drug that can be systemically administered to deliver cysteine to the brain [255, 359-362], acting as a precursor for glutathione synthesis as well as a stimulator of the cytosolic enzymes involved in glutathione regeneration. NAC can increase the Complex I activity in mitochondria isolated from pre-synaptic terminals of aged mice. These effects were proposed as evidence that NAC can cross the BBB having reparative effects on mitochondrial proteins of the brain and against age-associated memory decline [255, 257, 258]. Since glutathione levels become more depleted in the substantia nigra as the disease progress, NAC may contribute to GSH repletion, which in addition to its potent antioxidant effects by direct scavenging of ROS can make this antioxidant ideal for counteracts mitochondrial impairment in the substantia nigra of PD patients [270]. Furthermore, we have shown that NAC can prevent dopamine induced programmed cell death in cultured human cortical neurons [363], and also it can increase mitochondrial complex IV specific activity in presynaptic mitochondrial preparations from aged mice [360]. Effects on Complex I and IV activities can be related with the proven NAC beneficial action on the age-associated protein oxidation of mitochondrial proteins [361].

Systemic administration of NAC increases brain levels of glutathione in mice [255, 256, 370, 371], it reduces markers of oxidative damage [360], it increases brain synaptic [359] and non-synaptic [372] mitochondrial Complex I activities and it protects against MPTP toxicity [373, 374] and dopamine-induced cell death [363, 375]. Besides, dietary supplementation of NAC during 1 year was able to counteract age-related decrease in rat brain expressions of subunit 39 kDa and ND-1 of the mitochondrial respiratory Complex I, and other subunits of the mitochondrial oxidative phosphorylation [376]. In view of the above, there are sufficient scientific evidence that Complex I inhibition by prolonged glutathione depletion may be due, at least partially, to a reversible age-related event involving cysteine residues in the protein complex with impact on its enzymatic activity. This may be reversible by restoring glutathione to normal levels and it suggests that therapeutic strategies directed toward the maintenance of cellular glutathione concentration within dopaminergic neurons would be beneficial in PD. Moreover, *in vitro* studies have shown that NAC is able to restore Complex I age-related decreased activity, suggesting a direct role of NAC on cysteine residues of this mitochondrial protein complex [270, 359]. On the other hand, recent studies using positron emission tomography suggested that chronic use of methamphetamine (METH) causes a reduction of dopamine transporters in the human brain, suggesting that this is the mechanism of neurotoxicity in humans [377]. These findings are supported by a report that demonstrates that the densities of dopamine transporters are significantly decreased in the post-mortem striatum of chronic METH users [378]. Although the precise mechanisms of METH-induced neurotoxicity in dopaminergic nerve terminals are not fully known, NAC administration significantly attenuated the reduction of dopamine transporters in the striatum of the experimental animals, 3 weeks after the administration of METH [379], possibly by rescuing glutathione levels in the striatum [380]. Therefore, it is likely that NAC would be a suitable substance for the treatment of neurotoxicity in dopaminergic nerve terminals related to the chronic use of METH in humans.

Experiments in animal models have shown that oral NAC administration can protect against loss of dopaminergic terminals associated with over-expression of α-synuclein [381]. The results of this study showed that striatal tyrosine hydroxylase positive terminal density was increased in NAC-treated α-synuclein over-expressing mice compared to α-synuclein over-expressing mice with a control diet. This also correlated with a decrease in α-synuclein immunostaining in the brains of over-expressing mice treated with NAC.

Moreover, NAC supplementation significantly increased glutathione concentrations in the substantia nigra of transgenic mice over-expressing α-synuclein protein [381]. There is growing evidence that NAC plays a role against apoptosis in postmitotic cells and oligodendrocytes. *In vitro* studies have shown that NAC promotes survival of sympathetic neurons and pheochromocytoma (PC12) cells in the absence of trophic factors. This action of NAC seems to be attributable to the activation of the Ras-extracellular signal-regulated kinase (ERK) pathway. In fact, PC12 cell survival by NAC was totally blocked by an inhibitor of the ERK-activating MAP kinase/ERK kinase, suggesting a required role for ERK activation in the NAC mechanism of action. These findings support a key role of Ras-ERK activation pathway in the mechanism by which NAC prevents neuronal death [382, 383]. Moreover, since alterations in mitochondrial structure and function are early events in apoptosis [384] and NAC can prevent ROS accumulation, telomere shortening and cell death in an *in vitro* model that disrupt mitochondrial electron transport chain, it is conceivable that this thiolic antioxidant could act *in vivo* against apoptosis in PD [385]. NAC can also inhibit the expression of c-fos and c-jun genes and TGF β-1 mediated apoptosis in human ovarian carcinoma cells [386]. In addition, long-term treatment with NAC affected NF-kB signaling in the brain of mice by increasing cytoplasmic retention of NF-kB thus preventing its action as a transcription factor in the nucleus [381]. Since increased activation of NF-kB may contribute to the pathology in models of PD, it is possible that NAC actions against modification of cysteine thiol groups in the proteins involved in regulating cell survival and NF- kB pathway were linked to the reduced NF-kB activity in these models [387, 388]. NAC can also inhibit TNFα-induced apoptosis in human neuronal and U937 cells by the preservation of mitochondrial integrity and function since NAC was able to partially prevent the mitochondrial membrane depolarization induced by this cytokine [389, 390]. One study has shown that NAC is a potent direct scavenger of both hydrogen peroxide and toxic quinones derived from dopamine preventing dopamine-mediated inhibition of Na+, K+-ATPase activity. This action suggests another mechanism for the use of NAC in the treatment of PD [391]. Therefore, NAC may act against Na+, K+-ATPase inhibition, counteracting intracellular damage pathways that lead to death of dopaminergic neurons.

In summary, preclinical studies support the therapeutic potential of NAC in the treatment of PD. This thiolic antioxidant has been shown to counteract age-related mitochondrial damage, can prevent apoptosis, and can scavenge hydrogen peroxide and reactive quinones. Besides, NAC administration can protect against loss of dopaminergic terminals associated with over-expression of α-synuclein in animal models and it produces beneficial effects in mice treated with MPTP. Then, NAC is the unique known substance with the property to restore age-associated changes in cysteine modification of proteins and redox balance in brain aging and age-related neurodegenerative disorders.

However, the therapeutic efficacy of NAC in the treatment of PD disease has not been examined yet, although some clinical trials are currently going on. A beneficial role of NAC on the cystine-glutamate antiporter system is probable. Indeed, zonisamide has recently been shown to reduce disease severity in a clinical trial, possibly by the increase in cystine-glutamate antiporter system activity [392, 393]. In agreement, zonisamide administration to hemi-parkinsonian mice resulted in increased glutathione synthesis through upregulation of cystine-glutamate antiporter system [394, 395]. Numerous trials with NAC have provided inconsistent evidence for its benefit *in vivo* and potential toxicity *in vitro* [4]. However, evaluation of the efficacy of NAC administration through the extent to which NAC affects glutathione levels in humans is difficult to assess. Recent data have shown that GSH/GSSG and

Cys/CySS redox states follow the diurnal dietary intake of sulfur amino acids suggesting that careful timing and dosing are needed because NAC absorption and clearance are rapid [395]. Furthermore, homeostatic regulation of cysteine and glutathione declines with age, and this occurs at a younger age in men than in women [395]. Thus, dosing of NAC may be critical for chronic administration and the evaluation of its benefit to counteract the dysregulation of thiol redox homeostasis is not easy. On the other hand, the bioavailability of NAC is a major limitation for maximizing its effects on neurodegenerative disorders. To overcome this limitation some modifications of the molecule have been reported. Etherification of the carboxyl group of NAC to produce N-acetyl-cysteine ethyl ester (NACET) can increase its pharmacokinetics. In fact, NACET is rapidly absorbed after oral administration reaching low concentrations in plasma since NACET enters the cells and rapidly it is transformed in NAC and cysteine [396]. The ability to cross the BBB by a compound is critical for the treatment of brain aging and neurodegenerative diseases. Interestingly, a significant increase in the brain levels of NAC and cysteine were detected due to a rapid hydrolysis of NACET after its administration [343, 396].

An amide derivative of NAC, N-acetyl-cysteine amide (NACA) has been synthesized to improve its liposolubility and membrane permeability, demonstrating its BBB permeability and therapeutic potentials in neurological disorders [397]. However, it is important to keep in mind that it is not the same to treat acute acetaminophen overdose than use NAC chronically to restore disturbed redox homeostasis as the result of the dysregulation of the bioenergetics capacity and ROS overproduction associated to aging and age-related neurodegenerative diseases.

N-Acetylcysteine and Alzheimer's disease

AD is an age-associated multifactorial disease that has increased levels of lipid peroxides in the temporal and cerebral cortex and decreased glutathione concentrations in cortical areas and the hippocampus [398-400]. Most clinical trials of antioxidants for the treatment of AD have employed α-tocopherol or selegiline, an irreversible and selective MAO-B inhibitor. Besides, NAC has been tested in some murine models of AD [401, 402]. In a study on probable AD patients, treatment with NAC to drive cystine-glutamate antiporter system caused a beneficial trend in all measures tested as well as a significant improvement in some cognitive tasks [403]. The release of glutamate from cystine-glutamate antiporter system and subsequent activation of extra-synaptic NMDA receptors, may contribute to β-amyloid production and toxicity [404, 405]. In contrast, cystine uptake by cystine-glutamate antiporter system seems to exert beneficial effects by lowering β-amyloid stress, preventing apoptosis triggered by oxidative stress, and normalizing the activity of sodium dependent glutamate transporters [406, 407].

Increasing evidence suggests that AD is associated with mitochondrial disorders. A reduced level of cytochrome c oxidase (complex IV) activity has been observed in post-mortem cerebral cortex in AD patients [408-410]. Mitochondrial DNA mutations have also been demonstrated in AD patients [411-413], suggesting a role for mitochondrial DNA damage in the impairment of oxidative phosphorylation associated with the disease [414]. Two studies have demonstrated a decrease in the mitochondrial mRNA that encodes complex IV in the temporal cortex and hippocampus of AD patients [415, 416]. Additionally, ROS generation in cybrid cells transferred with mitochondria from AD platelets showed that complex IV defects are associated with the disease [417]. Pyruvate dehydrogenase complex is another mitochondrial enzyme that is decreased in AD brains [418, 419]. A study evaluated the effect of lipoic acid and NAC on oxidative and apoptotic markers in fibroblasts from patients with AD and age-matched and young controls. AD fibroblasts showed the highest levels of oxidative stress, and both antioxidants exerted a protective effect as evidenced by the decrease in oxidative stress and apoptotic markers. Moreover, the oxidative damage observed was associated with mitochondrial dysfunction, which were reversed or attenuated by both lipoic acid and NAC. These data suggest that mitochondria are important in the cellular oxidative damage that occurs in AD and also they show that antioxidant therapies based on lipoic acid and NAC supplementation are promising [420]. Finally, we have demonstrated that NAC is able to restore complex IV (cytochrome c oxidase) activity measured in synaptic mitochondria from aged mice, decreasing the age-related oxidative damage to mitochondrial lipids and proteins [255, 360, 361]. This effect of chronic NAC administration was associated with a delay in age-associated memory impairment in aged mice [257].

N-Acetyl-Cysteine and Amyotrophic Lateral Sclerosis

Familial amyotrophic lateral sclerosis (ALS) has mutations in the gene encoding superoxide dismutase (SOD1), supporting the role of free radicals in the progression of the disease [421]. Levels of SOD1 are reduced in patients with familial cases, but they are often normal in sporadic ALS. However, in some patients with sporadic ALS with normal SOD1 activity, glutathione peroxidase and glutathione reductase activities were markedly reduced. ALS has been associated to changes in glutamate signaling, glutathione disruption and disturbances in astrocytes [422-424]. NAC treatments of patients with ALS have modified the course of the disease [425]. In a randomized, double blind and controlled trial, NAC produced only a modest, non-significant increase in survival with no evidence of reduction in disease progression. However, the dose used was considerably lower than the dose used in clinical trials that have obtained a positive effect for other CNS disorders [426]. Likewise, the effect of NAC may depend on the type of ALS expressed since NAC was able to increase survival in subgroups of patients with disease of the limbs onset in comparison with those with bulbar onset [425]. Thus, the induction of glutamate release by NAC may be detrimental for some types of neurodegenerative disorders. However, NAC administration can reduce oxidative stress and mitochondrial dysfunction in human neuroblastoma cells (SH-SY5Y) expressing the G93A-SOD1 mutation and it can also delay the onset of motor impairments and survival in G93A-SOD1 mutated mice [427, 428].

N-Acetylcysteine and Huntington's Disease

A recent study showed a significant decrease in mitochondrial complex IV activity and cytochrome aa3 content in Huntington's disease (HD) brains [429]. Mitochondrial complex II-III and IV activities were reduced in the caudate of HD patients [430, 431]. Besides, the administration of an irreversible inhibitor of succinate dehydrogenase (mitochondrial complex II), produced lesions that closely replicate the neuropathology of HD [432, 433]. Experiments in the R6/2 transgenic mouse model of HD showed that complex IV deficiency and ROS generation precede neuronal death [434]. Moreover, huntingtin is able to bind the mitochondrial enzyme glyceraldehyde-3-phosphate dehydrogenase as a function of disease-related glutamine repeats [435], although the activity of this enzyme was not altered in human HD brains [436]. Thus, it is evident that mitochondrial dysfunction is a major event in the pathogenesis of HD.

NAC was evaluated recently for preventing mitochondrial dysfunction in a 3-nitropropionic acid (3-NP) induced HD model in rat [437]. Injection of rats with 3-NP, an irreversible inhibitor of complex II in the mitochondria, resulted in increased oxidative stress in both striatum and cortical synaptosomes of treated animals [438]. Treatment of these rats with a free-radical spin trap agent daily or with NAC starting 2 hours before 3-NP injection, protected against the oxidative damage with a significant reduction in the striatal lesion volumes. Moreover, it was found an increased ROS generation and lipid peroxidation in mitochondria of 3-NP-induced animals as well as decreased thiol levels and SOD activity in mitochondria of 3-NP-induced rat no treated with NAC. However, NAC treatment was capable of reversing 3-NP-induced mitochondrial dysfunction and behavioural deficits in this study, suggesting a potential beneficial effect of NAC in HD.

CONCLUSION AND FUTURE PERSPECTIVES

Reactive species, including ROS, RNS and RSS are fundamental molecules to the regulation of numerous physiological and pathological cellular processes. These reactive species modulate the delicate cellular redox homeostasis and metabolism by the activation or inactivation of a widespread number of protein-mediated pathways through posttranslational modification of cysteine residues. Although our understanding of the contribution that ROS, RNS and RSS intermediates have in the maintenance of redox homeostasis and cellular metabolism is incomplete, in the present review it has been proposed that thiolic groups of many cysteine-bearing proteins form a cellular network that is interconnected by reactive species and respond coherently by synchronizing metabolic pathways that in turn regulate the redox homeostasis and bioenergetic capacity of the cell (Fig. **6**). Indeed, ROS, RNS and RSS act as complex but highly integrated molecular systems that modulate with exquisite precision the proposed cellular "cysteinet". Mitochondria have essential cellular functions that are ultimately dependent on the balance between the production of ATP and reactive species. ATP is required to support most cellular processes while ROS, RNS and RSS can drive multitude of signaling pathways. Cellular viability may be viewed as the compromise between the ATP and ROS generation, which are finally dependent on the transfer of electrons from nutrients to O2 in the mitochondrial electron transport chain. Mitochondrial key proteins involved in nutrient oxidation, metabolism, oxidative phosphorylation, solute transport, ROS production and quenching, mitochondrial morphology and apoptosis are modulated principally by S-oxidation, S-glutathionylation and S-nitrosylation. All these complex reactions can converge on protein cysteine thiols resulting in redox changes in the function of many proteins. It is also important to develop computational models to help us understand how "cysteinet" works in physiological as well as pathological conditions as well as what is the ideal therapeutic approach to restore its age-associated dysregulation. In particular, a number of studies suggest that oxidative stress contributes to aging and the pathogenesis of age-associated neurodegenerative diseases by altering neuroprotective proteins. Protein oxidative modification can disrupt mitochondrial dynamics, protein folding, ubiquitination, synaptic transmission and signal transduction pathways, contributing to neuronal cell death and neurodegenerative disorders. For example, proteomic studies have already found thousands of proteins that are S-nitrosylated, which participate in many cellular pathways with important implications in the pathogenesis of aging and age-associated neurodegenerative diseases.

NAC has a broad spectrum of actions and possible applications in human aging and diseases across multiple conditions and systems. NAC is generally safe and well tolerated even at high doses and it represents the ideal substance for modulate redox homeostasis in many diseases. This ability is the result of its interaction with the "cysteinet" system and its capability of directly entering endogenous biochemical processes as a consequence of its own metabolism (Fig. **10**). For this reason, it would be an error to evaluate the possible beneficial effect of NAC in clinical trials only through the measurement of glutathione concentrations and/or cysteine pharmacokinetic parameters. In addition, NAC may cross the BBB and its potential use in neurological disorders is only at the beginning. They are necessary double blind randomized, placebo, well designed and controlled clinical studies to test the probable benefit of oral NAC administration in ameliorate the symptoms and slow down the progression of neurodegenerative disorders. Specifically, the age of onset of the symptoms can be important in order to classify the patients in the sub-groups that will receive NAC. Standardization of patients' variables, doses of NAC used and reporting results will facilitate inter-study comparisons. For example, given that most patients with PD may be under current treatments with L-dopa into, and given that an important treatment outcome is whether an additional drug allows for a decrease in L-dopa dose, data regarding actual L-dopa doses are quite important in the evaluation of NAC benefit. Besides, although PD is mainly a disease of the elderly, it does occur in young patients as well, and it would be inappropriate to assume that patients with early onset of PD should necessarily be treated the same as patients with older onset.

Similarly, patients with ALS of the limbs onset should be considered distinct to those with bulbar onset for NAC treatment purposes. For these reasons and for the possibility of variable and even contradictory results, it would be necessary to classify patients in homogeneous groups.

In conclusion, NAC should be considered a singular medication that can regulate many disturbed physiological processes acting principally on the bioenergetic capacity of the mitochondria as well as on the complex cellular redox homeostasis through the regulation and modulation of the proposed "cysteinet". Future research should focus on components possessing the properties of NAC with more adequate pharmacokinetic parameters.

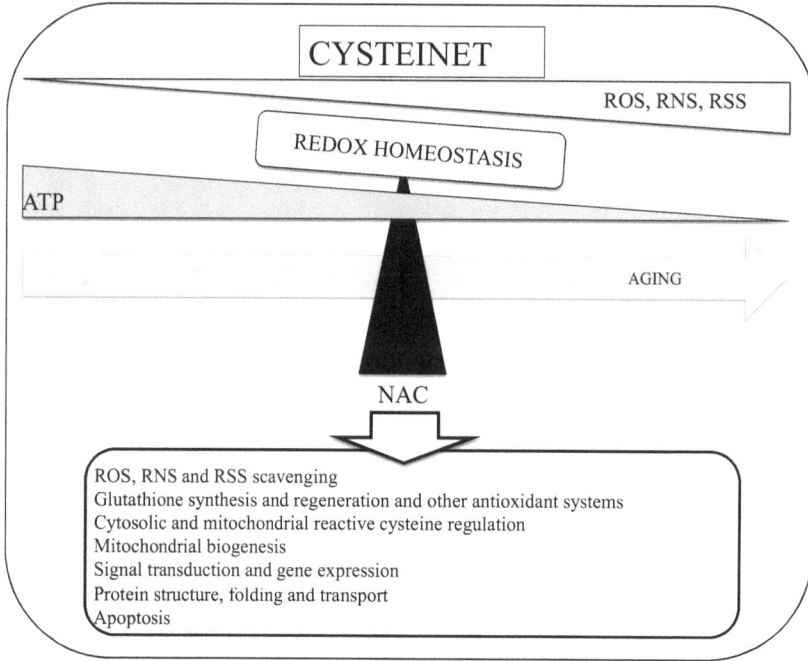

Fig. (10). Proposed effects of N-acetylcysteine (NAC) on CYSTEINET dysregulation associated with aging and neurodegenerative diseases.

CONFLICT OF INTEREST

The author confirms that author has no conflict of interest to declare for this publication.

ACKNOWLEDGEMENTS

Declared none.

REFERENCES

[1]   Claiborne A, Mallett TC, Yeh JI, Luba J, Parsonage D. Structural, redox, and mechanistic parameters for cysteine-sulfenic acid function in catalysis and regulation. Adv Protein Chem 2001; 58: 215-76.
[2]   Barford D. The role of cysteine residues as redox-sensitive regulatory switches. Curr Opin Struc Biol 2004; 14: 679-86.
[3]   Jacob C, Knight I, Winyard PG. Aspects of the biological redox chemistry of cysteine: From simple redox responses to sophisticated signalling pathways. Biol Chem 2006; 387: 1385-97.
[4]   Jones DP. Radical-free biology of oxidative stress. Am J Physiol-Cell Phys 2008; 295: C849-68.
[5]   Wood ZA, Schroder E, Harris JR, Poole LB. Structure, mechanism and regulation of peroxiredoxins. Trends Biochem Sci 2003; 28: 32-40.
[6]   Miseta A, Csutora P. Relationship between the occurrence of cysteine in proteins and the complexity of organisms. Mol Biol Evo 2000; 17: 1232-39.

[7]   Rutherfurd SM, Moughan PJ. Determination of sulfur amino acids in foods as related to bioavailability. J AOAC Int 2008; 91: 907-13.

[8]   Bradley H, Gough A, Sokhi RS, Hassell A, Waring R, Emery P. Sulfate metabolism is abnormal in patients with rheumatoid arthritis. Confirmation by *in vivo* biochemical findings. J Rheumatol 1994; 21: 1192-96.

[9]   Heafield MT, Fearn S, Steventon GB, Waring RH, Williams AC, Sturman SG. Plasma cysteine and sulphate levels in patients with motor neurone, Parkinson's and Alzheimer's disease. Neurosci Lett 1990; 110: 216-20.

[10]   Gordon C, Bradley H, Waring RH, Emery P. Abnormal sulphur oxidation in systemic lupus erythematosus. Lancet 1992; 339: 25-6.

[11]   Ozkan Y, Ozkan E, Simsek B. Plasma total homocysteine and cysteine levels as cardiovascular risk factors in coronary heart disease. Int J Cardiol 2002; 82: 269-77.

[12]   El-Khairy L, Vollset SE, Refsum H, Ueland PM. Plasma total cysteine, pregnancy complications and adverse pregnancy outcomes: the Hordaland Homocysteine Study. Am J Clin Nutr 2003; 77: 467-72.

[13]   Garcia RA, Stipanuk MH. The splanchnic organs, liver and kidney have unique roles in the metabolism of sulfur amino acids and their metabolites in rats. J Nutr 1992; 122: 1693-701.

[14]   Bagley PJ, Hirschberger LL, Stipanuk MH. Evaluation and modification of an assay procedure for cysteine dioxygenase activity: high-performance liquid chromatography method for measurement of cysteine sulfinate and demonstration of physiological relevance of cysteine dioxygenase activity in cysteine catabolism. Anal Biochem 1995; 227: 40-8.

[15]   Stipanuk MH, Coloso RM, Garcia RAG, Banks MF. Cysteine concentration regulates cysteine metabolism to glutathione, sulfate and taurine in rat hepatocytes. J Nutr 1992; 122: 420-7.

[16]   LoPachin RM, Gavin T, Geohagen BC, Das S. Neurotoxic mechanisms of electrophilic type-2 alkenes: soft interactions described by quantum mechanical parameters. Toxicol Sci 2007; 98: 561-70.

[17]   Benesch RE, Benesch R. The acid strength of the SH group in cysteine and related compounds. J Am Chem Soc 1955; 77: 5877-81.

[18]   Copley SD, Novak WRP, Babbitt PC. Divergence of function in the thioredoxin fold suprafamily: evidence for evolution of peroxiredoxins from a thioredoxin-like ancestor. Biochemistry 2004; 43: 13981-95.

[19]   Jacob C, Giles GI, Giles NM, Sies H. Sulfur and selenium: the role of oxidation state in protein structure and function. Angew Chem Int Ed 2003; 42: 4742-58.

[20]   Moriarty-Craige SE, Jones DP. Extracellular thiols and thiol/disulfide redox in metabolism. Ann Rev Nutr 2004; 24: 481-509.

[21]   Go YM, Jones DP. Redox compartmentalization in eukaryotic cells. Biochim Biophys Acta 2008; 1780: 1273-90.

[22]   Wade LA, Brady HM. Cysteine and cystine transport at the blood-brain barrier. J Neurochem 1981; 37: 730-34.

[23]   Knickelbein RG, Seres T, Lam G, Johnston RB Jr, Warshaw JB. Characterization of multiple cysteine and cystine transporters in rat alveolar type II cells. Am J Physiol 1997; 273: L1147-55.

[24]   McBean GJ. Cerebral cystine uptake: a tale of two transporters. Trends Pharmacol Sci 2002; 23: 299-302.

[25]   Bannai S, Kitamura E. Role of proton dissociation in the transport of cystine and glutamate in human diploid fibroblasts in culture. J Biol Chem 1981; 256: 5770-2.

[26]   Zerangue N, Kavanaugh MP. ASCT-1 is a neutral amino acid exchanger with chloride channel activity. J Biol Chem 1996; 27: 27991-4.

[27]   Kanai Y, Hediger MA. The glutamate/neutral amino acid transporter family SLC1: molecular, physiological and pharmacological aspects. Pflug Arch Eur J Phy 2004; 447: 469-479.

[28]   Zerangue N, Kavanaugh MP. Interaction of L-cysteine with a human excitatory amino acid transporter. J Physiol 1996; 493: 419-423.

[29]   Garcia-Garcia A, Zavala-Flores L, Rodriguez-Rocha H, Franco R. Thiol-redox signaling, dopaminergic cell death, and Parkinson's disease. Antioxid Redox Signal 2012; 17: 1764-84.

[30]   Chen Y, Swanson RA. The glutamate transporters EAAT2 and EAAT3 mediate cysteine uptake in cortical neuron cultures. J Neurochem 2003; 84: 1332-9.

[31]   Halliwell B. Oxidative stress and neurodegeneration: where are we now? J Neurochem 2006; 97: 1634-58.

[32]   Lin MT, Beal MF. Mitochondrial dysfunction and oxidative stress in neurodegenerative diseases. Nature 2006; 443: 787-95.

[33]   Dringen R, Pfeiffer B, Hamprecht B. Synthesis of the antioxidant glutathione in neurons: supply by astrocytes of CysGly as precursor for neuronal glutathione. J Neurosci 1999; 19: 562-9.

[34]   Dringen R, Gutterer JM, Hirrlinger J. Glutathione metabolism in brain metabolic interaction between astrocytes and neurons in the defense against reactive oxygen species. Eur J Biochem 2000; 267: 4912-6.

[35]   Wang XF, Cynader MS. Astrocytes provide cysteine to neurons by releasing glutathione. J Neurochem 2000; 74: 1434-42.

[36]   Conrad M, Sato H. The oxidative stress-inducible cystine/glutamate antiporter, system x (c) (-): cystine supplier and beyond. Amino Acids 2012; 42: 231-46.

[37]   Sagara JI, Miura K, Bannai S. Maintenance of neuronal glutathione by glial cells. J Neurochem 1993; 61: 1672-6.

[38]   Yudkoff M, Pleasure D, Cregar L, *et al.* Glutathione turnover in cultured astrocytes: studies with [15N] glutamate. J Neurochem 1990; 55: 137-45.

[39]   Anderson CL, Iyer SS, Ziegler TR, Jones DP. Control of extracellular cysteine/cystine redox state by HT-29 cells is independent of cellular glutathione. Am J Physiol-Reg I 2007; 293: R1069-75.

[40]   Banjac A, Perisic T, Sato H, *et al.* The cystine/cysteine cycle: a redox cycle regulating susceptibility *versus* resistance to cell death. Oncogene 2008; 27: 1618-28.

[41]   Janssen-Heininger YM, Mossman BT, Heintz NH, *et al.* Redox-based regulation of signal transduction: principles, pitfalls, and promises. Free Rad Biol Med 2008; 45: 1-17.

[42]   Trotti D, Rizzini BL, Rossi D, *et al.* Neuronal and glial glutamate transporters possess an SH-based redox regulatory mechanism. Eur J Neurosci 1997; 9: 1236-43.

[43]   Lipton SA, Choi YB, Takahashi H, *et al.* Cysteine regulation of protein function as exemplified by NMDA-receptor modulation. Trends Neurosci 2002; 25: 474-80.

[44]   Pan ZH, Zhang X, Lipton SA. Redox modulation of recombinant human GABA(A) receptors. Neuroscience 2000; 98: 333-8.

[45]   Calero CI, Calvo DJ. Redox modulation of homomeric rho1 GABA receptors. J Neurochem 2008; 105: 2367-74.

[46]   Pan ZH, Ba̅hring R, Grantyn R, Lipton SA. Differential modulation by sulfhydryl redox agents and glutathione of GABA and glycine-evoked currents in rat retinal ganglion cells. J Neurosci 1995; 15: 1384-91.

[47]   Song MY, Makino A, Yuan JX. Role of reactive oxygen species and redox in regulating the function of transient receptor potential channels. Antiox Redox Signal 2011; 15: 1549-65.

[48]   Nelson RA, Boyd SJ, Ziegelstein RC, *et al.* Effect of rate of administration on subjective and physiological effects of intravenous cocaine in humans. Drug Alcohol Depend 2006; 82: 19-24.

[49]   DiChiara TJ, Reinhart PH. Redox modulation of hslo Ca^{2+}-activated K^+channels. J Neurosci 1997; 17: 4942-55.

[50]   Barger SW, Goodwin ME, Porter MM, Beggs ML. Glutamate release from activated microglia requires the oxidative burst and lipid peroxidation. J Neurochem 2007; 101: 1205-13.

[51]   Marino SM, Gladyshev VN. Analysis and functional prediction of reactive cysteine residues. J Biol Chem 2012; 287: 4419-25.

[52]   Lutolf MP, Tirelli N, Cerritelli S, Cavalli L, Hubbell JA. Systematic modulation of Michael-type reactivity of thiols through the use of gharged amino acids. Bioconjugate Chem 2001; 12: 1051-56.

[53]   Roos G, Foloppe N, Messens J. Understanding the pK_a of redox cysteines: the key role of hydrogen bonding. Antioxid Redox Signal 2013; 18: 94-127.

[54]   Kortemme T, Creighton TE. Ionisation of cysteine residues at the termini of model alpha-helical peptides. Relevance to unusual thiol pKa values in proteins of the thioredoxin family. J Mol Biol 1995, 253: 799-812.

[55]   Hall A, Parsonage D, Poole LB, Karplus PA. Structural evidence that peroxiredoxin catalytic power is based on transition-state stabilization. J Mol Biol 2010, 402: 194-209.

[56]   Paulsen CE, Truong TH, Garcia FJ, *et al.* Peroxide-dependent sulfenylation of the EGFR catalytic site enhances kinase activity. Nat Chem Biol 2012; 8: 57-64.

[57] Winterbourn CC, Metodiewa D. Reactivity of biologically important thiol compounds with perodide and hydrogen peroxide. Free Rad Biol Med 1999; 27: 322-28.

[58] Paulsen CE, Carroll KS. Cysteine-mediated redox signaling: chemistry, biology, and tools for discovery. Chem Rev 2013; 113: 4633-79.

[59] Martindale JL, Holbrook NJ. Cellular response to oxidative stress: signaling for suicide and survival. J Cell Physiol 2002; 192: 1-15.

[60] Levonen AL, Landar A, Ramachandran A, *et al.* Cellular mechanisms of redox cell signalling: role of cysteine modification in controlling antioxidant defences in response to electrophilic lipid oxidation products. Biochem J 2004; 378: 373-82.

[61] Diers AR, Higdon AN, Ricart KC, *et al.* Mitochondrial targeting of the electrophilic lipid 15-deoxy-Delta 12, 14-prostaglandin J2 increases apoptotic efficacy *via* redox cell signalling mechanisms. Biochem J 2010; 426: 31-41.

[62] D'Autreaux B, Toledano MB. ROS as signalling molecules: mechanisms that generate specificity in ROS homeostasis. Nat Rev Mol Cell Biol 2007; 8: 813-24.

[63] Toledo JC Jr, Augusto O. Connecting the chemical and biological properties of nitric oxide. Chem Res Toxicol 2012; 25: 975-89.

[64] de Groot H, Littauer A. Hypoxia, reactive oxygen, and cell injury. Free Rad Biol Med 1989; 6: 541-51.

[65] Cross AR, Jones OT. Enzymic mechanisms of superoxide production. Biochim Biophys Acta 1991; 1057: 281-98.

[66] Finkel T. Oxygen radicals and signaling. Curr Opin Cell Biol 1998; 10: 248-53.

[67] Prosser BL, Ward CW, Lederer WJ. X-ROS signaling: rapid mechanochemo transduction in heart. Science 2011; 333: 1440-45.

[68] Brown DI, Griendling KK. Nox proteins in signal transduction. Free Rad Biol Med 2009; 47: 1239-53.

[69] Aguirre J, Lambeth JD. Nox enzymes from fungus to fly to fish and what they tell us about Nox function in mammals. Free Rad Biol Med 2010; 49: 1342-53.

[70] Chen K, Craige SE, Keaney JF Jr. Downstream targets and intracellular compartmentalization in Nox signaling. Antiox Redox Signal 2009; 11: 2467-80.

[71] Yazdanpanah B, Wiegmann K, Tchikov V, *et al.* Riboflavin kinase couples TNF receptor 1 to NADPH oxidase. Nature 2009; 460: 1159-63.

[72] Oshikawa J, Urao N, Kim HW, *et al.* Extracellular SOD-derived H_2O_2 promotes VEGF signaling in caveolae/lipid rafts and post-ischemic angiogenesis in mice. PLoS One 2010; 5: e10189.

[73] Smith SM, Min J, Ganesh T, *et al.* Ebselen and congeners inhibit NADPH oxidase 2-dependent superoxide generation by interrupting the binding of regulatory subunits. Chem Biol 2012; 19: 752-63.

[74] Hahn NE, Meischl C, Wijnker PJ, *et al.* NOX2, p22phox and p47phox are targeted to the nuclear pore complex in ischemic cardiomyocytes colocalizing with local reactive oxygen species. Cell Physiol Biochem 2011; 27: 471-8.

[75] Giorgio M, Trinei M, Migliaccio E, Pelicci PG. Hydrogen peroxide: a metabolic by-product or a common mediator of ageing signals? Nat Rev Mol Cell Biol 2007; 8: 722-8.

[76] Mailloux RJ, Jin X, Willmore WG. Redox regulation of mitochondrial function with emphasis on cysteine oxidation reactions. Redox Biol 2013; 2: 123-39.

[77] Guzy RD, Hoyos B, Robin E, *et al.* Mitochondrial complex III is required for hypoxia-induced ROS production and cellular oxygen sensing. Cell Metab 2005; 1: 401-8.

[78] Hamanaka RB, Chandel NS. Mitochondrial reactive oxygen species regulate cellular signaling and dictate biological outcomes. Trends Biochem Sci 2010; 35: 505-13.

[79] Anastasiou D, Poulogiannis G, Asara JM, et al. Inhibition of pyruvate kinase M2 by reactive oxygen species contributes to cellular antioxidant responses. Science 2011; 334: 1278-83.

[80] Grüning N-M, Rinnerthaler M, Bluemlein K, et al. Pyruvate kinase triggers a metabolic feedback loop that controls redox metabolism in respiring cells. Cell Metab 2011; 14: 415-27.

[81] Murphy MP. Mitochondrial thiols in antioxidant protection and redox signaling: distinct roles for glutathionylation and other thiol modifications, Antioxid Redox Signal 2012; 16: 476-95.

[82] Mailloux RJ, Lemire J, Appanna VD. Hepatic response to aluminum toxicity: dyslipidemia and liver diseases. Exp Cell Res 2011; 317: 2231-38.

[83]   Hurd TR, Costa NJ, Dahm CC, *et al.* Glutathionylation of mitochondrial proteins. Antioxid Redox Signal 2005; 7: 999-1010.

[84]   Piantadosi CA. Regulation of mitochondrial processes by protein S-nitrosylation. Biochim Biophys Acta 2012; 1820: 712-21.

[85]   Fomenko DE, Xing W, Adair BM, Thomas DJ, Gladyshev VN. High-throughput identification of catalytic redox-active cysteine residues. Science 2007; 315: 387-9.

[86]   Weerapana E, Wang C, Simon GM, *et al.* Quantitative reactivity profiling predicts functional cysteines in proteomes. Nature 2010; 468: 790-5.

[87]   Sanchez R, Riddle M, Woo J, Momand J. Prediction of reversibly oxidized protein cysteine thiols using protein structure properties. Protein Sci 2008; 17: 473-81.

[88]   Reddie KG, Carroll KS. Expanding the functional diversity of proteins through cysteine oxidation. Curr Opin Chem Biol 2008; 12: 746-54.

[89]   Tanner JJ, Parsons ZD, Cummings AH, Zhou H, Gates KS. Redox regulation of protein tyrosine phosphatases: structural and chemical aspects. Antioxid Redox Signal 2011; 15: 77-97.

[90]   Denu JM, Tanner GK. Specific and reversible inactivation of protein tyrosine phosphatases by hydrogen peroxide: evidence for a sulfenic acid intermediate and implications for redox regulation. Biochemistry 1998; 37: 5633-42.

[91]   Barrett WC, DeGnore JP, Keng YF, Zhang ZY, Yim MB, Chock PB. Roles of superoxide radical anion in signal transduction mediated by reversible regulation of protein-tyrosine phosphatase 1B. J Biol Chem 1999; 274: 34543-46.

[92]   Guttmann RP. Redox regulation of cysteine-dependent enzymes. J Animal Sci 2010; 88: 1297-1306.

[93]   Goll DE, Thompson VF, Li H, Wei W, Cong J. The calpain system. Physiol Rev 2003; 83: 731-801.

[94]   Bulteau Al, LundbergKC, Ikeda-Saito M, Isaya G, Szweda LI. Reversible redox-dependent modulation of mitochondrial aconitase and proteolytic activity during *in vivo* cardiac ischemia/reperfusion. P Natl Acad Sci USA 2005; 102: 5987-91.

[95]   Beer SM, Taylor ER, Brown SE, *et al.* Glutaredoxin 2 catalyzes the reversible oxidation and glutathionylation of mitochondrial membrane thiol proteins: implications for mitochondrial redox regulation and antioxidant defense. J Biol Chem 2004; 279: 47939-51.

[96]   Liesa M, Shirihai OS. Mitochondrial dynamics in the regulation of nutrient utilization and energy expenditure. Cell Metabol 2013; 17: 491-506.

[97]   Chan DC. Mitochondria: dynamic organelles in disease, aging, and development. Cell 2006; 125: 1241-52.

[98]   Shutt T, Geoffrion M, Milne R, McBride HM. The intracellular redox state is a core determinant of mitochondrial fusión. EMBO Rep 2012; 13: 909-15.

[99]   Redpath CJ, Bou Khalil M, Drozdzal G, Radisic M, McBride HM. Mitochondrial hyperfusion during oxidative stress is coupled to a dysregulation in calcium handling within a C2C12 cell model. PLoS One 2013; 8: e69165.

[100]   Kowaltowski AJ, Castilho RF, Vercesi AE. Mitochondrial permeability transition and oxidative stress. FEBS Lett 2001; 495: 12-5.

[101]   Halestrap AP, Brenner C. The adenine nucleotide translocase: a central component of the mitochondrial permeability transition pore and key player in cell death. Curr Med Chem 2003; 10: 1507-25.

[102]   Kowaltowski AJ, Castilho RF, Vercesi AF. Ca(2+)-induced mitochondrial membrane permeabilization: role of coenzyme Q redox state. Am J Physiol 1995; 269: C141-7.

[103]   Yin F, Sancheti H, Cadenas E. Mitochondrial thiols in the regulation of cell death pathways. Antioxid Redox Signal 2012; 17: 1714-27.

[104]   McStay GP, Clarke SJ, Halestrap AP. Role of critical thiol groups on the matrix surface of the adenine nucleotide translocase in the mechanism of the mitochondrial permeability transition pore. Biochem J 2002; 367: 541-8.

[105]   Kaukonen J, Juselius JK, Tiranti V, *et al.* Role of adenine nucleotide translocator 1 in mtDNA maintenance. Science 2000; 289: 782-5.

[106]   Queiroga CS, Almeida AS, Martel C, Brenner C, Alves PM, Vieira HL. Glutathionylation of adenine nucleotide translocase induced by carbon monoxide prevents mitochondrial membrane permeabilization and apoptosis. J Biol Chem 2010; 285: 17077-88.

[107]   Goshima Y, Nakamura F, Strittmatter P, Strittmatter S.M. Collapsin-induced growth cone collapse mediated by an intracellular protein related to UNC-33. Nature 1995; 376: 509-14.

[108]   Deo RC, Schmidt EF, Elhabazi A, Togashi H, Burley SK, Strittmatter SM. Structural bases for CRMP function in plexin-dependent semaphorin3A signaling. EMBO J 2004; 23: 9-22.

[109]   Stenmark P, Ogg D, Flodin S, *et al.* The structure of human collapsin response mediator protein 2, a regulator of axonal growth. J Neurochem 2007; 101: 906-17.

[110]   Luanpitpong S, Chanvorachote P, Stehlik C, *et al.* Regulation of apoptosis by Bcl-2 cysteine oxidation in human lung epithelial cells. Mol Biol Cell 2013; 24: 858-69.

[111]   Truong TH, Carroll KS. Redox regulation of EGFR signaling through cysteine oxidation. Biochemistry 2012; 51: 9954-65.

[112]   Lassing I, Schmitzberger F, Björnstedt M, *et al.* Molecular and structural basis for redox regulation of β-actin. J Mol Biolo 2007; 370: 331-48.

[113]   Lu M, Kim HE, Li CR, *et al.* Two distinct disulfide bonds formed in human heat shock transcription factor 1 actin opposition to regulate its DNA binding activity. Biochemistry 2008; 47: 6007-15.

[114]   Sekhar KR, Yan XX, Freeman ML. Nrf2 degradation by the ubiquitin proteasome pathway is inhibited by KIAA0132, the human homolog to INrf2. Oncogene 2002; 21: 6829-34.

[115]   Piccirillo S, Filomeni G, Brune B, Rotilio G, Ciriolo MR. Redox mechanisms involved in the selective activation of Nrf2-mediated resistance *versus* p53-dependent apoptosis in adenocarcinoma gastric cells. J Biol Chem 2009; 284: 27721-33.

[116]   Hall CN, Garthwaite J. What is the real physiological NO concentration *in vivo*? Nitric Oxide 2009; 21: 92-103.

[117]   Bogdan C. Nitric oxide and the immune response. Nat Immunol 2001; 2: 907-16.

[118]   Ferrer-Sueta G, Radi R. Chemical biology of peroxynitrite: kinetics, diffusion, and radicals. ACS Chem Biol 2009; 4: 161-77.

[119]   Winterbourn CC, Kettle AJ. Redox reactions and microbial killing in the neutrophil phagosome. Antioxid Redox Signal 2013; 18: 642-60.

[120]   Roman LJ, Martásek P, Masters BS. Intrinsic and extrinsic modulation of nitric oxide synthase activity. Chem Rev 2002; 102: 1179-90.

[121]   Lane P, Gross SS. Disabling a C-terminal autoinhibitory control element in endothelial nitric-oxide synthase by phosphorylation provides a molecular explanation for activation of vascular NO synthesis by diverse physiological stimuli. J Biol Chem 2002; 277: 19087-94.

[122]   Shiva S, Wang X, Ringwood LA, Xu X, *et al.* Ceruloplasmin is a NO oxidase and nitrite synthase that determines endocrine NO homeostasis. Nat Chem Biol 2006; 2: 486-93.

[123]   Nagababu E, Ramasamy S, Abernethy DR, Rifkind JM. Active nitric oxide produced in the red cell under hypoxic conditions by deoxyhemoglobin-mediated nitrite reduction. J Biol Chem 2003; 278: 46349-56.

[124]   Sarti P, Arese M, Forte E, Giuffrè A, Mastronicola D. Mitochondria and nitric oxide: chemistry and pathophysiology. Adv Exp Med Biol 2012; 942: 75-92.

[125]   Moran EE, Timerghazin QK, Kwong E, English AM. Kinetics and mechanism of S-nitrosothiol acid-catalyzed hydrolysis: sulfur activation promotes facile NO+ release. J Phys Chem B. 2011; 115: 3112-26.

[126]   Migaud M, Charlesworth P, Dempster M, *et al.* Enhanced long-term potentiation and impaired learning in mice with mutant postsynaptic density-95 protein. Nature 1998; 396: 433-9.

[127]   Foster MW, McMahon TJ, Stamler JS. S-Nitrosylation in health and disease. Trends Mol Med 2003; 9: 160-8.

[128]   Gow AJ, Chen Q, Hess DT, Day BJ, Ischiropoulos H, Stamler JS. Basal and stimulated protein S-nitrosylation in multiple cell types and tissues. J Biol Chem 2002; 277: 9637-40.

[129]   Wong A, Luth HJ, Deuther-Conrad W, *et al.* Advanced glycation endproducts colocalize with inducible nitric oxide synthase in Alzheimer's disease. Brain Res 2001; 920: 32-40.

[130]   Thippeswamy T, Morris R. The roles of nitric oxide in dorsal root ganglion neurons. Ann New York Acad Sci 2002; 962: 103-10.

[131]   Kligman D, Marshak DR. Purification and characterization of a neurite extension factor from bovine brain. Proc Natl Acad Sci USA 1985; 82: 7136-9.

[132] Zhukova L, Zhukov I, Bal W, Wyslouch-Cieszynska A. Redox modifications of the C-terminal cysteine residue cause structural changes in S100A1 and S100B proteins. Biochim Biophys Acta 2004; 1742: 191-201.

[133] Mishanina TV, Libiad M, Banerjee R. Biogenesis of reactive sulfur species for signaling by hydrogen sulfide oxidation pathways. Nat Chem Biol 2015; 11: 457-64.

[134] Giles GI, Tasker KM, Collins C, Giles NM, O'rourke E, Jacob C. Reactive sulphur species: an *in vitro* investigation of the oxidation properties of disulphide S-oxides. Biochem J 2002; 364: 579-85.

[135] Szabó C. Hydrogen sulphide and its therapeutic potential. Nat Rev Drug Disc 2007; 6: 917-35.

[136] Predmore BL, Lefer DJ, Gojon G. Hydrogen sulfide in biochemistry and medicine. Antioxid Redox Signal 2012; 17: 119-40.

[137] Kabil O, Banerjee R. Redox Biochemistry of Hydrogen Sulfide. J Biol Chem 2010; 285: 21903-7.

[138] Dombkowski RA, Russell MJ, Schulman AA, Doellman MM, Olson KR. Vertebrate phylogeny of hydrogen sulfide vasoactivity. Am J Physiol-Reg I 2005; 288: R243-52.

[139] Olson KR. Is hydrogen sulfide a circulating ''gasotransmitter'' in vertebrate blood? Biochim Biophis Acta 2009; 1787: 856-63.

[140] Tomaskova Z, Cacanyiova S, Benco A, *et al*. Lipids modulate H(2)S/ HS(-) induced NO release from S-nitrosoglutathione. Biochem Bioph Res Co 2009; 390: 1241-4.

[141] Kimura Y, Goto Y, Kimura H. Hydrogen sulfide increases glutathione production and suppresses oxidative stress in mitochondria. Antioxid Redox Signal 2010; 12: 1-13.

[142] Nagy P, Winterbourn CC. Rapid reaction of hydrogen sulfide with the neutrophil oxidant hypochlorous acid to generate polysulfides. Chem Res Toxicol 2010; 23: 1541- 3.

[143] Stasko A, Brezova V, Zalibera M, Biskupic S, Ondrias K. Electron transfer: A primary step in the reactions of sodium hydrosulphide, an H2S/HS(-) donor. Free Rad Res 2009; 43: 581-93.

[144] Li L, Rose P, Moore PK. Hydrogen sulfide and cell signaling. Annu Rev Pharmacol Toxicol 2011; 51: 169-87.

[145] Sen N, Paul BD, Gadalla MM, *et al*. Hydrogen sulfide-linked sulfhydration of NF-κB mediates its antiapoptotic actions. Mol Cell 2012; 45: 13-24.

[146] Whiteman M, Moore PK. Is hydrogen sulfide a regulator of nitric oxide bioavailability in the vasculature? In: Jacob C, Winyard PG, Eds. Redox signaling and regulation in biology and medicine. Weinheim, Germany: Wiley-VCH 2009; pp. 293–314.

[147] Whiteman M, Moore PK. Hydrogen sulfide and the vasculature: a novel vasculoprotective entity and regulator of nitric oxide bioavailability? J Cell Mol Med 2009; 13: 488–507.

[148] Whiteman M, Armstrong JS, Chu SH, *et al*. The novel neuromodulator hydrogen sulfide: an endogenous peroxynitrite 'scavenger'? J Neurochem 2004; 90: 765-8.

[149] Tyagi N, Moshal KS, Sen U, *et al*. H2S protects against methionine-induced oxidative stress in brain endothelial cells. Antioxid Redox Signal 2009; 11: 25–33.

[150] Giles GI, Jacob C. Reactive sulfur species: an emerging concept in oxidative stress. Biol Chem 2002; 383: 375–88.

[151] Baskar R, Bian J. Hydrogen sulfide gas has cell growth regulatory role. Eur J Pharmacol 2011; 656: 5-9.

[152] Gadalla MM, Snyder SH. Hydrogen sulfide as a gasotransmitter. J Neurochem 2010; 113: 14–26.

[153] Kimura H. Hydrogen sulfide: its production, release and functions.Amino Acids 2011; 41: 113–121.

[154] Jacob C, Battaglia E, Burkholz T, Peng D, Bagrel D, Montenarh M. Control of oxidative posttranslational cysteine modifications: from intricate chemistry to widespread biological and medical applications. Chem Res Toxicol 2012; 25: 588-604.

[155] Mikami Y, Shibuya N, Kimura Y, Nagahara N, Ogasawara Y, Kimura H. Thioredoxin and dihydrolipoic acid are required for 3-mercaptopyruvate sulfurtransferase to produce hydrogen sulfide. Biochem J 2011; 439: 479-85.

[156] Viscomi C, Burlina AB, Dweikat I, *et al*. Combined treatment with oral metronidazole and N-acetylcysteine is effective in ethylmalonic encephalopathy. Nat Med 2010; 16: 869-71.

[157] Abe K, Kimura H. The possible role of hydrogen sulfide as an endogenous neuromodulator. J Neurosci 1996; 16: 1066-71.

[158] Eto K, Ogasawara M, Umemura K, Nagai Y, Kimura H. Hydrogen sulfide is produced in response to neuronal excitation. J Neurosci 2002; 22: 3386-91.

[159]   Calvert JW, Jha S, Gundewar S, *et al.* Hydrogen sulfide mediates cardioprotection through Nrf2 signaling. Circ Res 2009; 105: 365-74.

[160]   Giustarini D, Del Soldato P, Sparatore A, Rossi R. Modulation of thiol homeostasis induced by H2S-releasing aspirin. Free Rad Biol Med 2010; 48: 1263-72.

[161]   Kimura Y, Kimura H. Hydrogen sulfide protects neurons from oxidative stress. FASEB J 2004; 18: 1165-7.

[162]   Nelson KC, Armstrong JS, Moriarty S, *et al.* Protection of retinal pigment epithelial cells from oxidative damage by oltipraz, a cancer chemopreventive agent. Invest Ophthalmol Vis Sci 2002; 43: 3550-4.

[163]   Osborne NN, Ji D, Abdul Majid AS, Fawcett RJ, Sparatore A, Del Soldato P. ACS67, a hydrogen sulfide-releasing derivative of latanoprost acid, attenuates retinal ischemia and oxidative stress to RGC-5 cells in culture. Invest Ophthalmol Vis Sci 2010; 51: 284-94.

[164]   Pryor WA, Houk KN, Foote CS, *et al.* Free radical biology and medicine: It's a gas, man! Am J Physiol-Reg I 2006; 291: R491-511.

[165]   Ballatori N, Krance SM, Notenboom S, Shi S, Tieu K, Hammond CL. Glutathione dysregulation and the etiology and progression of human diseases. Biol Chem 2009; 390: 191-214.

[166]   Ha MW, Ma R, Shun LP, Gong YH, Yuan Y. Effects of allitridi on cell cycle arrest of human gastric cancer cells. World J Gastroenterol 2005; 11: 5433-7.

[167]   Sparatore A, Santus G, Giustarini D, Rossi R, Del Soldato P. Therapeutic potential of new hydrogen sulfide-releasing hybrids. Expert Rev Clin Pharmacol 2011; 4: 109-21.

[168]   Tan BH, Wong PT, Bian JS. Hydrogen sulfide: A novel signaling molecule in the central nervous system. Neurochem Int 2010; 56: 3-10.

[169]   Kabil O, Vitvitsky V, Banerjee R. Sulfur as a Signaling nutrient through hydrogen sulfide. Ann Rev Nutr 2014; 34: 171-205.

[170]   Goubern M, Andriamihaja M, Nübel T, Blachier F, Bouillaud F. Sulfide, the first inorganic substrate for human cells. FASEB J 2007; 21: 1699-706.

[171]   Powell MA, Somero GN. Hydrogen sulfide oxidation is coupled to oxidative phosphorylation in mitochondria of solemya reidi. Science 1986; 233: 563-6.

[172]   Hildebrandt TM, Grieshaber MK. Three enzymatic activities catalyze the oxidation of sulfide to thiosulfate in mammalian and invertebrate mitochondria. FEBS J 2008; 275: 3352-61.

[173]   Mueller EG. Trafficking in persulfides: delivering sulfur in biosynthetic pathways. Nat Chem Biol 2006; 2: 185-94.

[174]   Kurihara T, Mihara H, Kato S, Yoshimura T, Esaki N. Assembly of iron-sulfur clusters mediated by cysteine desulfurases, IscS, CsdB and CSD, from *Escherichia coli.* Biochim Biophys Acta 2003; 1647: 303-9.

[175]   Li K, Tong WH, Hughes RM, Rouault TA. Roles of the mammalian cytosolic cysteine desulfurase, ISCS, and scaffold protein, ISCU, in iron-sulfur cluster assembly. J Biol Chem 2006; 281: 12344-51.

[176]   Rudolph TK, Freeman BA. Transduction of redox signaling by electrophile-protein reactions. Sci Signal 2009; 2: re7.

[177]   Schreier SM, Muellner MK, Steinkellner H, *et al.* Hydrogen sulfide scavenges the cytotoxic lipid oxidation product 4-HNE. Neurotox Res 2010; 17: 249-56.

[178]   Toohey JI. Sulfur signaling: is the agent sulfide or sulfane? Anal Biochem 2011; 413: 1-7.

[179]   Tiranti V, Viscomi C, Hildebrandt T, *et al.* Loss of ETHE1, a mitochondrial dioxygenase, causes fatal sulfide toxicity in ethylmalonic encephalopathy. Nat Med 2009; 15: 200-5.

[180]   Geng B, Cui Y, Zhao J, *et al.* Hydrogen sulfide downregulates the aortic L-arginine/nitric oxide pathway in rats. Am J Physiol-Reg I 2007; 293: R1608-18.

[181]   Kubo S, Kurokawa Y, Doe I, Masuko T, Sekiguchi F, Kawabata A. Hydrogen sulfide inhibits activity of three isoforms of recombinant nitric oxide synthase. Toxicology 2007; 241: 92-7.

[182]   Oh GS, Pae HO, Lee BS, *et al.* Hydrogen sulfide inhibits nitric oxide production and nuclear factor-kappaB *via* heme oxygenase-1 expression in RAW264.7 macrophages stimulated with lipopolysaccharide. Free Rad Biol Med 2006; 41: 106-19.

[183]   Sarti P, Arese M, Giuffrè A. The molecular mechanisms by which nitric oxide controls mitochondrial complex IV. Ital J Biochem 2003; 52: 37-42.

[184]   Muzaffar S, Shukla N, Bond M, *et al.* Exogenous hydrogen sulfide inhibits superoxide formation, NOX-1 expression and Rac1 activity in human vascular smooth muscle cells. J Vasc Res 2008; 45: 521-8.

[185]   Pacher P, Beckman JS, Liaudet L. Nitric oxide and peroxynitrite in health and disease. Physiol Rev 2007; 87: 315-424.

[186]   Klomsiri C, Karplus PA, Poole LB. Cysteine-based redox switches in enzymes. Antiox Redox Signal 2011; 14: 1065-77.

[187]   Hall A, Karplus, PA, Poole LB. Typical 2-Cys peroxiredoxins structures, mechanisms and functions. FEBS J 2009; 276: 2469-77.

[188]   Jones DP, Go YM. Mapping the cysteine proteome: analysis of redox-sensing thiols. Curr Opin Chem Biol 2011; 15: 103-12.

[189]   Damdimopoulos, AE, Miranda-Vizuete A, Pelto-Huikko M, Gustafsson JA, Spyrou G. Human mitochondrial thioredoxin. Involvement in mitochondrial membrane potential and cell death. J Biol Chem 2002; 277: 33249-57.

[190]   Ren X, Bjornstedt M, Shen B, Ericson ML, Holmgren A. Mutagenesis of structural half-cystine residues in human thioredoxin and effects on the regulation of activity by seleno-diglutathione. Biochemistry 1993; 32: 9701-8.

[191]   Holmgren A, Lu J. Thioredoxin and thioredoxin reductase: current research with special reference to human disease. Biochem Biophys Res Co 2010; 396: 120-4.

[192]   Ichijo H, Nishida E, Irie K, *et al.* Induction of apoptosis by ASK1, a mammalian MAPKKK that activates SAPK/ JNK and p38 signaling pathways. Science 1997; 275: 90-4.

[193]   Saitoh M, Nishitoh H, Fujii M, *et al.* Mammalian thioredoxin is a direct inhibitor of apoptosis signal-regulating kinase (ASK) 1. EMBO J 1998; 17: 2596-606.

[194]   Lillig CH, Berndt C, Holmgren A. Glutaredoxin systems. Biochim Biophys Acta 2008; 1780: 1304-17.

[195]   Pai HV, Starke DW, Lesnefsky EJ, Hoppel CL, Mieyal JJ. What is the functional significance of the unique location of glutaredoxin 1 (GRx1) in the intermembrane space of mitochondria? Antioxid Redox Signal 2007; 9: 2027-33.

[196]   Lu J, Holmgren A. Selenoproteins. J Biol Chem 2009; 284: 723-7.

[197]   Park JW, Mieyal JJ, Rhee SG, Chock PB. Deglutathionylation of 2-Cys peroxiredoxin is specifically catalyzed by sulfiredoxin. J Biol Chem 2009; 284: 23364-74.

[198]   Rhee SG, Kang SW, Netto LE, Seo MS, Stadtman ER. A family of novel peroxidases, peroxiredoxins. Biofactors 1999; 10: 207-9.

[199]   Rhee SG, Chae HZ, Kim K. Peroxiredoxins: a historical overview and speculative preview of novel mechanisms and emerging concepts in cell signaling. Free Rad Biol Med 2005; 38: 1543-52.

[200]   Stacey MM, Vissers MC, Win- terbourn, CC. Oxidation of 2-cys peroxiredoxins in human endothelial cells by hydrogen peroxide, hypochlorous acid, and chloramines. Antioxid Redox Signal 2012; 17: 411-21.

[201]   Chae HZ, Uhm TB, Rhee SG. Dimerization of thiol-specific antioxidant and the essential role of cysteine 47. Proc Natl Aca Sci USA. 1994; 91: 7022-6.

[202]   Chang TS, Jeong W, Woo HA, Lee SM, Park S, Rhee SG. Characterization of mammalian sulfiredoxin and its reactivation of hyperoxidized peroxiredoxin through reduction of cysteine sulfinic acid in the active site to cysteine. J Biol Chem 2004; 279: 50994-1001.

[203]   Rhee SG, Woo HA, Kil IS, Bae SH. Peroxiredoxin functions as a peroxidase and a regulator and sensor of local peroxides. J Biol Chem 2012; 287: 4403-10.

[204]   Monteiro G, Horta BB, Pimenta DC, Augusto O, Netto LE. Reduction of 1-Cys peroxiredoxins by ascor- bate changes the thiol-specific antioxidant paradigm, revealing another function of vitamin C. P Natl Acad Sci USA 2007; 104: 4886-91.

[205]   Bannai S, Christensen HN, Vadgama JV, Ellory JC, Englesberg E, Guidotti GG. Amino acid transport systems. Nature 1984; 311: 308.

[206]   Mullineaux PM, Creissen GP. Glutathione reductase: regulation and role in oxidative stress. Oxidative stress and the molecular biology of antioxidant defenses. Cold Spring Harbor Monograph Arch 1997; 34: 667-713.

[207]   Lind C, Gerdes R, Schuppe-Koistinen I, Cotgreave IA. Studies on the mechanism of oxidative modification of human glyceraldehyde-3 phosphate dehydrogenase by glutathione: catalysis by glutaredoxin. Biochem Biophys Res Co 1998; 247: 481-6.

[208]   Klatt P, Pineda-Molina E, Garcia de Lacoba M, Padilla A, Martinez-Galisteo E, Barcena JA. Redox regulation of c-Jun DNA binding by reversible S-glutathiolation. FASEB J 1999; 13: 1481-90.

[209]   Ward NE, Pierce DS, Chung SE, Gravitt KR, O'Brian CA. Irreversible inactivation of protein kinase C by glutathione. J Biol Chem 1998; 273: 12558-66.

[210]   Rainwater R, Parks D, Anderson ME, Tegtmeyer P, Mann K. Role of cysteine residues in regulation of p53 function. Mol Cell Biol 1995; 15: 3892-903.

[211]   Liu M, Pelling JC, Ju J, Chu E, Brash DE. Antioxidant action *via* p53-mediated apoptosis. Cancer Res 1998; 58: 1723-9.

[212]   Østergaard H, Tachibana C, Winther JR. Monitoring disulfide bond formation in the eukaryotic cytosol. J Cell Biol 2004; 166: 337-45.

[213]   Morgan B, Ezerina D, Amoako TN, Riemer J, Seedorf M, Dick TP. Multiple glutathione disulfide removal pathways mediate cytosolic redox homeostasis. Nat Chem Biol 2013; 9: 119-25.

[214]   Ziegler DM. Role of reversible oxidation-reduction of enzyme thiols- disulfides in metabolic regulation. Ann Rev Biochem 1985; 54: 305-29.

[215]   Hill BG, Bhatnagar A. Protein S-glutathiolation: redox-sensitive regulation of protein function. J Mol Cell Cardiol 2012; 52: 559-67.

[216]   Klatt P, Lamas S. Regulation of protein function by S-glutathiolation in response to oxidative and nitrosative stress. Eur J Biochem 2000; 267: 4928-44.

[217]   Gutscher M, Sobotta MC, Wabnitz GH, *et al.* Proximity-based protein thiol oxidation by H_2O_2-scavenging peroxidases. J Biol Chem 2009; 284: 31532-40.

[218]   Mandl J, Szarka A, Banhegyi G. Vitamin C: Update on physiology and pharmacology. Br J Pharmacol 2009; 157: 1097-110.

[219]   Barja G. Endogenous oxidative stress: relationship to aging, longevity and caloric restriction. Ageing Res Rev 2002; 1: 397-411.

[220]   Bokov A, Chaudhuri A, Richardson A. The role of oxidative damage and stress in aging. Mech Ageing Dev 2004; 125: 811-26.

[221]   Perez VI, Bokov A, Van Remmen H, *et al.* Is the oxidative stress theory of aging dead? Biochim Biophys Acta 2009; 1790: 1005-14.

[222]   Nuttall SL, Martin U, Sinclair AJ, Kendall JJ. Glutathione: in sickness and in health. Lancet 1998; 351: 645-6.

[223]   Chen TS, Richie JP, Hagasawa HT, Lang CA. Glutathione monoethyl ester protects against glutathione deficiencies due to aging and acetaminophen in mice. Mech Ageing Dev 2000; 120: 127-39.

[224]   De Benedictis G, Carrieri G, Varcasia O, Bonafe M, Franceschi C. Inherited variability of the mitochondrial genome and successful aging in humans. Ann New York Acad Sci 2000; 908: 208-18.

[225]   Hamilton ML, van Remmen H, Drake JA, *et al.* Does oxidative damage to DNA increase with age? Proc Natl Acad Sci USA 2001; 98: 10469-74.

[226]   Brochu M, Poehlman ET, Ades PA. Obesity, body fat distribution, and coronary artery disease. J Cardiopulm Rehab 2000; 20: 96-108.

[227]   Guo Z, Ersoz A, Butterfield DA, Mattson MP. Beneficial   effects of dietary restriction on cerebral cortical synaptic terminals: preservation of glucose transport and mitochondrial function after exposure to amyloid β-peptide and oxidative and metabolic insults. J Neurochem 2000; 75: 314-20.

[228]   Duan W, Mattson MP. Dietary restriction and 2-deoxyglucose administration improve behavioral outcome and reduce degeneration of dopaminergic neurons in models of Parkinson's disease. J Neurosci Res 1999; 57: 195-206.

[229]   Lowenstein DH, Chan PH, Miles MF. The stress protein response in cultured neurons: characterization and evidence for a protective role in excitotoxicity. Neuron 1991; 7: 1053-60.

[230]   Jones DP, Mody Jr VC, Carlson JL, Lynn MJ, Sternberg Jr P. Redox analysis of human plasma allows separation of pro-oxidant events of aging from decline in antioxidant defenses. Free Rad Biol Med 2002; 33: 1290-300.

[231]   Flint DH, Tuminello JF, Emptage MH. The inactivation of Fe-S cluster containing hydrolyases by superoxide. J Biol Chem 1993; 268: 22369-76.

[232]   Gardner PR, Fridovich I. Superoxide sensitivity of the *Escherichia coli* aconitase. J Biol Chem 1991; 266: 19328-33.

[233]   Mallis RJ, Hamann MJ, Zhao W, Zhang T, Hendrich S, Thomas JA. Irreversible thiol oxidation in carbonic anhydrase III: protection by S-glutathiolation and detection in aging rats. Biol Chem 2002; 383: 649-62.

[234]   Liedhegner EAS, Gao XH, Mieyal JJ. Mechanisms of altered redox regulation in neurodegenerative diseases—focus on S-glutathionylation. Antioxid Redox Signal 2012; 16: 543-66.

[235]   Ansari MA, Scheff SW. Oxidative stress in the progression of Alzheimer disease in the frontal cortex. J Neuropathol Exp Neurol 2010; 69: 155-67.

[236]   Akterin S, Cowburn RF, Miranda-Vizuete A, *et al.* Involvement of glutaredoxin-1 and thioredoxin-1 in beta-amyloid toxicity and Alzheimer's disease. Cell Death Diff 2006; 13: 1454-65.

[237]   Kenchappa RS, Ravindranath V. Glutaredoxin is essential for maintenance of brain mitochondrial complex I: studies with MPTP. FASEB J 2003; 17: 717-9.

[238]   Dinoto L, Deture MA, Purich DL. Structural insights into Alzheimer filament assembly pathways based on site-directed mutagenesis and S-glutathionylation of three-repeat neuronal Tau protein. Microsc Res Tech 2005; 67: 156-63.

[239]   Halliwell B, Gutteridge JM. Oxygen free radicals and iron in relation to biology and medicine: some problems and concepts. Arch Biochem Biophys 1986; 246: 501-14.

[240]   Koster JF, Sluiter W. Is increased tissue ferritin a risk factor for atherosclerosis and ischaemic heart disease? Br Heart J 1995; 73: 208-9.

[241]   Go YM, Jones DP. Cysteine/cystine redox signaling in cardiovascular disease. Free Rad Biol Med 2011; 50: 495-509.

[242]   Miquel J, Ferrándiz ML, De Juan E, Sevilla I, Martínez-Banaclocha M. N-acetylcysteine protecs against age-related decline of oxidative phosphorylation in liver mitochondria. Eur J Pharmacol 1995; 292: 333-5.

[243]   Musicco C, Capelli V, Pesce V, *et al.* Accumulation of overoxidized Peroxiredoxin III in aged rat liver mitocondria. Biochim Biophys Acta 2009; 1787: 890-6.

[244]   Velu CS, Niture SK, Doneanu CE, Pattabiraman N, Srivenugopal KS. Human p53 is inhibited by glutathionylation of cysteines present in the proximal DNA-binding domain during oxidative stress. Biochemistry 2007; 46: 7765-80.

[245]   Applegate MA, Humphries KM, Szweda LI. Reversible inhibition of alphaketoglutarate dehydrogenase by hydrogen peroxide: glutathionylation and protection of lipoic acid. Biochemistry 2008; 47: 473-8.

[246]   Nakamura T, Tu S, Akhtar MW, Sunico CR, Okamoto S, Lipton SA. Aberrant protein s-nitrosylation in neurodegenerative diseases. Neuron 2013; 78: 596-14.

[247]   Bouillaud F, Blachier F. Mitochondria and sulfide: a very old story of poisoning, feeding, and signaling? Antioxid Redox Signal 2011; 15: 379-91.

[248]   Youdim MB. Understanding Parkinson's disease. Scientific American 1997; 276: 52-9.

[249]   Forno LS. Neuropathology of Parkinson's disease. J Neuropathol Exp Neurol 1996; 55: 259-72.

[250]   Veldman BA, Wijn AM, Knoers N, *et al.* Genetic and environmental risk factors in Parkinson's Disease. Clin Neurol Neurosurg 1998; 100: 15-26.

[251]   Williams AC, Smith ML, Waring RH, *et al.* Idiopathic Parkinson's disease: a genetic and environmental model. Adv Neurol 1999; 80: 215-8.

[252]   Bowling AC, Mutisya EM, Walker LC, Price DL, Cork LC, Beal MF. Age-dependent impairment of mitochondrial function in primate brain. J Neurochem 1993; 60: 1964-7.

[253]   Curti D, Giangare MD, Redolfi ME, Fugaccia I, Benzi G. Age-related modification of cytochrome C oxidase activity in discrete brain regions. Mech Ageing Dev 1990; 55: 171-80.

[254]   Ferrándiz ML, Martínez-Banaclocha M, De Juan E, Díez A, Bustos G, Miquel J. Impairment of mitochondrial oxidative phosphorylation in the brain of aged mice. Brain Res 1994; 644: 335-8.

[255]   Martínez-Banaclocha M, Ferrándiz ML, De Juan E, Miquel J. Age-related changes in glutathione and lipid peroxide content in mouse synaptic mitochondria: relationship to cytochrome c oxidase decline. Neurosci Lett 1994; 170: 121-4.

[256]   Martínez-Banaclocha M, Ferrándiz ML, Diez A, Miquel J. Depletion of cytosolic GSH decreases the ATP levels and viability of synaptosomes from aged mice but not from young mice. Mech Ageing Dev 1995; 84: 77-81.

[257]   Martínez-Banaclocha M, Hernández AI, Martínez N. N-acetylcysteine delays age-associated memory impairment in mice: role in synaptic mitochondria. Brain Res 2000; 855: 100-6.

[258]   Martínez-Banaclocha M, Hernández AI, Martínez N, Ferrándiz ML. Age-related increase in oxidized proteins in mouse synaptic mitochondria. Brain Res 1996; 731: 246-8.

[259]   Langston JW. Parkinson's disease: current and future challenges. Neurotoxicology 2002; 23: 443-50.

[260]   Gilks WP, Abou-Sleiman PM, Gandhi S, *et al*. A common LRRK2 mutation in idiopathic Parkinson's disease. Lancet 2005; 365: 415-6.

[261]   Langston JW, Ballard P, Tetrud JW, Irwin I. Chronic parkinsonism in humans due to a product of meperidine-analog synthesis. Science 1983; 219: 979-80.

[262]   Sayre LM. Biochemical mechanism of action of the dopaminergic neurotoxin MPTP. Toxicol Lett 1989; 48: 121-49.

[263]   Langston JW, Forno LS, Tetrud J, Reeves AG, Kaplan JA, Karluk D. Evidence of active nerve cell degeneration in the substantia nigra of humans years after 1-methyl-4-phenyl-1,2,3,6-tetrahydropyridine exposure. Ann Neurol 1999; 46: 598-605.

[264]   Mizuno Y, Ikebe S, Hattori N, *et al*. Role of mitochondria in the etiology and pathogenesis of Parkinson's disease. Biochim Biophys Acta 1995; 1271: 265-74.

[265]   Swerdlow RH, Parks JK, Miller SW, *et al*. Origin and functional consequences of the complex I defect in Parkinson's disease. Ann Neurol 1996; 40: 663-71.

[266]   Schapira AH, Mann VM, Cooper JM, *et al*. Anatomic and disease specificity of NADH CoQ1 reductase (complex I) deficiency in Parkinson's disease. J Neurochem 1990; 55: 2142-5.

[267]   Sofic ELK, Jellinger K, Riederer P. Reduced and oxidized glutathione in the substantial nigra of patients with Parkinson's disease. Neurosci Lett 1992; 142: 128-30.

[268]   Adams Jr JD, Klaidman LK. Parkinson's disease-redox mechanisms. Curr Med Chem 2001; 8: 809-14.

[269]   Sechi GDM, Bua G, Satta WM, Deiana GA, Pes GM, Rosati G. Reduced intravenous glutathione in the treatment of early Parkinson's disease. Prog Neuro-Psychopharmacol 1996; 20: 1159-70.

[270]   Martínez-Banaclocha M. N-acetylcysteine in the treatment of Parkinson's disease. What are we waiting for? Med Hypoth 2012; 79: 8-12.

[271]   Perry TL, Godin DV, Hansen S. Parkinson's disease: a disorder due to nigral glutathione deficiency? Neurosci Lett 1982; 33: 305-10.

[272]   Perry TL, Yong VW. Idiopathic Parkinson's disease, progressive supranuclear palsy and glutathione metabolism in the substantia nigra of patients. Neurosci Lett 1986; 67: 269-74.

[273]   Meister A. Glutathione metabolism and its selective modification. J Biol Chem 1988; 263: 17205-8.

[274]   Jenner P. Oxidative mechanisms in nigral cell death in Parkinson's disease. Mov Disord 1998; 13: 24-34.

[275]   Cassarino DS, Fall CP, Swerdlow RH, *et al*. Elevated reactive oxygen species and antioxidant enzyme activities in animal and cellular models of Parkinson's disease. Biochim Biophys Acta 1997; 1362: 77-86.

[276]   Jenner P. Oxidative stress in Parkinson's disease. Ann Neurol 2003; 53: S26-36. [discussion S36-38].

[277]   Clementi E, Brown GC, Feelisch M, Moncada S. Persistent inhibition of cell respiration by nitric oxide: crucial role of S-nitrosylation of mitochondrial complex I and protective action of glutathione. P Natl Acad Sci USA 1998; 95: 7631-6.

[278]   Hsu M, Srinivas B, Kumar J, Subramanian R, Andersen J. Glutathione depletion resulting in selective mitochondrial complex I inhibition in dopaminergic cells is *via* an NO-mediated pathway not involving peroxynitrite: implications for Parkinson's disease. J Neuroci 2005; 92: 1091-103.

[279]   Jha N, Jurma O, Lalli G, *et al*. Glutathione depletion in PC12 results in selective inhibition of mitochondrial complex I activity. Implications for Parkinson's disease. J Biol Chem 2000; 275: 26096-7001.

[280]   Chinta SJ, Andersen JK. Reversible inhibition of mitochondrial complex I activity following chronic dopaminergic glutathione depletion *in vitro*: implications for Parkinson's disease. Free Rad Biol Med 2006; 41: 1442-8.

[281]   Uehara T, Nakamura T, Yao D, *et al*. S-Nitrosylated protein-disulphide isomerase links protein misfolding to neurodegeneration. Nature 2006; 441; 513-7.

[282]   Nakamura T, Lipton SA. Redox modulation by S-nitrosylation contributes to protein misfolding, mitochondrial dynamics, and neuronal synaptic damage in neurodegenerative diseases. Cell Death Differ 2011; 18: 1478-86.

[283]   Van Laar VS, Berman SB. Mitochondrial dynamics in Parkinson's disease. Exp Neurol 2009; 218: 247-56.

[284]   Chen H, Chan DC. Mitochondrial dynamics fusion, fission, movement, and mitophagy in neurodegenerative diseases. Hum Mol Genet 2009; 18: R169-76.

[285] Cho DH, Nakamura T, Fang J, *et al*. S-nitrosylation of Drp1 mediates beta-amyloid-related mitochondrial fission and neuronal injury. Science 2009; 324: 102-5.

[286] Fang J, Nakamura T, Cho DH, Gu Z, Lipton SA. S-nitrosylation of peroxiredoxin 2 promotes oxidative stress- induced neuronal cell death in Parkinson's disease. Proc Natl Acad Sci USA 2007; 104: 18742-7.

[287] Bonifati V, Rizzu P, Van Baren MJ, *et al*. Mutations in the DJ-1 gene associated with autosomal recessive early- onset parkinsonism. Science 2003; 299: 256-9.

[288] Ito G, Ariga H, Nakagawa Y, Iwatsubo T. Roles of distinct cysteine residues in S-nitrosylation and dimerization of DJ-1. Biochem Biophys Rese Commun 2006; 339: 667-72.

[289] Canet-Avilés RM, Wilson MA, Miller DW, *et al*. The Parkinson's disease protein DJ-1 is neuroprotective due to cysteine-sulfinic acid-driven mitochondrial localization. Proc Natl Acad Sci USA 2004; 101: 9103-8.

[290] Stefanis L. α-Synuclein in Parkinson's Disease. Cold Spring Harbor Perspect Med 2012, 2: a009399.

[291] Yoshii SR, Kishi C, Ishihara N, Mizushima N. Parkin mediates proteasome-dependent protein degradation and rupture of the outer mitochondrial membrane. J Biol Chem 2011; 286: 19630-40.

[292] Arduíno DM, Esteves AR, Cardoso SM. Mitochondrial fusion/fission, transport and autophagy in Parkinson's disease: when mitochondria get nasty. Parkinson Dis 2011; 2011: 767230.

[293] Da Costa CA, Sunyach C, Giaime E, *et al*. Transcriptional repression of p53 by parkin and impairment by mutations associated with autosomal recessive juvenile Parkinson's disease. Nat Cell Biol 2009; 11: 1370-5.

[294] Yao D, Gu Z, Nakamura T, *et al*. Nitrosative stress linked to sporadic Parkinson's disease: S-nitrosylation of parkin regulates its E3 ubiquitin ligase activity. Proc Natl Acad of Sci USA 2004; 101: 10810-4.

[295] Chung KKK, Thomas B, Li X, *et al*. S-nitrosylation of parkin regulates ubiquitination and compromises parkin's protective function. Science 2004; 304: 1328- 31.

[296] Hara MR, Agrawal N, Kim SF, *et al*. S-nitrosylated GAPDH initiates apoptotic cell death by nuclear translocation following Siah1 binding. Nat Cell Biol 2005; 7: 665-74.

[297] Eckelman BP, Salvesen GS, Scott FL. Human inhibitor of apoptosis proteins: why XIAP is the black sheep of the family. EMBO Reports 2006; 7: 988-94.

[298] Salvesen GS, Duckett CS. IAP proteins: blocking the road to death's door. Nat Rev Mol Cell Biol 2002; 3: 401-10

[299] Nakamura T, Wang L, Wong CCL *et al*. Transnitrosylation of XIAP regulates caspase-dependent neuronal cell death. Mol Cell 2010; 39: 184-95.

[300] Tsang AHK, Lee YIL, Ko HS *et al*. S-nitrosylation of XIAP compromises neuronal survival in Parkinson's disease. Proc Natl Acad Sci USA 2009; 106: 4900-5.

[301] Gouras GK, Tampellini D, Takahashi RH, Capetillo-Zarate E. Intraneuronal beta-amyloid accumulation and synapse pathology in Alzheimer's disease. Acta Neuropathol 2010; 119: 523-41.

[302] Wakabayashi K, Honer WG, Masliah E. Synapse alterations in the hippocampal-entorhinal formation in Alzheimer's disease with and without Lewy body disease. Brain Res 1994; 667: 24-32.

[303] Lacor PN, Buniel MC, Furlow PW, *et al*. Abeta oligomer- induced aberrations in synapse composition, shape, and density provide a molecular basis for loss of connectivity in Alzheimer's disease. J Neurosci 2007; 27: 796-807.

[304] Akhtar MW, Sunico C R, Nakamura T, Lipton SA. Redox regulation of protein function *via* cysteine S-nitrosylation and its relevance to neurodegenerative diseases. Intern J Cell Biol 2012; 2012: 463756.

[305] Devi L, Prabhu BM, Galati DF, Avadhani NG, Anandatheerthavarada HK. Accumulation of amyloid precursor protein in the mitochondrial import channels of human Alzheimer's disease brain is associated with mitochondrial dysfunction. J Neurosci 2006; 26: 9057-68.

[306] Dragicevic N, Mamcarz M, Zhu Y, *et al*. Mitochondrial amyloid-beta levels are associated with the extent of mitochondrial dysfunction in different brain regions and the degree of cognitive impairment in Alzheimer's transgenic mice. J Alzheimer's Dis 2010; 20: S535-50.

[307] Du H, Guo L, Fang F, *et al*. Cyclophilin D deficiency attenuates mitochondrial and neuronal perturbation and ameliorates learning and memory in Alzheimer's disease. Nat Med 2008; 14: 1097-105.

[308]   Du H, Guo L, Yan S, Sosunov AA, McKhann GM, Yan SS. Early deficits in synaptic mitochondria in an Alzheimer's disease mouse model. Proc Nat Acad Sci USA 2010; 107: 18670-5.

[309]   Du H, Guo L, Zhang W, Rydzewska M, Yan S. Cyclophilin D deficiency improves mitochondrial function and learning/memory in aging Alzheimer disease mouse model. Neurobiol Aging 2011; 32: 398-406.

[310]   Du H, Yan SS. Mitochondrial permeability transition pore in Alzheimer's disease: cyclophilin D and amyloid beta. Biochim Biophys Acta 2010; 1802: 198-204.

[311]   Qu J, Nakamura T, Cao G, Holland EA, McKercher SR, Lipton SA. S-Nitrosylation activates Cdk5 and contributes to synaptic spine loss induced by beta- amyloid peptide. P Nat Acad Sci USA 2011; 108: 14330-5.

[312]   Smith JD. Apolipoprotein E4: an allele associated with many diseases. Ann Med 2000; 32: 118-27.

[313]   Heijmsns BT, Westendorp RG, Slagboom PE. Common gene variants, mortality and extreme longevity in humans. Exp Gerontol 2000; 35: 865-77.

[314]   Katzman R. Apolipoprotein E and Alzheimer's disease. Curr Op Neurobiol 1994; 4: 703-7.

[315]   Pedersen WA, Chan SL, Mattson MP. A mechanism for the neuroprotective effect of apolipoprotein E: isoform-specific modification by the lipid peroxidation product 4-hydroxynonenal. J Neurochem 2000; 74: 1426-33.

[316]   Abrams AJ, Farooq A, Wang G. S-nitrosylation of ApoE in Alzheimer's disease. Biochemistry 2011; 50: 3405-7.

[317]   Hosoki R, Matsuki N, Kimura H. The possible role of hydrogen sulfide as an endogenous muscle relaxant in synergy with nitric oxide. Biochem Biophys Res Co 1997; 237: 527-31.

[318]   Eto K, Asada T, Arima K, Makifuchi T, Kimura H. Brain hydrogen sulfide is severely decreased in Alzheimer's disease. Biochem Biophys Res Commun. 2002; 293:1485-8.

[319]   Clarke R, Smith D, Jobst KA, Refsum H, Sutton L, Ueland PM. Folate, vitamin B12, and serum total homocysteine levels in confirmed Alzheimer disease. Arch Neurolo 1998; 55: 1449-55.

[320]   Lyles MM, Gilbert HF. Catalysis of the oxidative folding of ribonuclease A by protein disulfide isomerase: pre-steady-state kinetics and the utilization of the oxidizing equivalents of the isomerase. Biochemistry 1991; 30: 619-25.

[321]   Akiyama Y, Kamitani S, Kusukawa N, Ito K. *In vitro* catalysis of oxidative folding of disulfide-bonded proteins by the *Escherichia coli* dsbA (ppfA) gene product. J Biol Chem 1992; 267: 22440-5.

[322]   Mattson MP, Chan SL, Duan W. Modification of brain aging and neurodegenerative disorders by genes, diet, and behavior. Physiol Rev 2002; 82: 637-72.

[323]   Miquel J, Martínez-Banaclocha M, Diez A, *et al.* Effects of turmeric on blood and liver lipoperoxide levels of mice: lack of toxicity. Age 1995; 18: 171-4.

[324]   Yu ZF, Mattson MP. Dietary restriction and 2-deoxyglucose administration reduce focal ischemic brain damage and improve behavioral outcome: evidence for a preconditioning mechanism. J Neurosci Res 1999; 57: 830-9.

[325]   Culmsee C, Zhu Z, Yu QS, *et al.* A synthetic inhibitor of p53 protects neurons against death induced by ischemic and excitotoxic insults, and amyloid beta-peptide. J Neurochem 2001; 77: 220-8.

[326]   Duan W, Zhu X, Ladenheim B, *et al.* p53 inhibitors preserve dopamine neurons and motor function in experimental parkinsonism. Ann Neurol 2002; 52: 597-606.

[327]   Fassbender K, Simons M, Bergmann C, *et al.* Simvastatin strongly reduces levels of Alzheimer's disease β-amyloid peptides A β 42 and A β 40 *in vitro* and *in vivo*. P Nat Acad Sci USA 2001; 98: 5856-61.

[328]   Harman D. The biological clock: the mitochondria. J Am Geriatr Soc 1972; 20: 145-7.

[329]   Fraga CG, Shigenaga MK, Park JW, Degan P, Ames BN. Oxidative damage to DNA during aging: 8-hydroxy-2′-deoxyguanosine in rat organ DNA and urine. P Natl Acad Sci USA 1990; 87: 4533-37.

[330]   Stadtman ER. Protein oxidation and aging. Science 1992; 257: 1220-4.

[331]   Balaban RS, Nemoto S, Finkel T. Mitochondria, oxidants, and aging. Cell 2005; 120: 483-95.

[332]   Van Zandwijk N. N-Acetylcysteine for Lung Cancer Prevention. Chest 1995; 107: 1437-41.

[333]   De Flora S, Astengo M, Serra D, Bennicelli C. Inhibition of urethan-induced lung tumors in mice by dietary N-acetylcysteine. Cancer Lett 1986; 32: 235-41.

[334]   Rotstein JB, Slaga TJ. Anticarcinogenic mechanisms, as evaluated in the multistage mouse skin model. Mut Res 1988; 202: 421-7.

[335] Wilpart M, Speder A, Roberfroid M. Anti-initiation activity of N-acetylcysteine in experimental colonic carcinogenesis. Cancer Lett 1986; 31: 319-24.
[336] Dawson T, Dawson V. Protection of the brain from ischemia. Cerebrovas Dis1997; 7: 349-52.
[337] Cuzzocrea S, Mazzon E, Costantino G, *et al.* Beneficial effects of N-acetylcysteine on ischaemic brain injury. Brit J Pharmacol 2000; 130: 1219-26.
[338] Sen O, Caner H, Aydin MV, *et al.* The effect of mexiletine on the level of lipid peroxidation and apoptosis of endothelium following experimental subarachnoid hemorrhage. Neurol Res 2006; 28: 859-63.
[339] Hoffer E, Baum Y, Nahir AM. N-Acetylcysteine enhances the action of anti-inflammatory drugs as suppressors of prostaglandin production in monocytes. Mediat Inflamm 2002; 11: 321-3.
[340] Chen G, Shi J, Hu Z, Hang C. Inhibitory effect on cerebral inflammatory response following traumatic brain injury in rats: a potential neuroprotective mechanism of N-acetylcysteine. Mediat Inflamm 2008; 2008: 716458.
[341] Kelly GS. Clinical applications of N-acetylcysteine. Alter Med Rev 1998; 3: 114-27.
[342] Gere-Paszti E, Jakus J. The effect of N-acetylcysteine on amphetamine mediated dopamine release in rat brain striatal slices by ion-pair reversed-phase high performance liquid chromatography. Biomed Chromatograph 2009; 23: 658-64.
[343] Samuni Y, Goldstein S, Dean OM, Berk M. The chemistry and biological activities of N-acetylcysteine. Biochim Biophys Acta 2013; 1830: 4117-29.
[344] Caro L, Ghizzi A, Costa R, Longo A, Ventresca GP, Lodola E. Pharmacokinetics and bioavailability of oral acetylcysteine in healthy volunteers. Arzneimittelforschung 1989; 39: 382-6.
[345] Cotgreave IA, Moldeus P. Lung protection by thiol-containing antioxidants. Bull Eur Physiopathol Resp 1987; 23: 275-7.
[346] Borgström L1, Kågedal B. Dose dependent pharmacokinetics of N-acetylcysteine after oral dosing to man. Biopharm Drug Disp 1990; 11: 131-6.
[347] Prescott L, Donovan J, Jarvie D, Proudfoot A. The disposition and kinetics of intravenous N-acetylcysteine in patients with paracetamol overdosage. Eur J Clin Pharmacol 1989; 37: 501-6.
[348] Sarker KP, Abeyama K, Nishi J, *et al.* Inhibition of thrombin-induced neuronal cell death by recombinant thrombomodulin and E5510, a synthetic thrombin receptor signaling inhibitor. Thromb Haemost 1999; 82: 1071-7.
[349] Harada D, Naito S, Otagiri M. Kinetic analysis of covalent binding between N-acetyl-L-cysteine and albumin through the formation of mixed disulfides in human and rat serum *in vitro*. Pharmacol Res 2002; 19: 1648-54.
[350] Ziment I. Acetylcysteine: a drug that is much more than a mucokinetic. Pharmacotherapy 1988; 42: 513-9.
[351] Martínez-Banaclocha M. Therapeutic potential of N-acetyl-cysteine in age-related mitochondrial neurodegenerative diseases. Med Hypoth 2001; 56: 472-7.
[352] Ishige K, Tanaka M, Arakawa M, Saito H, Ito Y. Distinct nuclear factor-kappaB/Rel proteins have opposing modulatory effects in glutamate-induced cell death in HT22 cells. Neurochem Intern 2005; 47: 545-55.
[353] Henze A, Raila J, Scholze A, Zidek W, Tepel M, Schweigert FJ. Does N-acetylcysteine modulate post-translational modifications of transthyretin in hemodialysis patients? Antiox Redox Sign 2013; 19: 1166-1672.
[354] Kondratov RV, Vykhovanets O, Kondratova AA, Antoch MP. Antioxidant N-acetyl-L-cysteine ameliorates symptoms of premature aging associated with the deficiency of the circadian protein BMAL1. Aging 2009; 1: 979-87.
[355] Aoyama K, Suh SW, Hamby AM, *et al.* Neuronal glutathione deficiency and age-dependent neurodegeneration in the EAAC1 deficient mouse. Nat Neurosci 2006; 9: 119-26.
[356] Kerksick C, Willoughby D. The Antioxidant role of glutathione and N-acetyl-cysteine supplements and exercise-induced oxidative stress. J Int Soc Sports Nutr 2005; 2: 38-44.
[357] Medved I, Brown MJ, Bjorksten AR, Leppik JA, Sostaric S, McKenna MJ. N-acetylcysteine infusion alters blood redox status but not time to fatigue during intense exercise in humans. J Appl Physiol 2003, 94: 1572-82.
[358] Sen CK. Antioxidant and redox regulation of cellular signaling: introduction. Med Sci Sports Exerci 2001, 33: 368-70.

[359]   Martínez-Banaclocha M. N-acetylcysteine increase in complex I activity in synaptic mitochondria from aged mice: implications for treatment of Parkinson's disease. Brain Res 2000; 859: 173-5.

[360]   Martínez-Banaclocha M, Martínez N. N-acetylcysteine elicited increase in cytochrome c oxidase activity in mice synaptic mitochondria. Brain Res 1999; 842: 249-51.

[361]   Martínez-Banaclocha M, Hernández AI, Martínez N, Ferrándiz ML. N-acetylcysteine protects against age-related increase in oxidized proteins in mouse synaptic mitochondria. Brain Res 1997; 762: 256-8.

[362]   Martínez-Banaclocha M, Martínez N, Hernández AI, Ferrándiz ML. Hypothesis: can N-acetylcysteine be beneficial in Parkinson's disease? Life Sci 1999; 64: 1253-7.

[363]   Medina S, Martínez-Banaclocha M, Hernanz A. Antioxidants inhibit the human cortical neuron apoptosis induced by hydrogen peroxide, tumor necrosis factor alpha, dopamine and beta-amyloid peptide 1-42. Free Rad Res 2002; 36: 1179-84.

[364]   Reynaud E. Protein misfolding and degenerative diseases. Nat Edu 2010; 3: 28.

[365]   Shults CW, Beal MF, Fontaine D, Nakano K, Haas RH. Absorption, tolerability, and effects on mitochondrial activity of oral coenzyme Q10 in parkinsonian patients. Neurology 1998; 50: 793-5.

[366]   Parkinson Study Group. Effects of tocopherol and deprenyl on the progression of disability in early Parkinson's disease. New Eng J Med 1993; 328: 176-83.

[367]   Etminan M, Gill SS, Samii A. Intake of vitamin E, vitamin C, and carotenoids and the risk of Parkinson's disease: a meta-analysis. Lancet Neurol 2005; 4: 362-5.

[368]   Sechi G, Deledda MG, Bua G, et al. Reduced intravenous glutathione in the treatment of early Parkinson's disease. Prog Neuro-Psychop 1996; 20: 1159-70.

[369]   Wang XF, Cynader MS. Pyruvate released by astrocytes protects neurons from copper-catalyzed cysteine neurotoxicity. J Neurosci 2001; 21: 3322-31.

[370]   Pocernich CB, La Fontaine M, Butterfield DA. In-vivo glutathione elevation protects against hydroxyl free radical- induced protein oxidation in rat brain. Neurochem Intern 2000; 36: 185-91.

[371]   Vina J, Romero FJ, Saez GT, Pallardó FV. Effects of cysteine and N-acetyl cysteine on GSH content of brain of adult rats. Experientia 1983; 39: 164-5.

[372]   Coccoa T, Sgobbo P, Clemente M, et al. Tissue-specific changes of mitochondrial functions in aged rats: effect of a long-term dietary treatment with N-acetylcysteine. Free Rad Biol Med 2005; 38: 796-805.

[373]   Perry TL, Yong VW, Clavier RM, et al. Partial protection from the dopaminergic neurotoxin Nmethyl-4-phenyl-1,2,3,6-tetrahydropyridine by four different antioxidants in the mouse. Neurosci Lett 1985; 60: 109-14.

[374]   Sharma A, Kaur P, Kumar V, Gill KD. Attenuation of 1-methyl-4-phenyl-1, 2,3,6- tetrahydropyridine induced nigro-striatal toxicity in mice by N-acetyl cysteine. Cell Mol Biol 2007; 53: 48-55.

[375]   Offen D, Ziv I, Sternin H, Melamed E, Hochman A. Prevention of dopamine-induced cell death by thiol antioxidants: possible implications for treatment of Parkinson's disease. Exp Neurol 1996; 141: 32-9.

[376]   Nicoletti VG, Marino VM, Cuppari C, et al. Effect of antioxidant diets on mitochondrial gene expression in rat brain during aging. Neurochem Res 2005; 30: 737-52.

[377]   McCann UD, Wong DF, Yokoi F, Villemagne V, Dannais RF, Ricaurte GA. Reduced striatal dopamine transporter density in abstinent methamphetamine and methcathinone users: evidence from positron emission tomography studies with [11C]WIN-35,428. J Neurosci 1998; 18: 8417-22.

[378]   Wilson JM, Kalasinsky KS, Levey AI, et al. Striatal dopamine nerve terminal markers in human, chronic methamphetamine users. Nat Med 1996; 2: 699-703.

[379]   Hashimoto H, Tsukada H, Nishiyama S, et al. Protective effects of N-acetyl-L-cysteine on the reduction of dopamine transporters in the striatum of monkeys treated with methamphetamine. Neuropsychopharmacology 2004; 29: 2018-23.

[380]   Xu J, Kao SY, Lee FJ, Song W, Jin LW, Yankner BA. Dopamine-dependent neurotoxicity of alpha-synuclein: a mechanism for selective neurodegeneration in Parkinson disease. Nat Med 2002; 8: 600-6.

[381]   Clark J, Clore EL, Zheng K, et al. Oral N-acetyl-cysteine attenuates loss of dopaminergic terminals in a-synuclein over-expressing mice. PLoS One 2010; 5: e12333.

[382]   Mayer M, Noble M. N-acetyl-L-cysteine is a pluripotent protector against cell death and enhancer of trophic factor-mediated cell survival *in vitro*. P Natl Acad Sci USA 1994; 91: 7496-7500.

[382]   Yan CYI, Greene LA. Prevention of PC12 cell death by *N*-acetylcysteine requires activation of the Ras pathway. J Neurosci 1998; 18: 4042-49.

[384]   Green DR, Reed JC. Mitochondria and apoptosis. Science 1998; 281: 1309-12.

[385]   Ratan RR, Murphy TH, Baraban JM. Macromolecular synthesis inhibitors prevent oxidative stress-induced apoptosis in embryonic cortical neurons by shunting cysteine from protein synthesis to glutathione. J Neurosci 1994; 17: 4385-92.

[386]   Lafon C, Mathiew C, Guerrin M, Pierre O, Vidal S, Valette A. Transforming growth factor beta 1-induced apoptosis in human ovarian carcinoma cells: protection by the antioxidant N-acetylcysteine and bcl-2. Cell Grow Diff 1996; 7: 1095-104.

[387]   Aoki E, Yano R, Yokoyama H, Kato H, Araki T. Role of nuclear transcription factor kappa B (NF-kappaB) for MPTP (1-methyl-4-phenyl-1,2,3,6- tetrahyropyridine)-induced apoptosis in nigral neurons of mice. Exp Mol Pathol 2009; 86: 57-64.

[388]   Sha D, Chin LS, Li L. Phosphorylation of parkin by Parkinson disease linked kinase PINK1 activates parkin E3 ligase function and NF-kappaB signaling. Hum Mol Genet 2010; 19: 352-63.

[389]   Cossarizza A, Franceschi C, Monti D, *et al*. Protective effects of N-acetylcysteine in tumor necrosis factor-alpha-induced apoptosis in U937 cells: the role of mitochondria. Exp Cell Res 1996; 220: 232-40.

[390]   Talley AK, Dewhurst S, Perry SW, *et al*. Tumor necrosis factor-alpha-induced apoptosis in human neuronal cells: protection by the antioxidant N-acetylcysteine and the genes bcl-2 and crmA. Mol Cell Biol 1995; 15: 2359-66.

[391]   Bagh MB, Maiti AK, Jana S, Banerjee K, Roy A, Chakrabarti S. Quinone and oxyradical scavenging properties of N-acetylcysteine prevent dopamine mediated inhibition of Na+, K+-ATPase and mitochondrial electron transport chain activity in rat brain: implications in the neuroprotective therapy of Parkinson's disease. Free Rad Res 2008; 42: 574-81.

[392]   Murata M, Hasegawa K, Kanazawa I, Japan Zonisamide on PD Study Group. Zonisamide improves motor function in Parkinson disease: a randomized, double-blind study. Neurology 2007; 6845-50.

[393]   Murata M, Horiuchi E, Kanazawa I. Zonisamide has beneficial effects on Parkinson's disease patients. Neurosci Res 2001; 41: 397-9.

[394]   Asanuma M, Miyazaki I, Diaz-Corrales FJ, *et al*. Neuroprotective effects of zonisamide target astrocyte. Ann Neurol 2010; 67: 239-49.

[395]   McBean GJ, Aslan M, Griffiths HR, Torrão RC. Thiol redox homeostasis in neurodegenerative disease. Redox Biol. 2015; 5:186-94.

[396]   Giustarini D, Milzani A, Dalle-Donne I, Tsikas D, Rossi R. N-Acetylcysteine ethyl ester (NACET): a novel lipophilic cell-permeable cysteine derivative with an unusual pharmacokinetic feature and remarkable antioxidant potential. Biochem Pharmacol 2012; 84: 1522-33.

[397]   Sunitha K, Hemshekhar M, Thushara RM, *et al*. N-Acetylcysteine amide: a derivative to fulfill the promises of N-Acetyl cysteine. Free Rad Res 2013; 47: 357-67.

[398]   Adams JD Jr, KlaidmanLK, Odunze IN, Shen HC, Miller CA. Alzheimer's and Parkinson's disease. Brain levels of glutathione, glutathione disulfide, and vitamin E. Molecular and Chemical Neuropathology 1991; 14: 213-26.

[399]   Jenner P. Oxidative damage in neurodegenerative disease. Lancet 1994; 344: 796-8.

[400]   Lohr JB, Browning JA. Free radical involvement in neuropsychiatric illness. Psychopharmacol Bull 1995; 31: 159-65.

[401]   Tchantchou F, Graves M, Rogers E, Ortiz D, Shea TB. N-acteyl cysteine alleviates oxidative damage to central nervous system of ApoE-deficient mice following folate and vitamin E-deficiency. J Alzh Dis 2005; 7: 135-8.

[402]   Tucker S, Ahl M, Bush A, Westaway D, Huang X, Rogers JT. Pilot study of the reducing effect on amyloidosis *in vivo* by three FDA pre-approved drugs *via* the Alzheimer's APP 5' untranslated region. Curr Alzh Res 2005; 2: 249-54.

[403]   Adair JC, Knoefel JE, Morgan N. Controlled trial of N-acetylcysteine for patients with probable Alzheimer's disease. Neurology 2001; 57: 1515-17.

[404]   Qin S, Colin C, Hinners I, Gervais A, Cheret C, Mallat M. System xc- and apolipoprotein E expressed by microglia have opposite effects on the neurotoxicity of amyloid-beta peptide 1-40. J Neurosci 2006; 26: 3345-56.

[405]   Bordji K, Becerril-Ortega J, Nicole O, Buisson A. Activation of extrasyn- aptic, but not synaptic, NMDA receptors modifies amyloid precursor protein ex- pression pattern and increases amyloid-ss production. J Neurosci 2010; 30: 15927-42.

[406]   Olivieri G, Baysang G, Meier F, *et al.* N-Acetyl-L-cysteine protects SHSY5Y neuroblastoma cells from oxidative stress and cell cytotoxicity: effects on beta-amyloid secretion and tau phosphorylation. J Neurochem 2001; 76: 224-33.

[407]   Tanel A, Averill-Bates DA. Inhibition of acrolein-induced apoptosis by the antioxidant N-acetylcysteine. J Pharmacol Exp Ther 2007; 321: 73-83.

[408]   Mutisya EM, Bowling AC, Beal MF. Cortical cytochrome oxidase activity is reduced in Alzheimer's disease. J Neurochem 1994; 63: 2179-218.

[409]   Parker WD Jr, Filley CM, Parks JK. Cytochrome oxidase deficiency in Alzheimer's disease. Neurology 1990; 40: 1302-3.

[410]   Parker WD Jr, Parks JK, Filley CM. Kleinschmidt-DeMaster B. K. Electron transport chain defects in Alzheimer's disease brain. Neurology 1994; 44: 1090-6.

[411]   Blanchard BJ, Park T, Fripp WJ, Lerman LS, Ingram VM. A mitochondrial DNA deletion in normaly aging and in Alzheimer brain tissue. Neuroreport 1993; 4: 799-802.

[412]   Hutchin T, Cortopassi G. A mitochondrial DNA clone is associated with increased risk for Alzheimer's disease. P Natl Acad Sci USA 1995; 92: 6892-5.

[413]   Soffner JM, Brawn MD, Torroni A, Lott MT, Cabell MF, Mirra SS. Mitochondrial DNA variants observed in Alzheimer disease and Parkinson disease patients. Genomics 1993; 17: 171-84.

[414]   Beal MF. Mitochondria, free radicals, and neurodegeneration. Curr Op Neurobiol 1996; 6: 661-6.

[415]   Chandrasekaran K, Giordano T, Brady DR, Stoll J, Martin LJ, Rapoport SI. Impairment in mitochondrial cytochrome oxidase gene expression in Alzheimer's disease. Brain Research Mol Brain Res 1994; 24: 336-40.

[416]   Simonian NA, Hyman BT. Functional alterations in Alzheimer's disease: selective loss of mitochondrial-encoded cytochrome oxidase mRNA in the hyppocampal formation. J Neuropathol Exp Neurol 1994; 53: 508-12.

[417]   Swerdlow BJ, Parks JK, Cassarino SW, Maguire DJ, Maguire RS, Bennet JP. Cybrids in Alzheimer's disease: a cellular model of the disease? Neurology 1997; 49: 918-25.

[418]   Perry EK, Perry RH, Tomlinson BE. Coenzyme-A acetylating enzymes in Alzheimer's disease: possible cholinergic compartment of pyruvate dehydrogenase. Neurosci Lett 1980; 18: 105-10.

[419]   Sorbi S, Bird ED, Blass JP. Decreased pyruvate dehydrogenase complex activity in Huntington and Alzheimer diseases. Ann Neurol 1983; 13: 72-8.

[420]   Moreira PI, Harris PL, Zhu X, *et al.* Lipoic acid and N-acetyl cysteine decrease mitochondrial-related oxidative stress in Alzheimer disease patient fibroblasts. J Alzh Dis 2007; 12: 195-206.

[421]   Rosen DR, Siddique T, Patterson D, *et al.* Mutations in Cu/Zn superoxide dismutase gene are associated with familial amyotrophic lateral sclerosis. Nature 1993; 362: 59-62.

[422]   Cova E, Bongioanni P, Cereda C, *et al.* Time course of oxidant markers and antioxidant defenses in subgroups of amyotrophic lateral sclerosis patients. Neurochem Int 2010; 56: 687-93.

[423]   D'Alessandro G, Calcagno E, Tartari S, Rizzardini M, Invernizzi RW, Cantoni L. Glutamate and glutathione interplay in a motor neuronal model of amyo- trophic lateral sclerosis reveals altered energy metabolism. Neurobiol Dis 2011; 43: 346-55.

[424]   Boille e S, Vande Velde C, Cleveland DW. ALS: a disease of motor neurons and their nonneuronal neighbors. Neuron 2006; 52: 39-59.

[425]   Louwerse ES, Weverling GJ, Bossuyt PM, Meyjes F, de Jong J. Randomized, double-blind, controlled trial of acetylcysteine in amyotrophic lateral sclerosis. Arch Neurol 1995; 52: 559-64.

[426]   Grant JE, Odlaug BL, Kim SW. N-acetylcysteine, a glutamate modulator, in the treatment of trichotillomania: a double-blind, placebo-controlled study. Arch Gen Psych 2009; 66: 756-63.

[427]   Beretta S, Sala G, Mattavelli L, *et al.* Mitochondrial dysfunction due to mutant copper/zinc superoxide dismutase associated with amyotrophic lateral sclerosis is reversed by N-ace- tylcysteine. Neurobiol Dis 2003; 13: 213-21.

[428]   Andreassen OA, Dedeoglu A, Klivenyi P, Beal MF, Bush AI. N-acetyl-L-cysteine improves survival and preserves motor performance in an animal model of familial amyotrophic lateral sclerosis. Neuroreport 2000; 11: 2491-3.

[429]   Brennan WA, Bird ED, Aprille JR. Regional mitochondria respiratory activity in Huntington's disease brain. J Neurochemi 1985; 44: 1948-50.

[430]   Gu M, Gash MT, Mann VM, Javoy-Agid F, Cooper JM,   Schapira AHV. Mitochondrial defect in Huntington's disease caudate nucleus. Ann Neurol 1996; 39: 385-9.

[431]   Mann VM, Copper JM, Jvoy-Agid Y, Jenner Y, Schapira AHV. Mitochondrial function and parental sex effect in Huntington's disease. Lancet 1990; 336: 749.

[432]   Borlongan CV, Koutouzis TK, Freeman TB, Cahill DW, Sanberg PR. Behavioral pathology induced by repeated systemic injection of 3-nitropropionic acid mimics the motor symptoms of Huntington's disease. Brain Res 1995; 697: 254-7.

[433]   Brouillet E, Hantraye P, Ferrante RJ. *et al.* Chronic mitochondrial energy impairment produces selective striatal degeneration and abnormal choreiform movements in primates. P Natl Acad Sci USA 1995; 92: 7105-9.

[434]   Tabrizi SJ, Workman J, Hart PE. *et al.* Mitochondrial dysfunction and free radical damage in the Huntington R6/2 transgenic mouse. Ann Neurol 2000; 47: 80-6.

[435]   Burke JR, Enghild JJ, Martin ME, *et al.* Huntington and DRPLA proteins selectively interact with the enzyme GAPDH. Nat Med 1996; 2: 347-50.

[436]   Browne SE, Bowling AC, MacGarvey U, *et al.* Oxidative damage and metabolic dysfunction in Huntington's disease: selective vulnerability of the basal ganglia. Ann Neurol 1997; 41: 646-53.

[437]   Sandhir R, Sood A, Mehrotra A, Kamboj SS. N-Acetylcysteine reverses mitochondrial dysfunctions and behavioral abnormalities in 3-nitropropionic acid-induced Huntington's disease. Neurodeg Dis 2012; 9: 145-57.

[438]   La Fontaine MA, Geddes JW, Banks A, Butterfield DA. Effect of exogenous and endogenous antioxidants on 3-nitropropionic acid-induced *in vivo* oxidative stress and striatal lesions: insights into Huntington's disease. J Neurochem 2000; 75: 1709-15.

CHAPTER 3

Non-motor Symptoms in Parkinson's Disease and Drug Therapies

Kazuo Abe[*]

Department of Community Health Medicine, Graduate School of Hyogo College of Medicine, Division of Neurology, Hospital of Hyogo College of Medicine, 1-1 Mukogawa-Cho, Nishinomiya-City, Hyogo 663-8131, Japan

Abstract: Parkinson's disease (PD) is typically characterized by its motor symptoms, namely rigidity, resting tremor, bradykinesia and postural instability. However, non-motor symptoms (NMS) such as sleep disturbance, pain, constipation, urinary problems and fatigue are integral to PD and are the leading cause of poor quality of life for both people with PD and their caregivers. Although NMS affect almost every patient, they remain under-recognized and under-treated. An evaluation of the treatment consequences of NMS in over 60% of patients revealed that NMS such as apathy, pain, sexual difficulties, bowel incontinence and sleep disorders may not be revealed to health care professionals because the patients are either embarrassed or unaware that the symptoms are linked to PD. This mini-review provides an overview of NMS in PD along with possible drug therapies.

Keywords: Behavior, depression, digestive disturbance, dopamine agonist, fatigue, levodopa, non-motor symptoms (NMS), Parkinson's disease (PD), quality of life (QOL), sensory disturbance, sleep disturbances, zonisamide.

1. INTRODUCTION

In his historical essay on "shaking palsy", James Parkinson described an "involuntary tremulous motion, with lessened muscular power, in parts not in action and even when supported; with a propensity to bend the trunk forwards and to pass from a walking to a running pace: the senses and intellects being uninjured" [1]. Since then, Parkinson's disease (PD) has been classified as a "movement disorder" with a clinical diagnosis contingent upon the presence of tremor, akinesia, muscle rigidity and loss of the right reflex. The treatment focus of PD has therefore been to primarily improve these "motor syndromes" [1-3].

*Corresponding author Kazuo Abe: Department of Community Health Medicine, Graduate School of Hyogo College of Medicine, Division of Neurology, Hospital of Hyogo College of Medicine, 1-1 Mukogawa-Cho, Nishinomiya-City, Hyogo 663-8131, Japan; Tel: +81-798-45-6598; Fax: +81-798-45-6597; E-mails: abe-neur@hyo-med.ac.jp; abe_neurology@85.alumni.u-tokyo.ac.jp

The advent of advanced drug therapies for PD has led to improvements in motor function and activities of daily living (ADL) in the early stages of the disease [4-7]. For motor complications (especially dyskinesia), the reported risk factors include higher dosage and longer duration of levodopa treatment, longer duration and severity of PD and younger age at PD onset [8]. Some researchers consider non-motor syndromes as initial sign and symptom of PD [10-13].

This mini-review provides an overview of the non-motor symptoms and signs of PD as well as possible treatments for the same.

2. NON-MOTOR SYMPTOMS AND SIGNS IN PD

The non-motor fluctuations seen during long-term levodopa usage in PD have been categorized as dysautonomic, cognitive/psychiatric and sensory [9]. Although their causes remain obscure, dopaminergic dysfunction appears to be involved. Neuroimaging data have demonstrated dopaminergic dysfunction in the hypothalamus of the parkinsonian brain. It is important to note that in advanced PD, deep brain stimulation of the subthalamic nucleus seems best at alleviating the non-motor fluctuations affecting sensory, autonomic and cognitive function [9].

The amount of non-motor symptoms and signs in PD constantly increases with the accumulation of information on PD clinical pathogenesis. In order to investigate the non-motor symptoms scale for Parkinson's disease (NMSSPD), we investigated the overall frequency of non-motor symptoms and signs using a questionnaire [11, 12]. We investigated neuropsychological signs (e.g., depression, apathy, anxiety, anhedonia, cognitive dysfunction, attention deficit, hallucination, delusion, dementia, delirium, panic disorder), sleep disorders (restless leg syndrome, abnormal behavioral disorders in REM period, insomnia, excessive day time sleep, sleep apnea), dysautonomia (orthostatic hypotension, dysuria, impotent), digestion disorders (drooling, incontinence, dysphagia, dysosmia, visual disorder), behavioral disorders (hypersexuality, compulsive behavior, pathological shopping, pathological gambling) and others (fatigue, weight loss, weight gain) [11-16].

These foregoing investigations identified non-motor signs and symptoms in PD that have been known to affect motor syndromes and quality of life (QOL) [12].

2.1. Neuropsychological Syndromes

PD patients may develop neuropsychiatric syndromes such as hallucinations [17, 18] or depression [19], which can interfere with treatment, care and ADL.

If hallucinations affect ADL, a full explanation to patients and their caregivers may lighten their burdens. If hallucinations in PD require immediate integrative treatment with psychosocial interventions, antipsychotic medications are prescribed in addition to an adjustment of current anti-PD medications. Nonetheless, an adequate L-dopa dose must be maintained to prevent severe off-states [18].

For individuals with PD, depression is quite common. Research suggests that PD itself causes chemical changes in dopamine, serotonin and norepinephrine in the limbic system that may lead to depression. Dopamine and dopamine agonists are used for the treatment of depression [19-22]. In addition, depression in PD has a close relationship with motor dysfunction. Improvement of motor syndromes is therefore a key factor in the improvement of depression in PD.

Anxiety and apathy syndromes are dependent on depression. These signs and symptoms may be found with depression, but may also be found without depression if a full drug therapy for motor syndromes is administered [23, 24].

James Parkinson had initially claimed that PD patients never developed dementia. However, we now know that PD patients may develop frontal lobe dysfunctions such as forgetfulness [25-27]. PD patients have difficulty planning and carrying out tasks. When facing a task or situation on their own, an individual with PD may feel overwhelmed by having to make choices. They may also have difficulty remembering information, or have trouble finding the right words when speaking. To some degree, cognitive impairment affects most PD patients. The same chemical changes that lead to motor symptoms can also result in slowness in thinking [28]. Cognitive impairment in PD is distinct from dementia, which is a more severe loss of intellectual abilities that interferes with daily living so much that it may not be possible for a person to live independently. The causes of cognitive changes in PD remain unclear, but a possible cause is the decreased level of dopamine. However, the cognitive changes associated with dopamine declines are typically mild and circumscribed. It should be noted that PD patients are deeply stigmatized by their disorder and this impairs their cognitive function and increase the burden for caregivers.

2.2. Sleep Disturbance

PD patients have been reported to claim sleep disturbances, as they may have difficulty turning over in their sleep due to increased muscle tonus, may be disturbed in their sleep by pollakisuria, or may have depression [29]. Although non-ergot dopamine agonists such as ropinirole, pramipexole and rotigotine have been known to induce daytime sleepiness or sudden insomnia [30], PD itself induces sleep disturbances such as restless leg syndrome (RLS), abnormal behavioral disorders during REM periods, insomnia, excessive daytime sleep and sleep apnea [31-35]. PD patients reportedly develop RBD at a high frequency (up to 40%) and experience these sleep disturbances during the early stages. Diagnosis and treatment of sleep disturbances is therefore a key factor in PD treatment [9, 36].

2.3. Autonomic Failure (Dysautonomia)

PD patients frequently show dysautonomias such as orthostatic hypotension [37] and bowel disorders [38]. Unlike patients with multiple systemic atrophy (MSA), PD patients rarely lose consciousness due to orthostatic hypotension. However, light-headedness or dizziness due to orthostatic hypotension may be a risk factor for accidental falls and injuries after falls. Since PD patients are tended to have disused syndrome that impair their ADL and QOL, rehabilitation using tilt table and elastic stockings are needed [15, 39, 40]. If their environments require, reform of houses are also planned [41-42].

PD patients may also show pollakisuria, dysuria, feeling of residual urine, or incontinence of urine. A single individual may develop a wide variety of urinary disorders. PD patients may have sleep disorders due to nocturia [9, 33]. These urinary signs generally improve after adequate dopamine therapy, although anticholinergic agonists such as oxybutynin hydrochloride can also cause improvement [38, 43].

2.4. Digestive Disorders

PD patients frequently develop constipation without anti-parkinsonian agents; however, anti-parkinsonian agents themselves may cause patients to develop incontinence [44, 45]. In a pathological study, Braak *et al.* found Lewy bodies in the Auerbach neural plexus that may be responsible for incontinence in PD [46]. Incontinence may eventually lead to ileus and should be treated with drugs, diet and exercise [47].

PD patients increasingly exhibit drooling as the disease progresses, despite the fact that these patients do not have an increased amount of saliva sputum. Slowing of swallowing may therefore be causing the drooling [48]. Although some drugs can reduce the amount of saliva, which increases tooth decay and periodontal disease, wiping or postural drainage therapy may be useful for preventing aspiration pneumonia.

To treat dysphagia in PD, a nutrition support team (NST) may be useful in daily clinical practice [49, 50].

2.5. Sensory Disturbance

Although James Parkinson wrote that PD patients never show sensory disturbances [1], many patients may experience pain or dysesthesia [51]. PD patients are also known to demonstrate dysosmia [52, 53] and dysgeusia [52] in early stages. Dopamine is an agent that modulates pain in the spinal cord, thalamus, basal ganglia and cingulate gyrus [54, 55]. Hence, PD patients show sensory disturbances because of dopamine dysregulation. Up to 29% of PD patients complain of pain that impedes their QOL [11].

2.6. Behavior Disorders

Dis-inhibition including hypersexuality, compulsive behavior, pathological shopping and pathological gambling have also been reported as behavioral disturbances in PD [11, 12]. In our study in Japan, we rarely encountered these behavioral disturbances and proposed that this was because a relatively small amount of anti-parkinsonian drugs is used in Japan. Amantadine may be effective in treating disinhibitions such as those that cause pathological gambling [56]. However, further research is needed to more precisely define how amantadine may influence the development and treatment of impulsive compulsive disorders (ICDs) in PD [57].

2.7. Fatigue

Fatigue is the most troubling but least elucidated syndrome in PD [58-61]. Fatigue can be distinguished from depression and although depression in PD has some relationship with motor syndromes, fatigue does not [58, 59]. While depression also causes trouble for patients' caregivers [62, 63], fatigue has been reported as the most troublesome non-motor symptom [60]. Although the pathogenesis of fatigue in PD remains uncertain, we hypothesize that it may be related to cognitive dysfunction, especially that caused by frontal lobe dysfunction [58-60, 62, 64]. Although the drug therapy for fatigue has not been established, Pavese proposed that anti-serotoninergic drugs might be helpful [64].

2.8. Others

PD patients show respiratory dysfunction that result in speech disturbances, dysphagia and aspiration pneumonia [65, 66]. We have developed an exercise regime for PD involving respiratory exercises that could improve respiratory dysfunction.

3. DRUG THERAPIES FOR NON-MOTOR SYMPTOMS IN PD PATIE-NTS

In general, the published management recommendations for non-motor PD symptoms have not attempted to differentiate between symptoms intrinsic to PD and those arising as complications of PD therapy. A contributory problem may be that in clinical studies of dopaminergic options, improvement of non-motor PD symptoms (levodopa-related or not) has seldom been either a primary endpoint or a means for comparing treatments with potentially different abilities to produce continuous drug delivery (CDD). In a double-blind, 12-week, parallel-group study of patients with early PD, treatment with levodopa/carbidopa/entacapone resulted in significantly greater improvements in PDQ-8 scores compared with treatment with levodopa/carbidopa. Statistically significant improvements were predominantly observed in non-motor domains (e.g., depression, personal relationships, communication and stigma) [67]. Troeung conducted a meta-analysis of randomized placebo-controlled trials for depression and/or anxiety in PD. While the pooled effects of antidepressant therapies in PD were not significant, the moderate to large magnitude of each pooled effect was promising [68].

Impulse control disorders (ICDs) in PD are typically (but not exclusively) associated with dopamine agonist use. A history of obsessive-compulsive disorder, impulsive personality, or addictive behaviors increases the likelihood of ICDs [69]. Dopamine agonist dose reduction or discontinuation is generally an effective treatment of ICDs [70].

We attempted to quantify the various aspects of nocturnal sleep problems in PD using the PD sleep scale (PDSS) [32]. Sixty-four patients with PD and 60 age- and sex-matched controls completed the PDSS. The PDSS scores in the PD group were significantly different from those in controls. Individual items of the scale showed good discriminatory power between PD and controls. The overall tendencies among individuals in Japan and the UK were the same, however certain points, especially the absence of refreshing quality of sleep, were exclusive to Japan (Fig. 1). Zolpidem, a short-acting hypnotic drug, was reported to improve motor symptoms in patients with Parkinson's disease. This gamma-aminobutyric acid (GABA-ergic) drug is a selective agonist of the benzodiazepine receptor subtype BZ1 [35, 71]. The highest density of this receptor is in the output

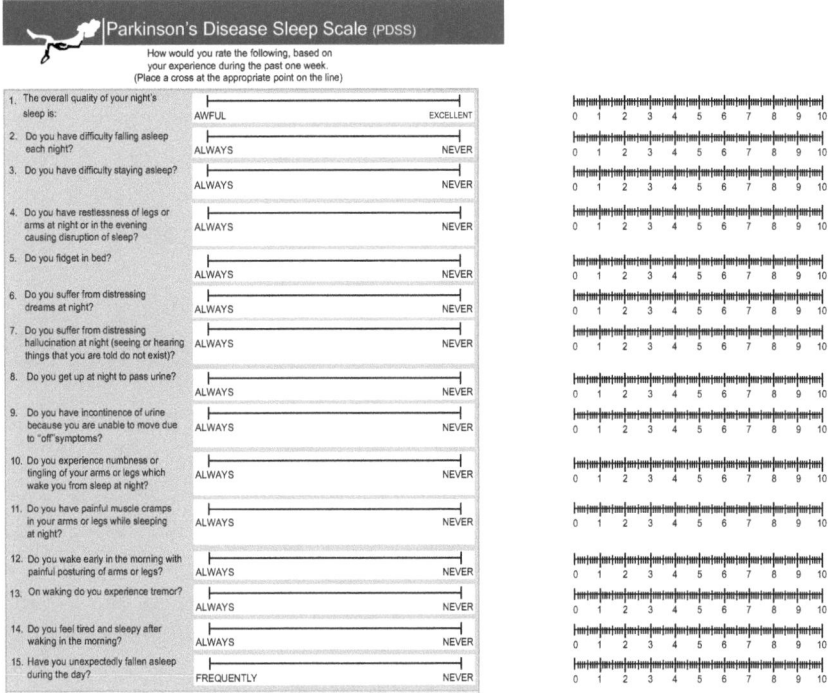

Fig. (1a): Parkinson's disease sleep scale (PDSS).

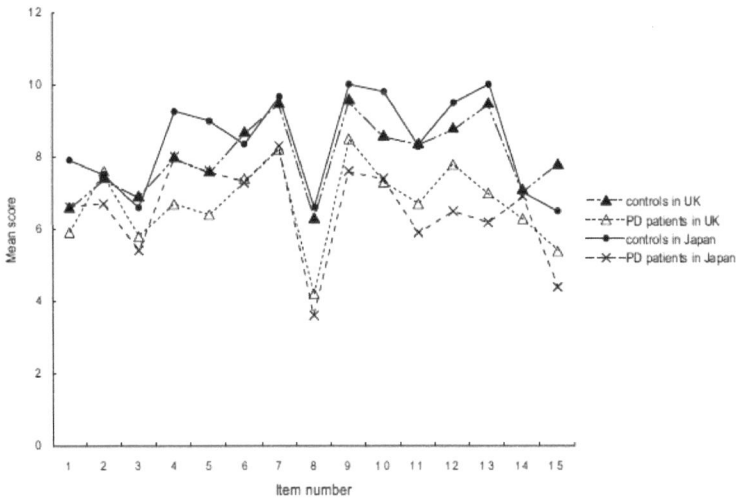

Fig. (1b): Profiles of mean PDSS in Japan and UK. Controls in Japan showed tended to exhibit a higher mean score for item 4 and item 5. PD patients in Japan tended to have a lower mean score for item 2, but had higher mean scores for items 4, 5 and 14. Patients with PD showed a mean score of 99, while that for controls was 122.7. There were highly significant differences between the patients and controls for most items except item 14. The differences in scores for items 1, 4 and 6 were borderline.

structures of the basal ganglia (*e.g.*, the ventral globus pallidus and the substantia nigra pars reticulata). Since these structures are involved in the pathology of many movement disorders, investigators have identified zolpidem as a possible drug to treat the neurological signs and symptoms of movement disorders. We proposed a new therapeutic approach to movement disorders such as Parkinson's disease that could be beneficial for patients with complications related to sleep disorders and other autonomic disorders [71].

Transdermal rotigotine has recently been assessed in a large, double blind placebo trial involving PD patients with unsatisfactory early morning motor-symptom control [72]. On the PDSS-2 [73], the mean 12-week change in total score was significantly improved in the rotigotine group compared with the placebo group from baseline to the end of maintenance. Difficulty falling asleep and feeling tired and sleepy in the morning were among the ten parameters that showed significant improvement. A significant improvement was also documented in the mean change in total score on the NMSS [14, 15] from baseline to end of treatment, with significant changes in the sleep/fatigue and mood/cognition domains. Based on the RECOVER (Randomized Evaluation of the 24-hour Coverage: Efficacy of Rotigotine; Clintrials.gov: NCT00474058) study, Kassubek concluded that pain was improved in patients with PD treated with rotigotine. This may be partly attributable to the improvements in motor function and sleep disturbances. Prospective studies are warranted to investigate this potential benefit and the clinical relevance of these findings [73]. In a small prospective trial, we assessed the effect of transdermal rotigotine on fatigue in 55 PD patients using an analogue fatigue scale (scale; 0-100). At approximately 6 weeks, we found improvements in fatigue symptoms following rotigotine prescription (Fig. **2**).

In a small prospective open-label trial, improvement in non-motor PD symptoms through subcutaneous apomorphine infusion has been a primary endpoint [74]. Seventeen patients with advanced PD received subcutaneous apomorphine infusion and 17 received conventional therapy. At approximately 6 months, patients in the apomorphine group showed a significant improvement in NMSS and PDQ-8 total scores from baseline, while no change was observed in the conventional-therapy group.

Emre investigated the long-term safety of rivastigmine and found effects on motor symptoms in patients with mild-to-moderately severe PD dementia [75]. Istradefylline, an adenosine A2A receptor antagonist, improves motor function in patients with PD. Recent findings in the areas of pharmacological manipulation and genetic ablation suggest that the adenosine A2A receptor is also related to cognitive functions, especially working memory. Istradefylline exerts antidepressant-like effects *via* modulation of A2A receptor activity, which is independent of monoaminergic transmission in the brain. Uchida has suggested that istradefylline might represent a novel treatment option for depression in PD as well as for the motor symptoms [76].

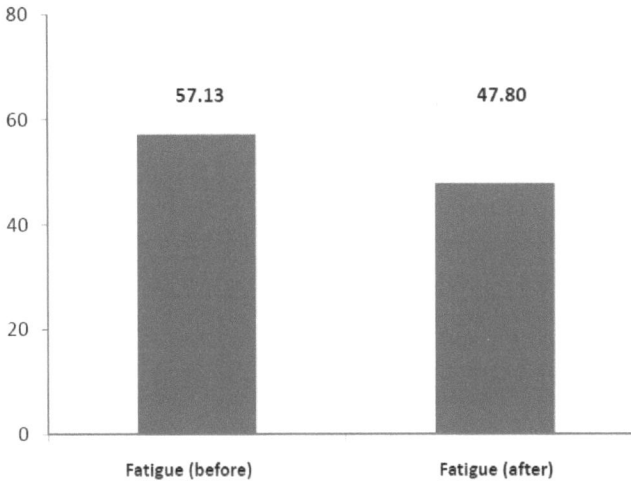

Figure 2: Mean value of Fatigue Severity in PD patients before and after prescribing transdermal rotigotine. Improvement was found after prescribing rotigotine.

CONCLUSION

Since PD patients are admitted to neurological clinics based on motor signs and symptoms such as bradykinesia, tremor, or dystonia, neurologists mainly focus on improving these motor syndromes. However, with the advent of more advanced PD therapies, non-motor syndromes have become increasingly recognized as a crucial factor for improving the QOL of patients and their caregivers [9, 14, 62, 63]. Many pathological examinations (mainly conducted by Braak *et al.*) have suggested that non-motor syndromes may result from the initial pathological changes in PD [46]. The non-motor syndromes of PD therefore require more attention, as they are extremely pertinent to the QOL of PD patients. Although drug therapies for the non-motor syndromes of PD remain under consideration, several promising treatments are now going to be developed.

CONFLICT OF INTEREST

The author confirms that author has no conflict of interest to declare for this publication.

ACKNOWLEDGMENTS

Declared none.

REFERENCES

[1]   Parkinson J. An essay on the shaking palsy. J Neuropsychiatry Clin Neurosci 2002;14:223-3.

[2]   Hoehn MM, Yahr MD. Parkinsonism: onset, progression and mortality. Neurology 1967;17:427-42.

[3]   Jankovic J. Parkinson's disease: clinical features and diagnosis. J Neurol Neurosurg. Psychiatry 2008;79:368-76.

[4]   Duvoisin RC, Sage J. Parkinson's disease. A guide for patient and family. 5th edition Lippincott Williams & Wilkins, Philadelphia, New York, 2001.

[5]   Abe K. Frontier in treatment. Parkinson disease-Role of rehabilitation- Brain Med 2002;14:45-54.

[6]   Coelho M, Ferreira JJ. Late-stage Parkinson disease. Nat Rev Neurol. 2012;8:435-42.

[7]   Schenkman M, Hall DA, Baron AE, Schwartz RS, Mettler P, Kohrt WM. Exercise for people in early- or mid-stage Parkinson Disease: A 16-month randomized controlled trial. Phys Ther. 2012 Jul 19. [Epub ahead of print]

[8]   Grandas Grandas huri KR, Healy DG, Schapira AHV. Non-motor symptoms of Parkinson's disease: diagnosis and management. Lancet Neurology 2009;8:464-74.

[9]   Gräber S, Liepelt-Scarfone I, Brüssel T, Schweitzer K, Gasser T, Berg D. Self estimated quality of life in monogenetic Parkinson's disease. Mov Disord. 2011;26(1):187-8.

[10]   Abe K, Takanashi M, Yanagihara T. Fatigue in patients with Parkinson's disease. Behav Neurol 2000;12;103-6.

[11]   Martinez-Martin P, Schapira AHV, Stocchi F, *et al.* Prevalence of nonmotor symptoms in Parkinson's disease in an international setting; Study using nonmotor symptoms questionnaire in 545 patients Mov Disord 2007;22:1623-9

[12]   Chaudhuri KR, Martinez-Martin P, Brown RG, *et al.* The metric properties of a novel non-motor symptoms scale for Parkinson's disease: Results from an international pilot study. Mov Disord 2007;22:1901-11.

[13]   Zanigni S, Calandra-Buonaura G, Grimaldi D, Cortelli P. REM behaviour disorder and neurodegenerative diseases. Sleep Med. 2011;12 Suppl 2:S54-8.

[14]   Martinez-Martin P, Rodriguez-Blazquez C, Abe K, *et al.* International validation of the non-motor symptoms scale: Comparison with the pilot study. Neurology 2009;73:1584-91.

[15]   Martinez-Martin P, Rodriguez-Blazquez C, Kurtis, MM, Chaudhuri KR. The impact of non motor symptoms on health-related quality of life of patients with Parkinson's disease. Mov Disord 2011;26(3):399-406.

[16]   Yahr MD, Duvoisin RC, Schear MJ, Barrett RE, Hoehn MM. Treatment of parkinsonism with levodopa. Arch Neurol. 1969; 21:343-354.

[17]   Fénelon G, Mahieux F, Huon R, *et al.* Hallucinations in Parkinson's disease. Prevalence, phenomenology and risk factors. Brain 2000;123:733-45.

[18]   Jenner J, Laar van T. 2012. Visual hallucinations in Parkinson's disease. In: JH Stone, M Blousing and editors. International Encyclopedia of Rehabilitation. Available online: http://cirrie.buffalo.edu/encyclopedia/en/article/147/

[19]   Cummings JL, Masterman DL. Depression in patients with Parkinson's disease. Int J Geriatr Psychiatry 1999;14 (9):711-18

[20]   Remy P, Doder M, Lees A, *et al.* Depression in Parkinson's disease. Loss of dopamine and noradrenaline innervations in the limbic system. Brain 2005;128:1314-22.

[21]   Rektorova I, Rektor I, Bares M, *et al.* Pramipexole and pergolide in the treatment of depression in Parkinson's disease: a national multicentre prospective randomized study. Eur J Neurol 2003;10:399-406.

[22]   Barone P, Scarzella L, Marconi R, *et al.* Pramipexole *versus* sertraline in the treatment of depression in Parkinson's disease. A national multicenter parallel group randomized study. J Neurol 2006;253:601-7.

[23]   Lemke MR, Brecht HM, Koester J, *et al.* Anhedonia, depression and motor functioning in Parkinson's disease during treatment with pramipexole. J Neuropsychiatry Clin Neurosci 2005;17:214-20.

[24]   Czernecki V, Pillon B, Houeto JL, *et al.* Motivation, reward and Parkinson's disease: influence of dopa therapy. Neuropsychologia 2002;40:2257-67.

[25]   Aarsland D, Andersen K, Larsen JP, Lolk A, Kragh-Sørensen P. Prevalence and characteristics of dementia in Parkinson disease. Arch Neurol 2003;60:387-92.

[26]   Williams-Gray CH, Foltynie T, Brayne CEG, *et al.* Evolution of cognitive dysfunction in an incident Parkinson's disease cohort. Brain 2007;130:1787-98.

[27]   Cooper JA, Sagar HJ, Jordan N, *et al.* Cognitive impairment in early, untreated Parkinson's disease and its relationship to motor disability. Brain 1991;2095-2122.

[28]   Fujiwara M, Yamauchi K, Abe K. Rehabilitation in Parkinson disease. Monthly Book Medical Rehabilitation 2007;76:44-52.

[29]   Chaudhuri KR, Pal S, DiMarco A, *et al.* The Parkinson's disease sleep scale: a new instrument for assessing sleep and nocturnal disability in Parkinson's disease. J Neurol Neurosurg Psychiatry 2002;73:629-35.

[30]   Homann CN, Wenzel K, Suppan K, Ivanic G, Kriechbaum N, Crevenna R, Ott E. Sleep attacks in patients taking dopamine agonists: review. BMJ 2002;324:1483-87.

[31]   International classification of sleeps disorders: Diagnostic and coding manual. Diagnostic Classification Steering Committee, Therapy MJ, Chairman. Rochester, Minnesota: American Sleep Disorders Association, 1990.

[32]   Abe K, Hikita T, Sakoda S. Sleep disturbances in Japanese patients with Parkinson's disease-- comparing with patients in the UK. J Neurol Science 234:73-8.

[33]   Abe K, Hikita T, Sakoda S. Zolpidem tartrate therapy for sleep disturbances in patients with Parkinson's disease. J Neurol Sci 2005;238:S350.

[34]   Christopher G. Goetz, Bichun Ouyang, Alice Negron, Glenn T. Stebbins. Hallucinations and sleep disorders in PD: Ten-year prospective longitudinal study. Neurology 2010 75:1773-9.

[35]   Abe K. Zolpidem therapy for movement disorders. Recent Patents on CNS Drug Discovery 2008;3:55-60.

[36]   Goetz CG, Ouyang B, Negron A, Stebbins GT. Hallucinations and sleep disorders in PD: Ten-year prospective longitudinal study. Neurology 2010;75:1773-9.

[37]   Medow MS, Stewart JM, Sanyal S, Mumtaz A, Sica D, Frishman WH. Pathophysiology, diagnosis and treatment of orthostatic hypotension and vasovagal syncope. Cardiol Rev 2008;16:4-20.

[38]   Fitzmaurice H, Fowler CJ, Rickards D, *et al.* Micturition disturbance in Parkinson's disease. Br J Urol 1985;57: 652-6.

[39]   Duncan RP, Leddy AL, Cavanaugh JT, *et al.* Accuracy of fall prediction in Parkinson disease: six-month and 12-month prospective analyses. Parkinsons Dis. 2012;2012:237673.

[40]   Johnson L, James I, Rodrigues J, Stell R, Thickbroom G, Mastaglia F. Clinical and posturographic correlates of falling in Parkinson's disease. Mov Disord. 2013 ;28:1250-6.

[41]   Clarke CE, Zobkiw RM, Gullaksen E. Quality of life and care in Parkinson's disease. Br J Clin Pract. 1995;49:288-93.

[42]   Kamata N, Abe K. Accidental falls and overestimation of stability limits in parkinsonian patients. In accidental falls: causes, prevention and interventions. Edited by Murray, Christine A. Nova Science Publishers and Inc. NY. pp200-43, 2008.

[43]   Jain S, Goldstein DS. Cardiovascular dysautonomia in Parkinson disease: from pathophysiology to pathogenesis. Neurobiol Dis 2012;46:572-80.

[44]   Singharam C, Ashraf W, Gaummitz EA, *et al.* Dopaminergic defect of enteric nervous system in Parkinson's disease patients with chronic constipation. Lancet 1995;346:861-64.

[45]   Abbott RD, Petrovitch H, White LR, *et al.* Frequency of bowel movements and the future risk of Parkinson's disease. Neurology 2001;57:456-62.

[46]   Braak H, Del Tredici K, Rüb U, de Vos RA, Jansen Steur EN, Braak E. Staging of brain pathology related to sporadic Parkinson's disease Neurobiology of Aging 2003;24:197-211.

[47]   Clark EC, Mulder DW, Erickson DJ, Clements BG, Maccaty CS. Parkinson's disease: therapeutic exercises in its management. Postgrad Med 1957;21:301-8.

[48]   Kalf JG, de Swart BJ, Borm GF, Bloem BR, Munneke M. Prevalence and definition of drooling in Parkinson's disease: a systematic review. J Neurol. 2009;56:1391-6.

[49]   Salat-Foix D, Suchowersky O. The management of gastrointestinal symptoms in Parkinson's disease. Expert Rev Neurother 2012;12:239-48.

[50]   Lyons KE, Pahwa R. The impact and management of nonmotor symptoms of Parkinson's disease. Am J Manag Care. 2011;17 S12: S308-14.

[51]   Quinn NP, Koller WC, Lang AE, Marsden CD. Painful Parkinson's disease. Lancet 1986;1:1366-9.

[52]   Shah M, Deeb J, Fernando M, Noyce A, Visentin E, Findley LJ, Hawkes CH. Abnormality of taste and smell in Parkinson's disease. Parkinsonism Relat Disord 2009;15:232-7.

[53]   Landis BN, Burkhard PR. Rhinorrhea and olfaction in Parkinson disease. Neurology. 2008;71:1041-2.

[54]   Chudler EH, Dong WK. The role of the basal ganglia in nociception and pain. Pain 1995;60:3-38.

[55]   Shyu BC, Kititsy-Roy JA, Morrow TJ. Neurophysiological, pharmacological and behavioral evidence for medial thalamic mediation of cocaine induced dopaminergic analgesia. Brain Res 1992;522:16-23

[56]   Thomas A, Bonanni L, Gambi F, Di Iorio A, Onofrj M. Pathological gambling in Parkinson disease is reduced by amantadine. Ann Neurol 2010;68:400-4.

[57]   Weintraub D, Sohr M, Potenza MN, *et al.* Amantadine use associated with impulse control disorders in Parkinson disease in cross-sectional study. Ann Neurol 2010;68:963-8.

[58]   Abe K. Tiredness and fatigue. Journal of International Society of Life Information Science 2006;24:60-2.

[59]   Abe K. Tiredness and fatigue. Brain 21 2001;4:35-9.

[60]   Abe K, Takanashi M, Yanagihara T, Sakoda S. Pergolide Mesilate may improve fatigue in patients with Parkinson's disease. Behav Neurol 2002;13:117-21.

[61]   Schrag A, Jahanshahi M, Quinn N. What contributes to quality of life in patients with Parkinson's disease? J Neurol Neurosurg Psychiatry 2000;69:308-12.

[62]   Martinez-Martine P, Benito-León J, Alonso F, *et al.* Quality of life of caregivers in Parkinson's disease. Qual Life Res 2005;14:463-72.

[63]   Krupp LB, Coyle PK, Doscher C, *et al.* Fatigue therapy in multiple svlerosis - results of a double-blind, randomized, parallel trial of amantadine, remoline and placebo. Neurol 1995;45:1956-61.

[64]   Pavese N, Metta V, Bose SK, Chaudhuri KR, Brooks DJ. Fatigue in Parkinson's disease is linked to striatal and limbic serotonergic dysfunction. Brain. 2010;133:3434-43.

[65]   Tamaki A, Matsuo Y, Abe K. Influence of thoracoabdominal movement on pulmonary function in patients with Parkinson's disease-comparison with healthy subjects. Neurorehabil Neural Repair 2000;14:43-7.

[66]   Matsuo Y, Kamata N, Abe K. Thoraco-abdominal movements during deep breathing in patients with Parkinson's disease may be reduced parallel to disease progression. Clin Neurophysiol 2006;117: S185.

[67]   Fung VS, Herawati L, Wan Y. Movement Disorder Society of Australia Clinical Research and Trials Group; QUEST-AP Study Group. Quality of life in early Parkinson's disease treated with levodopa/carbidopa/entacapone. Mov Disord 2009;24:25-31.

[68]   Troeung L, Egan SJ, Gasson N. A meta-analysis of randomized placebo-controlled treatment trials for depression and anxiety in Parkinson's disease. PLoS One 2013;8:e79510.

[69]   Weiss HD, Marsh L. Impulse control disorders and compulsive behaviors associated with dopaminergic therapies in Parkinson disease. Neurol Clin Pract 2012;2:267-74.

[70]   Connolly BS, Lang AE. Pharmacological treatment of Parkinson disease: a review. JAMA 2014;311:1670-83.

[71]   Depoortere H, Zivkovic B, Lloyd KG, *et al.* Zolpidem, a novel nonbenzodiazepine hypnotic. I. Neuropharmacological and behavioral effects. J Pharmacol Exp Ther 1986;237:649-58.

[72]   Trenkwalder C, Kies B, Rudzinska M, *et al.* Rotigotine effects on early morning motor function and sleep in Parkinson's disease: a double-blind, randomized, placebo-controlled study (RECOVER). Mov Disord. 2011;26:90-9.

[73]   Kassubek J, Chaudhuri KR, Zesiewicz T, *et al.* Rotigotine transdermal system and evaluation of pain in patients with Parkinson's disease: a post hoc analysis of the RECOVER study. BMC Neurol 2014;14:42.

[74]   Martinez-Martin P, Reddy P, Antonini A, *et al.* Chronic subcutaneous infusion therapy with apomorphine in advanced Parkinson's disease compared to conventional therapy: a real life study of nonmotor effect. J Parkinsons Dis 2011;1:197-203.

[75]   Emre M, Poewe W, De Deyn PP, *et al.* Long-term safety of rivastigmine in parkinson disease dementia: an open-label, randomized study. Clin Neuropharmacol 2014;37:9-16.

[76]   Uchida S, Kadowaki-Horita T, Kanda T. Effects of the adenosine A2A receptor antagonist on cognitive dysfunction in Parkinson's disease. Int Rev Neurobiol 2014;119:169-89.

CHAPTER 4

Alpha 7 Nicotinic Receptor Agonist Modulatory Interactions with Melatonin: Relevance not only to Cognition, but to Wider Neuropsychiatric and Immune Inflammatory Disorders

George Anderson[1] and Michael Maes[2]

[1]*CRC Scotland & London, Eccleston Square, London, UK and* [2]*Deakin University, Department of Psychiatry, Geelong, Australia*

Abstract: Recent clinical trials indicate the importance of the agonists at the alpha 7 nicotinic receptor (a7nAChR) in the modulation of cognitive deficits in both Alzheimer's disease and schizophrenia. Such benefits have been modelled on the effects of the a7nAChR agonists in neurons. However, it is of note that the a7nAChR is also a powerful immune and glia regulator, suggesting that some of a7nAChR agonist benefits may be mediated by the suppression of immune and glia reactivity, which is in line with more recent conceptualizations of these disorders that have emphasized the role of these cells. Notably, melatonin, also efficacious across an array of medical conditions, is a significant regulator of a7nAChR levels and activity, with the a7nAChR mediating many of the beneficial effects of melatonin, including in the regulation of mitochondrial functioning.

This chapter focuses on the wide-ranging benefits that may arise from melatonin interactions with the a7nAChR, especially as to how such interactions may impact on the cellular mechanisms of an array of medical conditions. Such interactions are likely to have relevance across a host of neurodegenerative and psychiatric conditions.

Keywords: Alpha 7 nicotinic, Alzheimer's disease, brain, breast milk, depression, glia, immune, melatonin, multiple sclerosis, Parkinson's disease, schizophrenia, treatment.

INTRODUCTION

Recent data shows the alpha 7 nicotinic acetylcholine receptor (a7nAChR) to have an important role in a number of physiological and pathophysiological processes, with clinical relevance to the regulation of cognition in Alzheimer's disease (AD)

*****Corresponding author George Anderson:** CRC Scotland & London, Eccleston Square, UK; Tel: +447 940745360; E-mail: anderson.george@rocketmail.com

and schizophrenia [1, 2]. Conceptualizations as to how this is achieved have focussed on the role of the a7nAChR in the regulation of neuronal activity [3]. However, the a7nAChR is also a significant regulator of immune cells, including macrophages [4] and T helper (Th) cells [5], as well as having inhibitory regulatory effects in glial cells, including astrocytes and microglia [6, 7]. Circadian melatonin is a significant positive regulator of a7nAChR levels and activity [8]. Although melatonin is associated with beneficial effects across a host of medical conditions [9-12], it is becoming increasingly clear that some of these melatonin benefits may be mediated by its regulation of the a7nAChR in a number of different cell types, *via* the regulation of mitochondrial functioning [13].

This chapter reviews such melatonin interactions with the a7nAChR in different cellular functions, and how such interactions may be important to the course and etiology of a host of medical conditions, including neurodegenerative and psychiatric disorders. Firstly, the a7nAChR is looked at in more detail.

The Alpha 7 Nicotinic Receptor (a7nAChR)

Although the a7nAChR predominantly forms homomeric, rather than heteromeric pentamers, it is able to form heteromeric pentamers [14]. The majority of central cell types express the a7nAChR, as do immune cells and an array of other cell types, including gut cells, the enteric nervous system and the vagal nerve [15, 16]. In neurons, presynaptic a7nAChR activation leads to calcium and sodium influx, which is generally associated with the positive modulation of neurotransmitter release, whilst it is also expressed post-synaptically suggesting a wider role in synaptic plasticity [17]. In astrocytes, activation of the a7nAChR reduces the levels of inflammatory response, including to Alzheimer's disease associated amyloid-beta [6]; whilst in microglia, a7nAChR activation inhibits the levels of pro-inflammatory cytokines induced by lipopolysaccharide (LPS) [18]. The a7nAChR may also play a role in the regulation of neurogenesis, partly by the modulation of astrocyte fluxes [19], as well as more directly *via* effects on the maturation of immature dendritic cells [20]. As such, the a7nAChR is present in all relevant central cell types and plays a significant role in the regulation of crucial central processes.

The a7nAChR is also present in systemic immune cells, including monocytes/macrophages and T cells [21, 22]. Given that recent data shows an important role for such immune cells in the regulation of central processing, including in the pathophysiological processes underlying Alzheimer's disease, Parkinson's disease, depression and multiple sclerosis [23, 24], the a7nAChR will also have indirect impacts on key central processes *via* such immune cell regulation. As well as directly modulating such systemic immune cells, the a7nAChR also acts *via* key processes involved in immune activation, including inhibiting the immune-activating consequences arising from increases in gut permeability [24]. Such impacts on gut permeability driven immune activation may be inhibited by a7nAChR stimulation, at least in part, *via* the modulation of vagal nerve activity [15]. Some of the effects of the a7nAChR in decreasing blood-brain barrier (BBB) permeability, seem mediated by its activation in the spleen [25], again highlighting the importance of systemic and immune effects on central processes.

Cognition and a7nAChR

Although associated with many clinically relevant processes, including analgesia [26], recent work on the a7nAChR has highlighted its importance in cognition. Preclinical studies have long shown a7nAChR activation to increase aspects of cognition, including recognition memory and cognitive flexibility [27, 28]. Recent clinical trials in Alzheimer's disease and schizophrenia patients have shown a7nAChR agonists to be cognitive enhancers and/or to inhibit neurodegenerative processes [1, 2]. The a7nAChR is also thought to modulate wider cognitive functioning in people with Down syndrome [29], as well as those on the autistic spectrum [30].

Overall, by virtue of its significant regulation of an array of cells, the a7nAChR is an important regulator of many physiological and pathophysiological processes, with consequences for an array of medical conditions. Some of the a7nAChR effects seem *via* its interaction with melatonin, which will be overviewed next.

MELATONERGIC PATHWAYS

Introduction

Melatonin (N-acetyl-5-methoxytryptamine) is a methoxyindole that is primarily known as the 'darkness hormone' when released by the pineal gland following stimulation by norepinephrine (NE). Consequently, melatonin is widely appreciated for its role in the regulation of the circadian rhythm. However, less well appreciated is the data showing melatonin to be released by many different cell types, including enterochromaffin cells of the gut, where melatonin release can be up to 400-fold higher than its maximal release levels by the pineal gland. A growing body of data suggests that melatonin may be released by all mitochondria-containing cells with consequences for a wide array of physiological and pathophysiological processes, reviewed in [31]. As well as its role in circadian rhythm regulation, melatonin: is a powerful anti-oxidant and inducer of nuclear factor erythroid 2 (NF-E2)-related factor 2 (*Nrf2*)-driven endogenous anti-oxidants; increases the longevity protein, sirtuin-1, thereby acting on mitochondrial functioning; is a significant immune cell regulator, including *via* both autocrine and paracrine effects; is a significant anti-inflammatory, primarily due to its immune cell regulating effects; is a clinically relevant antinociceptive; optimizes mitochondrial functioning; increases levels of neurogenesis [32].

Melatonin is derived from serotonin *via* the production of N-acetylserotonin (NAS). Serotonin is therefore the necessary precursor of NAS and melatonin, being enzymatically converted by aralkylamine N-acetyltransferase (AA-NAT) to NAS, with hydroxyindole O-methyltransferase (HIOMT) (also referred to as acetylserotonin methyltransferase) enzymatically converting NAS to melatonin. As such, melatonin availability is dependent on levels of tryptophan, which are necessary for serotonin synthesis. The cellular activation of the melatonergic pathways leads to the synthesis and efflux of both NAS and melatonin, which are both amphiphilic. NAS has many similar effects to melatonin, being a significant anti-oxidant, mitochondria regulator and inducer of neurogenesis. However, NAS is a brain derived neurotrophic factor (BDNF) mimic, given its activation of the BDNF receptor, tyrosine kinase receptor-beta (TrKB) [33]. As such, the NAS/melatonin ratio may be of some significance, including in pathophysiological processes underpinning glioblastoma [34] and bipolar disorder [9], given their differential regulation of the a7nAChR and TrkB.

As well as the protective effects afforded by melatonin precursors, many of the melatonin metabolites, including N1-acetyl-N2-formyl-5-methoxykynuramine (AFMK) and N1-Acetyl-5-Methoxykynuramine (AMK), also have anti-inflammatory, anti-oxidant and immune-regulatory effects, reviewed in [32]. This indicates that the activation of the melatonergic pathways is likely to produce an array of factors with common anti-oxidant effects, but with specific effects on the a7nAChR and the BDNF TrkB being determined by levels of melatonin and NAS, respectively. As such, wider cellular processes that act to differentially regulate the melatonergic pathway products may significantly determine the products and consequences of melatonergic pathway activation.

Melatonin effects are commonly modeled as being mediated by melatonin receptors (MT1r and MT2r) activation. However, due to the amphiphilic nature of melatonin and NAS, melatonin can readily cross cell membranes. Consequently, melatonin is commonly found to accumulate around intracellular organelles, particularly mitochondria. Melatonin also acts to regulate membrane fluidity, suggesting that it may have a role in the structural regulation, and therefore function, of membranes [35]. As well as its co-ordination of circadian genes, such structural membrane regulation may be another means by which melatonin acts to synchronize an array of cellular processes.

A growing number of factors have been shown to regulate the melatonergic pathways. Pineal melatonin is increased by leptin and 14-3-3, with 14-3-3 stabilizing AA-NAT and increasing its transcription. Tumor necrosis factor alpha (TNFa), along with other pro-inflammatory cytokines, decreases pineal melatonin synthesis [31, 32]. This is of some note in regard to the regulation of wider body processes, as the induction of pro-inflammatory cytokines, such as TNFa, interleukin(IL)-1b and interferon-gamma (IFNg), increases indoleamine 2,3-dioxygenase (IDO), thereby driving tryptophan down the kynurenine pathways and away from serotonin and melatonin synthesis. As such, inflammatory processes are intimately involved in the regulation of the melatonergic pathways, with significant consequences for the conceptualization and treatment of an array of medical conditions, including most neurodegenerative [10, 12, 36] and psychiatric conditions [37-39]. Overall, the melatonergic pathways are intimately linked to an array of cellular and systemic processes, with consequences for many medical conditions that are currently poorly managed.

The work of Regina Markus and colleagues has highlighted the presence of an immune-pineal axis (IPA) whereby pro-inflammatory processes and cytokines, particularly TNFa, switch off pineal melatonin synthesis, thereby preventing the generally immune-dampening effects of pineal melatonin during the course of infection and inflammation [40]. This suggests that integrating the regulation of the melatonergic pathways, in different organs, tissues and cell types, may be an integral aspect of fine-tuning the immune response.

Overall, the regulation of the melatonergic pathways is likely to be intimately associated with many core cellular and systemic processes. It is in this context the regulation of the a7nAChR by melatonin requires investigation.

MELATONIN AND a7nAChR INTERACTIONS

Mitochondria

As indicated above, circadian melatonin acts to regulate the a7nAChR [8], with some of the effects of melatonin being *via* the a7nAChR, as exemplified by the protection afforded by melatonin against ischaemia [41]. The latter study showed melatonin to protect against ischaemia in rodents, with effects dependent upon the activation of the a7nAChR and the induction of Nrf2 and one of its transcriptional products, heme oxygenase-1 (HO-1) [41]. Similar processes were shown to occur in organotypic hippocampal cultures [41]. Given the role of ischaemia in ageing and neurodegenerative conditions, such as Alzheimer's disease, such data suggests that the protection of melatonin in such conditions may be mediated, at least in part, by the regulation and activation of the a7nAChR.

With both melatonin and the a7nAChR activation known to have beneficial effects on mitochondrial functioning [13], it is not unlikely that the benefits arising from both factors involves their interaction in the optimization of mitochondrial functioning under challenge. The a7nAChR is expressed in the mitochondrial outer membrane, where it acts to regulate the voltage-dependent anion channel (VDAC)-mediated Ca(2+) transport and thereby the apoptosis-associated mitochondrial permeability transition [13]. Investigations utilizing melatonin have shown it to have a number of protective effects in mitochondria, including both directly and indirectly. Indirectly, melatonin increases sirtuin-1 which acts to regulate peroxisome proliferator-activated receptor gamma coactivator 1-alpha (PGC-1a) [42]. PGC-1a is known as the master mitochondria regulator [43]. More directly, the enzymes involved in driving the melatonergic pathway, including AANAT, have been shown to be directly expressed in mitochondria, suggesting that mitochondria are sites for the synthesis of melatonin [44], with this occurring in the absence of any known mitochondrial import motif or mechanism. Such melatonin synthesis urgently requires investigation in other cell types, including neurons, glia, immune and gut cells.

As to whether melatonin synthesis in mitochondria has any regulatory effect on a7nAChR levels and activity in mitochondria is unknown. It is also unknown as to whether any of the other products of the melatonergic pathway, including NAS, AMFK and AMK, would interact with such mitochondrial associated melatonin synthesis and a7nAChR regulation. Given that the a7nAChR requires to be in caveolin-associated lipid rafts for optimized function [45] and that melatonin regulates membrane fluidity [35], it should be investigated as to whether the interactions of the a7nAChR and melatonin, including when synthesized in mitochondria, are mediated to some degree by melatonin's effects on membrane fluidity. With mitochondrial functioning acting to regulate the activity and survival of all cells, including immune cells and glia, it is likely that such melatonin interactions with the a7nAChR will be an important determinant of reactivity processes underlying inflammatory conditions. As to whether a7nAChR and melatonin interactions in the regulation of inflammatory processes are mediated by changes in the fluidity of the plasma membrane, as well as mitochondrial membranes [46], especially in the modulation of the a7nAChR, remains to be determined. This may provide an important subcellular basis for the many overlapping effects of melatonin and the a7nAChR across different cell types and cellular processes.

Glia

Astrocytes and microglia are important determinants of neuronal activity in the central nervous system. Both are reactive cells, which are activated in most, if not all, neurodegenerative and psychiatric conditions [47, 48]. Recent thinking on these cells, suggests that neuronal activity may have evolved as a form of immune-to-immune communication, with glia being the brain's immune-type cells [48]. As such, the regulation of glia reactivity is likely to be a crucial determinant of the apoptosis and alterations in neuronal functioning and patterning that are common in CNS-associated disorders. The a7nAChR is present in both of these glia cell types, where its activation leads to a decrease in their reactivity [6, 18]. Such data parallels the effects of melatonin in astrocytes and microglia, with melatonin also decreasing the reactivity of these cells [49, 50].

As to whether such a7nAChR and melatonin effects in glia are mediated, at least in part, *via* their interaction in glia mitochondria requires investigation. Given that the protection afforded by melatonin against LPS-induced oxidative stress, Nrf2 inhibition, and neuroinflammation are dependent on sirtuin1 in microglia [51], it is likely that some of the anti-inflammatory effect is mediated by the indirect effects of melatonin on mitochondria functioning *via* sirtuin1-PGC1a. Mitochondria are powerful determinants of glia reactivity [52, 53].

Autophagy

Autophagy was originally conceived as a "self-eating" survival pathway, which allowed nutrient recycling during periods of starvation. However, work showing that autophagy can target intracellular bacteria for degradation led to autophagy being linked, *via* multiple cellular processes, to different aspects of immunity, including a role in antigen processing, lymphocyte homeostasis and thymic selection, as well as in the regulation of immunoglobulin and cytokine secretion. Autophagy is an important aspect of both central and systemic processing, including in the mucosal immune system of the gut [54]. Some of the protection afforded by melatonin across an array of pathophysiological conditions is driven by melatonin's regulation of autophagy [55, 56]. In prion-mediated mitochondrial damage the protection afforded by melatonin is mediated *via* the upregulation of a7nAChR signalling [57]. This again highlights the importance of the interactions of melatonin and the a7nAChR in the modulation of key cellular processes.

Mitophagy can be seen as a specialized form of autophagy, with mitophagy being the selective degradation of damaged mitochondria by autophagy. Melatonin affords protection in some brain injury models by increasing mitophagy, thereby decreasing the levels of oxidants produced by damaged mitochondria [58]. These authors showed that melatonin increased mitophagy through the mechanistic target of rapamycin (mTOR) pathway, leading to decreased levels of oxidants and pro-inflammatory cytokines [58]. Under conditions of cellular challenge, melatonin not only increases mitophagy but also increases mitochondrial biogenesis [59]. As to whether such effects of melatonin are mediated *via* the a7nAChR in mitophagy, as in autophagy more widely, requires urgent investigation, given the importance of mitochondrial functioning and regulation across all medical conditions.

Immunity

Melatonin is a significant regulator of the immune system, with its autocrine and paracrine effects generally decreasing the reactivity of immune cells, such as macrophages [60], whilst optimizing the more beneficial Th1 cell response and inhibiting the more prolonged and damaging Th17 cell response [61]. The alterations in the immune response driven by pro-inflammatory cytokine effects on pineal melatonin production [40] highlight the importance of melatonin in immune system regulation. As indicated above, the a7nAChR is a significant regulator of systemic immune cell responses [21, 22], with indirect immune effects mediated *via* its regulation of the vagal nerve [15] and BBB permeability [25]. As indicated above, the interactions of melatonin and the a7nAChR in the regulation of autophagy, mitophagy and mitochondrial functioning will modulate not only the cellular triggers to immune activation, but will also modulate the immune response itself.

A recent study shows that the alterations in offspring brain development arising from the injection of viral mimetic in a murine model of autism and schizophrenia can be prevented by a7nAChR activation [62]. These authors also showed that variations in the a7nAChR gene also modulate offspring outcome [62]. Given that alterations in the melatonergic pathways are also associated with pregnancy associated conditions such as autism and schizophrenia [63], it requires investigation as to how the melatonergic pathways interact with the a7nAChR in the early developmental etiology of childhood- and adult-onset conditions. Melatonin is highly produced by the placenta, with melatonin levels being decreased in preeclampsia [64]. The a7nAChR is also highly expressed in the placenta, with alterations in a7nAChR protein levels evident in severe preeclampsia [65]. This suggests that interactions of placental melatonin and the a7nAChR may be of some importance to the early developmental etiology of an array of childhood and adult medical conditions, likely in association with alterations in immune responses, both in the mother and foetus/offspring.

Gut-Brain Axis

The gut-brain axis is at the cutting edge of research across an array of medical conditions, including Alzheimer's disease, Parkinson's disease, multiple sclerosis and depression [10, 66, 67]. The primary changes in the gut seem to involve the specific bacteria present and variations in the levels of gut barrier permeability, with increased gut permeability leading to the crossing of gut bacteria and tiny fragments of partially digested food that drive the host's immune response. Many factors increase gut permeability, including alcohol, fatty acids and stress/cortisol [68]. Such increases in gut permeability can be prevented by melatonin, which is very highly expressed in the gut, primarily in enterochromaffin cells [68].

Given such melatonin benefits, it is important to note that the specific nature of gut bacteria is a major determinant of host tryptophan, and therefore of serotonin, availability [69], in turn determining the serotonin availability for melatonergic pathways. Given the interactions of the a7nAChR with melatonin, changes in the nature of the microbiome are likely to impact on the levels and/or activity of the a7nAChR. Both a7nAChr activation and melatonin are significant negative regulators of the inflammasome, which drives IL-1B and IL-18 synthesis and release. The inflammasome, by its induction of these pro-inflammatory cytokines, is an important regulator of gut homeostasis [70], including gut permeability [71]. Some of the protection afforded by melatonin against experimentally-induced increases in gut permeability is mediated by enhancing a7nAChr levels and activity [72]. As such, gut inflammasome regulation may be another important aspect of the biological underpinnings of an array of medical conditions, which can be modulated by the interactions of melatonin and the a7nAChR.

As well as inhibiting glia and macrophages, the a7nAChR also inhibits the activation of mast cells, which are important regulators of gut permeability [73]. Mast cell activation significantly enhances gut permeability in humans, following stress or corticotropin-releasing hormone-induced cortisol [74]. Work in rodents shows that chronic stress sensitizes the gut to subacute stress effects, leading to an increase in gut permeability [75]. However, these authors showed that the biological underpinnings of such stress-driven changes in the gut are different at different sites, with increased mast cell density being evident, following chronic plus subacute stress, in the colon, but not in the small intestine [75]. It is very likely that future research will show that other factors are differentially regulated in different subsections of the gut, and it will require investigation as to whether variations in melatonin and/or the a7nAChR are differentially important in such gut subsections.

Overall changes in mitochondria, autophagy, mitophagy, glia and immune cells as well as in the regulation of gut microbiota and gut permeability are likely to be important sites and processes for the interactions of melatonin and the a7nAChR in the etiology, course and treatment of an array of medical conditions. An important pathway in such pathophysiological processes is the tryptophan catabolites (TRYCATs) pathway, which is briefly described below.

Tryptophan Catabolites (TRYCATs)

TRYCATs are produced when pro-inflammatory cytokines, including IL-1B, IL-18 and TNFa but especially interferon-gamma, induce indoleamine 2,3-dioxygenase (IDO) or when stress-associated cortisol induces tryptophan 2,3-dioxygenase (TDO). The activation of IDO and TDO takes tryptophan away from serotonin and melatonin synthesis and drives it down the kynurenine pathway to the production of TRYCATs, including the neuroregulatory kynurenic acid (KYNA) [31]. As such, stress and pro-inflammatory conditions can drive alterations in serotonergic and the melatonergic pathways, as well as increasing the production of KYNA, which is an inhibitor of the a7nAChR. Consequently, TRYCATs activation and KYNA synthesis will decrease melatonin as well as inhibiting the a7nAChR, with significant implications for pathophysiological processes underlying a host of medical conditions. Indicants of TRYCATs pathway activation, including increased pro-inflammatory cytokines and oxidative stress, as well as increased levels in the ratio of kynurenine/tryptophan are evident in many neuropsychiatric conditions, including depression, bipolar disorder, schizophrenia, Alzheimer's disease and multiple sclerosis [9, 10, 12]. All of these conditions are associated with decreased levels of melatonin and increased levels of TRYCATs, such as KYNA. As such, TRYCATs pathway activation is likely to be associated with significant changes in the levels of melatonin and a7nAChR with consequences for the etiology, course and management of these conditions.

Although inflammatory factors and oxidative stress are seen as significant treatment targets in most neuropsychiatric conditions, it is likely that some of the symptomatology is driven by decreases in melatonin and the inhibition of the a7nAChR [31]. In this way, decreases in local and pineal melatonin as well as in a7nAChR levels and activity are co-ordinated not only with wider oxidative and inflammatory stress, but also with changes in mitochondrial functioning, autophagy, mitophagy and increased gut permeability, as well as with glia and immune cell reactivity (Fig. **1**).

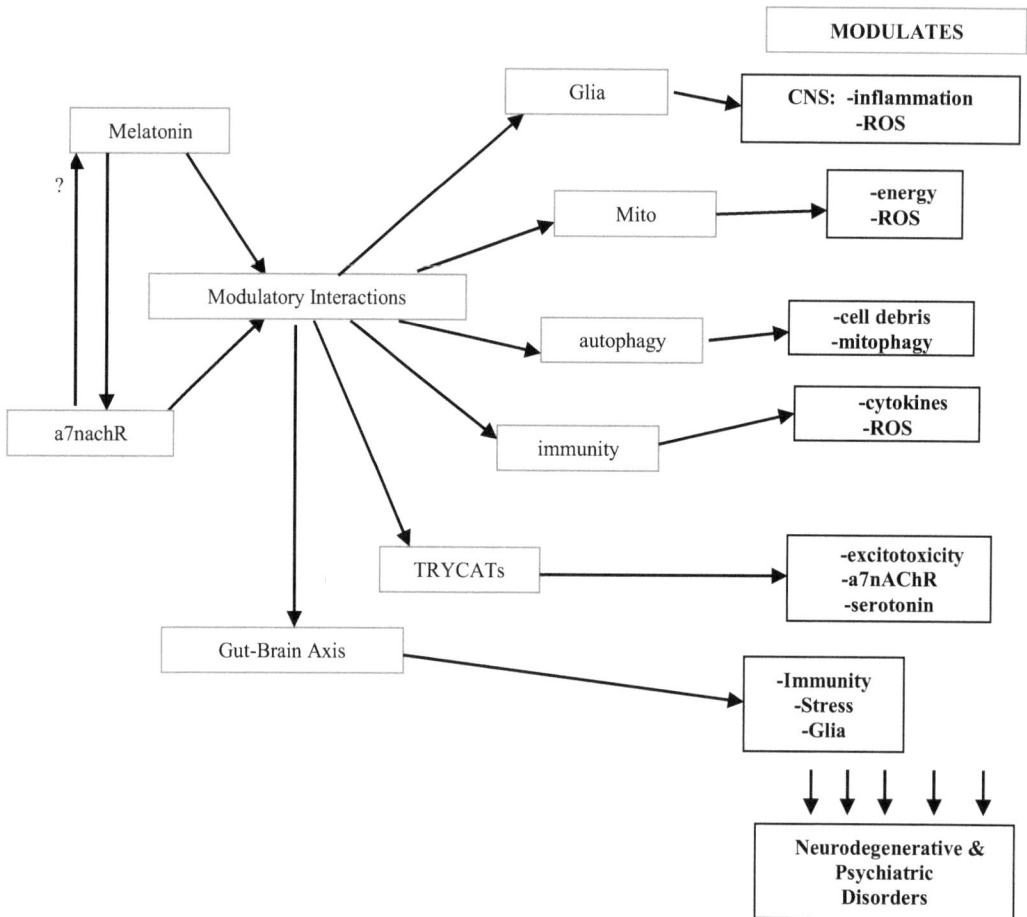

Fig. (1). The modulation of the a7nAChR by melatonin and the likely a7nAChR impacts on the melatonergic pathways interact to impact on many processes that are fundamental to cellular and systemic system functioning, including mitochondria, gut-brain axis and glia/immune reactivity. Decreases in a7nAChR and local melatonin at different cellular sites, tissues and organs will impact on important pathophysiological processes, including inflammation and ROS, across a host of neurodegenerative and psychiatric conditions. Such processes modulated by the interaction of the a7nAChR (and its agonists) with melatonin are highlighted in bold on the right-hand side of the figure.

Breastfeeding *versus* Formula-Feeding

Given the significant benefits that are derived from breastfeeding versus formula-feeding, it may be interesting to look at the role of melatonin and a7nAChR in this context. Breastfeeding has a plethora of benefits, including: lower hospital admissions rates for respiratory infections [76] and neonatal fever [77]; decreased obesity levels at age 2 years [78]; lower offspring cancer levels [79], type 1 and 2 diabetes, hypertension, hyperlipidemia, obesity and cardiovascular disease [80]. Given such health benefits, breastfeeding can provide substantial financial benefits to countries, driven by the improved health, IQ and cognition that breastfeeding provides [81, 82]. Such offspring benefits are likely to be mediated, at least in part, by breastfeeding lowering levels of obesity and metabolic dysregulation [83].

Recently, we proposed that formula feed could be improved by the addition of melatonin, perhaps to a night-time only feed [84], thereby better replicating the melatonin content in breast milk. Coupled to data showing that levels of choline, a natural agonist at the a7nAChR, in breast milk also positively correlate with offspring cognitive ability [85], it may be that melatonin interactions with the a7nAChR play a significant role in the benefits conferred by breast-feeding. It is not unlikely that some of the benefits of breast milk are mediated *via* changes in the infant's microbiome that drive an early influence of the gut-brain axis on CNS development. This requires investigation, but highlights the wide ranging benefits that may arise from a better understanding of the importance of melatonin interactions with the a7nAChR.

CONCLUSION

The overlapping benefits of melatonin and the a7nAChR across a host of medical conditions and pathophysiological processes seem to arise from melatonin's regulation of the a7nAChR, with many of melatonin's effects seemingly mediated by the a7nAChR. This interaction has important consequences for crucial cellular processes, including mitochondrial biogenesis, autophagy and mitophagy as well as for more systemic processes, including glia reactivity, immune cell patterning/reactivity and the regulation of the gut-brain axis. The interactions of melatonin and the a7nAChR with stress/cortisol, pro-inflammatory cytokines and oxidative stress may be mediated, in part, by the activation of TRYCATs, especially KYNA. Consequently, this has implications for the etiology, course and treatment of a host of neuropsychiatric conditions. The growing appreciation of the role of early development in the susceptibility to a wide array of childhood-

and adult-onset disorders would suggest that such melatonin interactions with the a7nAChR should be taken into account in the development of formula feed, thereby allowing formula feed to be closer to the developmental needs of the infant, as exemplified by breast milk. The importance of melatonin interactions with the a7nAChR in such an array of physiological processes would also suggest that the a7nAChR agonists, currently in phase III trials as cognitive enhancers in Alzheimer's disease and schizophrenia patients, should take account of patient melatonin levels, perhaps optimizing melatonin by its adjunctive use.

CONFLICT OF INTEREST

The author confirms that author has no conflict of interest to declare for this publication.

ACKNOWLEDGMENTS

Declared None

REFERENCES

[1]   Deardorff WJ, Shobassy A, Grossberg GT. Safety and clinical effects of EVP-6124 in subjects with Alzheimer's disease currently or previously receiving an acetylcholinesterase inhibitor medication. Expert Rev Neurother. 2015;15(1):7-17.
[2]   Marder SR. Alpha-7 nicotinic agonist improves cognition in schizophrenia. Evid Based Ment Health. 2016;19(2):60.
[3]   Kunii Y, Zhang W, Xu Q, *et al.* CHRNA7 and CHRFAM7A mRNAs: co-localized and their expression levels altered in the postmortem dorsolateral prefrontal cortex in major psychiatric disorders. Am J Psychiatry. 2015;172(11):1122-30.
[4]   Lee RH, Vazquez G. Evidence for a prosurvival role of alpha-7 nicotinic acetylcholine receptor in alternatively (M2)-activated macrophages. Physiol Rep. 2013;1(7):e00189.
[5]   Liu Z, Han B, Li P, Wang Z, Fan Q. Activation of α7nAChR by nicotine reduced the Th17 response in CD4(+)T lymphocytes. Immunol Invest. 2014;43(7):667-74.
[6]   Wang Y, Zhu N, Wang K, Zhang Z, Wang Y. Identification of α7 nicotinic acetylcholine receptor on hippocampal astrocytes cultured *in vitro* and its role on inflammatory mediator secretion. Neural Regen Res. 2012;7(22):1709-14.
[7]   Suzuki T, Hide I, Matsubara A, *et al.* Microglial alpha7 nicotinic acetylcholine receptors drive a phospholipase C/IP3 pathway and modulate the cell activation toward a neuroprotective role. J Neurosci Res. 2006;83(8):1461-70.
[8]   Markus RP, Silva CL, Franco DG, Barbosa EM Jr, Ferreira ZS. Is modulation of nicotinic acetylcholine receptors by melatonin relevant for therapy with cholinergic drugs? Pharmacol Ther. 2010;126(3):251-62.
[9]   Anderson G, Jacob A, Bellivier F, Geoffroy PA. Bipolar disorder: the role of the kynurenine and melatonergic pathways. Curr Pharm Des. 2016;22(8):987-1012.
[10]   Maes M, Anderson G. Overlapping the tryptophan catabolite (TRYCAT) and melatoninergic pathways in Alzheimer's disease. Curr Pharm Des. 2016; 22(8):1074-85.
[11]   Anderson G, Maes M, Markus RP, Rodriguez M. Ebola virus: melatonin as a readily available treatment option.J Med Virol. 2015; 87(4):537-43.

[12]   Anderson G, Rodriguez M. Multiple sclerosis: the role of melatonin and N-acetylserotonin. Mult Scler Relat Disord. 2015;4(2):112-23.

[13]   Gergalova G, Lykhmus O, Kalashnyk O, *et al.* Mitochondria express α7 nicotinic acetylcholine receptors to regulate Ca2+ accumulation and cytochrome c release: study on isolated mitochondria. PLoS One. 2012;7(2):e31361.

[14]   Liu Q, Huang Y, Shen J, Steffensen S, Wu J. Functional α7β2 nicotinic acetylcholine receptors expressed in hippocampal interneurons exhibit high sensitivity to pathological level of amyloid β peptides. BMC Neurosci. 2012;13:155.

[15]   Kalkman HO, Feuerbach D. Modulatory effects of α7 nAChRs on the immune system and its relevance for CNS disorders. Cell Mol Life Sci. 2016;73(13):2511-30.

[16]   Costantini TW, Krzyzaniak M, Cheadle GA, *et al.* Targeting α-7 nicotinic acetylcholine receptor in the enteric nervous system: a cholinergic agonist prevents gut barrier failure after severe burn injury. Am J Pathol. 2012;181(2):478-86.

[17]   Yakel JL. Nicotinic ACh receptors in the hippocampal circuit; functional expression and role in synaptic plasticity. J Physiol. 2014;592(19):4147-53.

[18]   Noda M, Kobayashi AI. Nicotine inhibits activation of microglial proton currents *via* interactions with α7 acetylcholine receptors. J Physiol Sci. In press.

[19]   Anderson G, Maes M. Reconceptualizing adult neurogenesis: role for sphingosine-1-phosphate and fibroblast growth factor-1 in co-ordinating astrocyte-neuronal precursor interactions.CNS Neurol Disord Drug Targets. 2014;13(1):126-36.

[20]   John D, Shelukhina I, Yanagawa Y, Deuchars J, Henderson Z. Functional alpha7 nicotinic receptors are expressed on immature granule cells of the postnatal dentate gyrus. Brain Res. 2015;1601:15-30.

[21]   Rosas-Ballina M, Goldstein RS, Gallowitsch-Puerta M, *et al.* The selective alpha7 agonist GTS-21 attenuates cytokine production in human whole blood and human monocytes activated by ligands for TLR2, TLR3, TLR4, TLR9, and RAGE. Mol Med. 2009;15(7-8):195-202.

[22]   De Rosa MJ, Dionisio L, Agriello E, Bouzat C, Esandi Mdel C. Alpha 7 nicotinic acetylcholine receptor modulates lymphocyte activation. Life Sci. 2009;85(11-12):444-9.

[23]   Slyepchenko A, Maes M, Köhler CA, *et al.* T helper 17 cells may drive neuroprogression in major depressive disorder: Proposal of an integrative model. Neurosci Biobehav Rev. 2016;64:83-100.

[24]   Anderson G, Seo M, Carvalho A, Berk M, Maes M. Gut permeability and Parkinson's disease. Curr Pharm Des. In press.

[25]   Dash PK, Zhao J, Kobori N, *et al.* Activation of Alpha 7 Cholinergic nicotinic receptors reduce blood-brain barrier permeability following experimental traumatic brain injury. J Neurosci. 2016;36(9):2809-18.

[26]   Damaj MI, Meyer EM, Martin BR. The antinociceptive effects of alpha7 nicotinic agonists in an acute pain model. Neuropharmacology. 2000;39(13):2785-91.

[27]   Nikiforuk A, Kos T, Potasiewicz A, Popik P. Positive allosteric modulation of alpha 7 nicotinic acetylcholine receptors enhances recognition memory and cognitive flexibility in rats.Eur Neuropsychopharmacol. 2015;25(8):1300-13.

[28]   Van Kampen M, Selbach K, Schneider R, Schiegel E, Boess F, Schreiber R. AR-R 17779 improves social recognition in rats by activation of nicotinic alpha7 receptors. Psychopharmacology (Berl). 2004;172(4):375-83.

[29]   Deutsch SI, Rosse RB, Mastropaolo J, Chilton M.Progressive worsening of adaptive functions in Down syndrome may be mediated by the complexing of soluble Abeta peptides with the alpha 7 nicotinic acetylcholine receptor: therapeutic implications. Clin Neuropharmacol. 2003;26(5):277-83.

[30]   Deutsch SI, Burket JA, Urbano MR, Benson AD. The α7 nicotinic acetylcholine receptor: A mediator of pathogenesis and therapeutic target in autism spectrum disorders and Down syndrome. Biochem Pharmacol. 2015;97(4):363-77.

[31]   Anderson G, Maes M. Local melatonin regulates inflammation resolution: a common factor in neurodegenerative, psychiatric and systemic inflammatory disorders. CNS Neurol Disord Drug Targets. 2014;13(5):817-27.

[32]   Hardeland R, Cardinali DP, Srinivasan V, Spence DW, Brown GM, Pandi-Perumal SR. Melatonin--a pleiotropic, orchestrating regulator molecule. Prog Neurobiol. 2011;93(3):350-84.

[33]   Jang SW, Liu X, Pradoldej S, *et al.* N-acetylserotonin activates TrkB receptor in a circadian rhythm. Proc Natl Acad Sci U S A. 2010;107(8):3876-81.

[34]   Beischlag TV, Anderson G, Mazzoccoli G. Glioma: tryptophan catabolite and melatoninergic pathways link microRNA, 14-3- 3, chromosome 4q35, epigenetic processes and other glioma biochemical changes. Curr Pharm Des. 2016;22(8):1033-48.

[35]   Drolle E, Kučerka N, Hoopes MI, Choi Y, Katsaras J, Karttunen M, Leonenko Z. Effect of melatonin and cholesterol on the structure of DOPC and DPPC membranes. Biochim Biophys Acta. 2013;1828(9):2247-54.

[36]   Anderson G, Maes M. TRYCAT pathways link peripheral inflammation, nicotine, somatization and depression in the etiology and course of Parkinson's disease. CNS Neurol Disord Drug Targets. 2014;13(1):137-49.

[37]   Anderson G and Maes M. Co-enzyme Q10, depression and depression associated conditions. Chapter in Book: Co-Enzyme Q10: from fact to fiction. Ed: Hargreaves IP. 2015.

[38]   Anderson G, Kubera M, Duda W, Lason W, Berk M, Maes M. Increased IL-6 trans-signaling in depression: focus on the tryptophan catabolite pathway, melatonin and neuroprogression. Pharmacol Rep. 2013;65(6):1647-54.

[39]   Anderson G, Maes M, Berk M. Schizophrenia is primed for an increased expression of depression through activation of immuno-inflammatory, oxidative and nitrosative stress, and tryptophan catabolite pathways. Prog Neuropsychopharmacol Biol Psychiatry. 2013;42:101-14.

[40]   Markus RP, Cecon E, Pires-Lapa MA. Immune-pineal axis: nuclear factor κB (NF-kB) mediates the shift in the melatonin source from pinealocytes to immune competent cells. Int J Mol Sci. 2013;14(6):10979-97.

[41]   Parada E, Buendia I, León R, Negredo P, Romero A, Cuadrado A, López MG, Egea J. Neuroprotective effect of melatonin against ischemia is partially mediated by alpha-7 nicotinic receptor modulation and HO-1 overexpression. J Pineal Res. 2014;56(2):204-12.

[42]   Traboulsi H, Davoli S, Catez P, Egly JM, Compe E. Dynamic partnership between TFIIH, PGC-1α and SIRT1 is impaired in trichothiodystrophy. PLoS Genet. 2014;10(10):e1004732.

[43]   Puigserver P. Tissue-specific regulation of metabolic pathways through the transcriptional coactivator PGC1-alpha. Int J Obes (Lond). 2005;29 Suppl 1:S5-9.

[44]   He C, Wang J, Zhang Z, et al. Mitochondria synthesize melatonin to ameliorate its function and improve mice oocyte's quality under in vitro conditions. Int J Mol Sci. 2016;17(6).

[45]   Brusés JL, Chauvet N, Rutishauser U. Membrane lipid rafts are necessary for the maintenance of the (alpha)7 nicotinic acetylcholine receptor in somatic spines of ciliary neurons. J Neurosci. 2001;21(2):504-12.

[46]   García JJ, Piñol-Ripoll G, Martínez-Ballarín E, et al. Melatonin reduces membrane rigidity and oxidative damage in the brain of SAMP8 mice. Neurobiol Aging. 2011;32(11):2045-54.

[47]   Wes PD, Sayed FA, Bard F, Gan L. Targeting microglia for the treatment of Alzheimer's Disease. Glia. 2016;64(10):1710-32.

[48]   Anderson G. Neuronal-immune interactions in mediating stress effects in the etiology and course of schizophrenia: role of the amygdala in developmental co-ordination. Med Hypotheses. 2011;76(1):54-60.

[49]   Babaee A, Eftekhar-Vaghefi SH, Asadi-Shekaari M, et al. Melatonin treatment reduces astrogliosis and apoptosis in rats with traumatic brain injury. Iran J Basic Med Sci. 2015;18(9):867-72.

[50]   Ding K, Wang H, Xu J, Lu X, Zhang L, Zhu L. Melatonin reduced microglial activation and alleviated neuroinflammation induced neuron degeneration in experimental traumatic brain injury: Possible involvement of mTOR pathway. Neurochem Int. 2014 Oct;76:23-31.

[51]   Shah SA, Khan M, Jo MH, Jo MG, Amin FU, Kim MO. Melatonin stimulates the SIRT1/Nrf2 signaling pathway counteracting lipopolysaccharide (LPS)-induced oxidative stress to rescue postnatal rat brain. CNS Neurosci Ther. In press.

[52]   Ye J, Jiang Z, Chen X, Liu M, Li J, Liu N. Electron transport chain inhibitors induce microglia activation through enhancing mitochondrial reactive oxygen species production. Exp Cell Res. 2016;340(2):315-26.

[53]   Sarafian TA, Montes C, Imura T, et al. Disruption of astrocyte STAT3 signaling decreases mitochondrial function and increases oxidative stress in vitro.PLoS One. 2010;5(3):e9532.

[54]   Kabat AM, Pott J, Maloy KJ. The mucosal immune system and its regulation by autophagy. Front Immunol. 2016;7:240.

[55]   Ding K, Xu J, Wang H, Zhang L, Wu Y, Li T. Melatonin protects the brain from apoptosis by enhancement of autophagy after traumatic brain injury in mice. Neurochem Int. 2015;91:46-54.

[56]   Carloni S, Favrais G, Saliba E, *et al*. Melatonin modulates neonatal brain inflammation through ER stress, autophagy and miR-34a/SIRT1 pathway. J Pineal Res. 2016;61(3):370-80.

[57]   Jeong JK, Park SY. Melatonin regulates the autophagic flux *via* activation of alpha-7 nicotinic acetylcholine receptors. J Pineal Res. 2015;59(1):24-37.

[58]   Lin C, Chao H, Li Z, *et al*. Melatonin attenuates traumatic brain injury-induced inflammation: a possible role for mitophagy. J Pineal Res. 2016;61(2):177-86.

[59]   Kang JW, Hong JM, Lee SM. Melatonin enhances mitophagy and mitochondrial biogenesis in rats with carbon tetrachloride-induced liver fibrosis. J Pineal Res. 2016;60(4):383-93.

[60]   Muxel SM, Pires-Lapa MA, Monteiro AW, *et al*. NF-κB drives the synthesis of melatonin in RAW 264.7 macrophages by inducing the transcription of the arylalkylamine-N-acetyltransferase (AA-NAT) gene. PLoS One. 2012;7(12):e52010.

[61]   Lee JS, Cua DJ. Melatonin Lulling Th17 Cells to Sleep. Cell. 2015;162(6):1212-4.

[62]   Wu WL, Adams CE, Stevens KE, Chow KH, Freedman R, Patterson PH. The interaction between maternal immune activation and alpha 7 nicotinic acetylcholine receptor in regulating behaviors in the offspring. Brain Behav Immun. 2015;46:192-202.

[63]   Anderson G, Maes M. Redox regulation and the autistic spectrum: role of tryptophan catabolites, immuno-inflammation, autoimmunity and the amygdala. Curr Neuropharmacol. 2014;12(2):148-67.

[64]   Lanoix D, Guérin P, Vaillancourt C. Placental melatonin production and melatonin receptor expression are altered in preeclampsia: new insights into the role of this hormone in pregnancy. J Pineal Res. 2012;53(4):417-25.

[65]   Kwon JY, Kim YH, Kim SH, *et al*. Difference in the expression of alpha 7 nicotinic receptors in the placenta in normal versus severe preeclampsia pregnancies. Eur J Obstet Gynecol Reprod Biol. 2007;132(1):35-9.

[66]   Anderson G, Seo M, Berk M, Carvalho AF, Maes M. Gut permeability and microbiota in Parkinson's disease: role of depression, tryptophan catabolites, oxidative and nitrosative stress and melatoninergic pathways. Curr Pharm Des. In press.

[67]   Rodriguez M, Wootla B, Anderson G. Multiple sclerosis, gut microbiota and permeability: role of tryptophan catabolites, depression and the driving down of local melatonin. Curr Pharm Des. In press.

[68]   Anderson G, Maes M. The Gut-Brain Axis: The role of melatonin in linking psychiatric, Inflammatory and neurodegenerative conditions. Advances In Integrative Medicine. 2015;2(1):31-37.

[69]   Wikoff WR, Anfora AT, Liu J, *et al*. Metabolomics analysis reveals large effects of gut microflora on mammalian blood metabolites. Proc Natl Acad Sci U S A. 2009;106(10):3698-703.

[70]   Zambetti LP, Mortellaro A. NLRPs, microbiota, and gut homeostasis: unravelling the connection. J Pathol. 2014;233(4):321-30.

[71]   Xiao YT, Li GX, Wang XM. Effect mechanism of NOD like receptor signaling pathway on intestinal mucosal barrier of rat during early phase of acute intra-abdominal infection. Zhonghua Wei Zhong Bing Ji Jiu Yi Xue. 2013;25(9):527-32.

[72]   Sommansson A, Nylander O, Sjöblom M. Melatonin decreases duodenal epithelial paracellular permeability *via* a nicotinic receptor-dependent pathway in rats in vivo. J Pineal Res. 2013;54(3):282-91.

[73]   Mishra NC, Rirsimaah J, Boyd RT, *et al*. Nicotine inhibits Fc epsilon RI-induced cysteinyl leukotrienes and cytokine production without affecting mast cell degranulation through alpha 7/alpha9/alpha 10-nicotinic receptors. J Immunol 2010; 185(1): 588-96.

[74]   Vanuytsel T, van Wanrooy S, Vanheel H, *et al*. Psychological stress and corticotropin-releasing hormone increase intestinal permeability in humans by a mast cell-dependent mechanism. Gut. 2014;63(8):1293-9.

[75]   Lauffer A, Vanuytsel T, Vanormelingen C, *et al*. Subacute stress and chronic stress interact to decrease intestinal barrier function in rats. Stress. 2016;19(2):225-34.

[76]   Dagvadorj A, Ota E, Shahrook S, *et al*. Hospitalization risk factors for children's lower respiratory tract infection: A population-based, cross-sectional study in Mongolia. Sci Rep. 2016;6:24615.

[77]   Netzer-Tomkins H, Rubin L, Ephros M. Breastfeeding is associated with decreased hospitalization for neonatal fever. Breastfeed Med. 2016;11:218-21.

[78]   Bider-Canfield Z, Martinez MP, Wang X, *et al*. Maternal obesity, gestational diabetes, breastfeeding and childhood overweight at age 2 years. Pediatr Obes. In press.

[79]   Blackadar CB. Historical review of the causes of cancer. World. J. Clin. Oncol. 2016;7(1):54-86.

[80]   Binns C, Lee M, Low WY. The long-term public health benefits of breastfeeding. Asia Pac J Public Health. 2016;28(1):7-14.

[81]   Walters D, Horton S, Siregar AY, *et al*. The cost of not breastfeeding in Southeast Asia. Health Policy Plan. 2016;31(8):1107-16.

[82]   Straub N, Grunert P, Northstone K, Emmett P. Economic impact of breast-feeding-associated improvements of childhood cognitive development, based on data from the ALSPAC. Br J Nutr. 2016:1-6.

[83]   Moser VA, Pike CJ. Obesity and sex interact in the regulation of Alzheimer's disease. Neurosci Biobehav Rev. 2016;67:102-18.

[84]   Anderson G, Vaillancourt C, Maes M, Reiter RR. Breastfeeding and melatonin: implications for improving perinatal health. J Breast-feeding Biology. 2016; 1: 8-20.

[85]   Cheatham CL, Sheppard KW. Nutrients.synergistic effects of human milk nutrients in the support of infant recognition memory: An Observational Study. 2015;7(11):9079-95.

CHAPTER 5

Multi-modal Pharmacological Treatments for Major Depressive Disorder: Testing the Hypothesis

Trevor R. Norman*

Department of Psychiatry, University of Melbourne, Austin Hospital, Heidelberg, Australia

Abstract: Antidepressant medications have been available for more than fifty years, yet the proportion of patients helped by the various classes of these agents has hardly changed at all. This despite greater insight into the disorder at a biological level based on knowledge derived from contemporary advances in neuroscience. Two fundamental issues may lie at the heart of this apparent paradox. First, depression is a highly heterogeneous disorder and almost certainly is not a single 'disease' entity. The diagnosis is based on the clustering of a discrete set of symptoms, within a defined time frame and as such is best described as a syndrome or disorder. The cause(s) of depression remain unknown and are, in all probability, multi-factorial. The identification of bio-marker defined depression sub-groups may aid both more precise diagnoses and better treatment outcomes. Secondly, the development of pharmacological treatments has been limited by the prevailing aetiological hypothesis of the disorder, the monoamine hypothesis. While it is now well recognised that such a postulate is limited in its explanatory power, both for aetiology and treatment, it has, nevertheless, driven drug development for well over five decades. Clearly, while monoamines surely are important in mediating responses to antidepressant medication, they are almost certainly not the only important driver. Thus the shortcomings in such agents have been well recognised almost since their inception. Principal among these drawbacks has been speed of onset of action with full recovery and remission taking several weeks if not months. Additionally, the relative lack of efficacy of medications in all likelihood reflects the heterogeneity of the disorder and the inability to define predictive factors such as symptom patterns, personality variables or biomarkers, which are responsive to particular pharmacological properties of individual medications. Serious adverse events, side effects, cardiovascular safety and multiple potential drug interactions have also been cited as drawbacks of the existing plethora of antidepressant medications. While few medications have yet emerged into clinical practice based on the insights gleaned from recent basic studies, the notion of targeting multiple, relevant sites in the central nervous system to improve treatment outcomes in major depression has produced some new agents. These so called multi-modal antidepressants simultaneously interact with several different receptors and transporter molecules thought to contribute to antidepressant responses. Expanding the effects on central

***Corresponding author Trevor R. Norman:** Department of Psychiatry, University of Melbourne, Austin Hospital, Heidelberg, 3084, Victoria, Australia; Tel: ++613-94965680; Fax: ++613-94590821; E-mail: trevorrn@unimelb.edu.au

monoamine activity through the development of so-called triple reuptake inhibitors has been one multi-modal approach to the treatment of depression. Amitifiadine is the first triple reuptake exemplar to reach early clinical trials. To date clinical outcomes could be described as mixed. Vilazodone and vortioxetine are multi-modal agents which target reuptake mechanisms as well as other neurotransmitter receptors. This review examines the pharmacological properties of these new agents and critically evaluates their efficacy in clinical trials with a particular emphasis on whether the multi-modal approach obviates some of the perceived shortcomings of existing medications. Achievement of high remission rates in depression is increasingly recognised as the bench mark of an efficacious drug. The extent to which these new agents achieve better remission rates can be regarded as a measure of the extent to which the multi-modal hypothesis is realized and may guide treatment approaches into the future.

Keywords: Depression, monoamines, multimodal medications, remission, response, triple reuptake inhibitors, vilazodone, vortioxetine.

INTRODUCTION

Antidepressant medications have been available from the 1950's. Since the introduction of the tricyclic antidepressants and monoamine oxidase inhibitors there has been an evolution of multiple classes of drug each with putatively different mechanisms of action. While the efficacy of antidepressants has at times been questioned, with some analyses suggesting that the drugs are no more effective than placebo [1, 2], they remain one of the mainstays of management of depressive conditions. Antidepressant response depends on the initial severity of the presenting complaints: moderate to severe forms are more responsive than milder forms. The number of patients who achieve adequate remission of symptoms however remains low. In addition to relatively low remission rates antidepressants have some clinically important shortcomings. Among the impediments to use is the inability to match individual patients to a medication *a priori* with the certainty of improvement of the condition. While some progress has been made with respect to personalised medicine in psychiatry [3], choices of medication for the most part remain a 'trial and error' process. Furthermore, despite the diversity of agents the proportion of patients who recover when treated with members of the various classes has hardly changed at all. For example the STAR*D (Sequenced Treatment Alternatives to Relieve Depression) trial found that more than half of all patients treated were considered 'nonresponsive' to first line administration of a selective serotonin reuptake inhibitor (SSRI) [4]. Efficacy does not appear to be particularly different between the different classes of antidepressants: response rates to the first course of medication are relatively similar. At the clinical coal face choices between drugs are often idiosyncratically based on supposed differences in side effect profiles or presenting patient symptom patterns, which might be perceived as responding differentially to one drug or another. Such clinical practices persist in the face of intensive research into the aetiology of depression at both a psychosocial and neurobiological level. Clearly, research findings so far have not been particularly helpful for improving clinical outcomes for individual patients.

These ambiguities, in all probability, reflect an uncertainty about the nature of depression itself. Most particularly, clinical trials (and indeed most neurobiological studies) regard major depression (MDD) as a unitary disorder. Diagnostic manuals, such as the DSM-V reinforce the unitary notion of disorder whereas earlier version of the manual distinguished melancholia [5]. With respect to the latter, some would regard DSM defined melancholia as a more severe form of MDD lacking the subtleties of the original meaning of the term [6]. Purists would argue that melancholia is a distinct sub-group of depression with characteristic symptoms typified by movement disorder (either psychomotor retardation or agitation) and a neurobiological basal ganglia dopamine dysfunction [7, 8]. Irrespective of these more nuanced approaches it is clear that depression is a highly heterogeneous disorder and almost certainly is not a single 'disease' entity. The diagnosis is based on the clustering of a discrete set of symptoms, within a defined time frame and as such is best described as a syndrome or disorder. The cause(s) of depression therefore remain unknown and are, in all probability, multi-factorial even within apparently clinically distinct sub-groups of patients. The identification of sub-groups of depression defined by reproducible bio-markers would provide uniformity of objective diagnoses and arguably lead to improved medication treatment outcomes within such groups.

Further, the development of pharmacological treatments has been limited by the prevailing aetiological hypothesis of the disorder, the monoamine hypothesis [9]. Succinctly stated in its earliest iteration, the hypothesis posits that depression arises due to a deficiency of noradrenaline (and / or serotonin) at critical synapses in the central nervous system. It was soon recognised that such a postulate was of limited explanatory power, both for aetiology and for treatment. Modifications of the hypothesis to account for the discrepancy between the delayed onset of clinical effects and the almost immediate effects of medications on neurotransmitter concentrations in brain, invoked adaptation of critical central receptors. Principally this involved down-regulation of α2- and β1-adrenoceptors following chronic drug administration but later developments have envisaged a role for serotonergic receptors (5HT2A and 5HT1A) to account for the action of SSRI agents [10]. Neurotrophic factors, pro-inflammatory cytokines and antioxidants are more recent mechanistic elaborations to account for the therapeutic actions of antidepressants [11]. Clearly, while monoamines surely are important in mediating responses to antidepressant medication, they are almost certainly not the only important driver. Despite the proliferation of additional explanatory hypotheses, thinking grounded in mono-amine accounts have continued to provide the impetus for drug development for well over five decades.

The clinical shortcomings of monoamine based agents has been well recognised almost since their introduction into clinical practice [12]. Principal among these are slow onset of action with full recovery and remission taking several weeks if not months, unwanted side effects leading to non-compliance, relatively poor cardiovascular safety on overdose and multiple potential drug interactions having been cited as drawbacks of existing agents. Furthermore, a significant number of patients are described as 'treatment resistant' usually defined as two or more adequate courses of medication which have failed to bring about sufficient clinical improvements [13, 14]. This less than ideal situation has stimulated the search for improved antidepressants or to the adoption of pharmacological strategies where outcomes might be improved. Combining medications has been advocated as one potential method for enhancing drug responses, although this is not a recommended first line approach [13]. Such a method brings with it increased potential for drug-drug interactions (at both a pharmacokinetic and pharmacodynamic level) and an increased risk for toxicity. Nevertheless, it has been claimed that combining antidepressants with putatively different mechanisms of action is successful in cases of treatment resistant depression [15-17]. The extent to which such a strategy is successful begs the question of whether combining selective targets within a single molecule would not be as successful without the burden of potential interaction effects as well as reduced toxicity.

COMBINATION TREATMENTS: THE ORIGIN OF THE MULTI-MODAL HYPOTHESIS

The primacy of monoamine targets for the symptomatic relief of depression is based on the pharmacological effects of the majority of extant antidepressant drugs which principally target monoamine transporter molecules. Combination treatments for major depression are based on the rationale that targeting different pharmacological actions at monoamine receptors or transporters should produce greater efficacy than the single agent alone [18]. Indeed the combination of a tricyclic antidepressant and an SSRI has arguably demonstrated more effective treatment of depression than either drug alone [19, 20]. Attendant side effect issues, not the least of which has been the demonstration of increased tricyclic antidepressant plasma concentrations and the potential for associated cardiac arrhythmias, has limited the use of the combination as a first line treatment. A rapid, robust clinical effect was demonstrated for the combination of desipramine (predominantly noradrenergic) and fluoxetine (predominantly serotonergic) in an open evaluation [21]. A later study showed that combining fluoxetine with desipramine in non-treatment-resistant in-patients with a major depressive episode was significantly more likely to result in remission than was fluoxetine alone or desipramine alone [22]. It could be questioned to what extent does combining a medication with predominant effects on noradrenaline with a drug with predominant effects on serotonin (a so-called serotonin-noradrenaline reuptake inhibitor; SNRI) produce an advantage over a drug such as venlafaxine, which at sufficiently high doses already combines these actions? Few studies have addressed this issue directly; nevertheless comparative studies and meta-analysis suggest a modest efficacy advantage and a slightly faster onset of antidepressant effect for SNRIs compared to SSRIs, but with potentially lower tolerability [23].

Proof of the multi-target concept was apparently vindicated based on the results of a double-blind comparative study examining the combination of mirtazapine with paroxetine in patients with unipolar major depression [24]. A greater proportion of patients treated with the combination achieved remission of symptoms after 6 weeks than patients treated with monotherapies. In a follow-up study various antidepressants were combined from the commencement of treatment and the outcomes were compared with fluoxetine given alone [25]. Patients met DSM-IV criteria for major depressive disorder and were randomly assigned to receive fluoxetine monotherapy; mirtazapine combined with fluoxetine, venlafaxine or bupropion for 6 weeks. The proportion of patients achieving remission of symptoms was statistically significantly higher for the comparison of the fluoxetine monotherapy (25%) and both the fluoxetine (52%) and venlafaxine combinations (58%), but not the bupropion combination (46%).

The success of these trials was not repeated in a larger single blind study comparing remission rates in 665 patients with major depression treated with escitalopram or one of two combination treatments: escitalopram and bupropion or venlafaxine and mirtazapine [26]. Remission rates and most secondary outcome measures were not different among treatment groups at 12 weeks. The remission rates were 38.8% for escitalopram-placebo, 38.9% for bupropion-escitalopram, and 37.7% for venlafaxine-mirtazapine. Neither medication combination was significantly different from the monotherapy. The results are not strictly comparable with the two earlier trials as the doses used were different.

RATIONALE FOR MULTI-MODAL TREATMENTS

The term 'multimodal' was coined for medications which have two or more separate pharmacological actions that are complementary in terms of either efficacy or tolerability [27, 28]. Almost all extant antidepressants have more than one pharmacological action, particularly at supra-therapeutic doses. In most cases, these additional mechanisms are a potential cause of unwanted effects rather than a means of increasing efficacy. Such agents are not regarded as multimodal. On the other hand, multi-modal agents are described as compounds where multiple actions exist and also contribute to therapeutic effects. By such definitions the tricyclic antidepressants could be regarded as multi-modal on the one hand as they block the noradrenaline transporter molecule (NET) and/or the serotonin transporter (SERT). However their additional actions at cholinergic, histaminergic and adrenergic receptors make them poorly tolerated. On overdose the tricyclics interact with sodium channels which may partly account for their effects on cardiovascular function [29]. Certainly the extra actions do not appear to contribute to the therapeutic effects of the drugs. Similarly SNRI agents have serotonergic and noradrenergic (and to some extent cortical dopaminergic) actions; bupropion is a noradrenaline-dopamine reuptake inhibitor [30], while mirtazapine has antagonist actions at α2-adrenoceptors, 5-HT2C, 5-HT2A and histamine H1 receptors [31].

Multi-functionality of a molecule may be contingent upon the dose and potency at the receptors responsible for the multiple pharmacologic actions exhibited. Thus doxepin, a tricyclic antidepressant is a selective H1 antagonist when administered at a fraction of its antidepressant dose. With increasing doses it becomes an antidepressant, as additional pharmacologic actions are recruited [32]. Trazodone is another dose dependent multifunctional agent exhibiting hypnotic but not an antidepressant effect at low doses [33]. Sedative effects are probably related to actions of the drug at low doses on 5-HT2A receptors, H1 receptors and α1 receptors. By increasing the dose a 3-5 fold higher blockade of the SERT is recruited and trazodone exhibits antidepressant effects.

The pharmacological rationale for the development multi-modal antidepressants has been elaborated by Millan [34]. Major depression is thought to arise from a multiplicity of genetic, epigenetic, environmental and developmental influences with no single predominant factor. Numerous neural networks have been implicated in the development of a depressive disorder and almost certainly involve several interacting neurotransmitter pathways within those networks [35]. A system has been described that links the medial prefrontal cortex to the amygdala, the ventral striatum and pallidum, the medial thalamus, the hypothalamus, and the periaqueductal gray and other parts of the brainstem [35]. Data from human functional and structural imaging studies, as well as analysis of lesions and histological material indicates that this system is centrally involved in the generation of mood disorders. Serotonergic, dopaminergic, noradrenergic, GABA and glutamate pathways have all been implicated in the development of depression [36]. Consequently it is argued that treatment strategies which broadly influence these cortico-limbic circuits implicated in depression are more likely than highly selective agents to be effective for the majority of patients [36]. Furthermore, patients with MDD often suffer from associated symptoms such as anxiety, pain and cognitive dysfunction. Agents with broader, complementary components of action should, in theory at least, have a greater chance of controlling mood disturbances of depression and associated symptoms [34]. These factors coupled with the efficacy of combined medication described previously argue for a multi-modal approach to the treatment of MDD.

While few medications have yet emerged into clinical practice based on the insights gleaned from basic neurobiological findings, the notion of targeting multiple, relevant sites in the central nervous system to improve treatment outcomes in major depression has produced some new agents. These so named multi-modal antidepressants simultaneously interact with several different receptors and transporter molecules thought to contribute to antidepressant efficacy.

HOW CAN THE MULTI-MODAL HYPOTHESIS BE EVALUATED?

Evaluation of the success of the multi-modal hypothesis of antidepressant action encompasses several potential factors by which to judge a new medication. The ideal antidepressant should be effective for the maximum number of patients with few or no side effects, minimal drug interactions, lack of toxicity and a rapid onset of action. None of the extant agents has ever achieved such ideals and it would seem unlikely, given the lack of fundamental knowledge of the causes of MDD, that any new medication would achieve such goals. Therefore, any significant improvements are likely to be incremental.

Based on the current management guidelines for MDD remission of symptoms, rather than response, is regarded as the most desirable treatment outcome and the targeted goal [37]. The multimodal hypothesis would imply that higher remission rates should be achieved since these treatments are directed to interact with more appropriate (therapeutically relevant) neuronal receptors. The extent to which multi-modal agents achieve higher remission rates than existing compounds at therapeutically equivalent doses administered over the same time frame can be taken as the 'gold' standard by which to judge success of this hypothesis.

It can be contended that success might be judged by the ability of multi-modal agents to alleviate secondary targets in MDD. For example there has been a focus in recent times on anxiety symptoms associated with depression [38]. Complaints of pain are a frequent accompaniment in patients presenting with MDD while cognitive symptoms have also been identified as a barrier to restoration of function in patients with MDD [39]. While these are legitimate concerns of treatment there are relatively few studies where such secondary symptoms have been addressed on a consistent basis in clinical trials. Furthermore, few registration clinical trial databases include functional measures of depression outcome or if they do, these are not the primary outcome goals. Direct comparison of newer agents with the old is therefore problematic. Similarly comparisons of side effect profiles and speed of onset of action is unlikely to have the same methodological rigour as comparison of efficacy outcomes based on the use of well validated, widely recognised rating scales for depressive symptomatology.

TESTING THE HYPOTHESIS

Triple Reuptake Inhibitors

These compounds typically combine inhibition of the serotonin, noradrenaline and dopamine transporters. The therapeutic rationale for simultaneously targeting the three transporter molecules is that activity in the three pathways is specific to the main symptoms of depression. Thus, it has been postulated that activity in noradrenaline pathways is related to alertness, energy, anxiety, attention and interest in life [40]. Serotonin is related to anxiety, obsessions and compulsions while dopamine is related to attention, pleasure, cognition, motivation and interest in life. Increasing activity in these three pathways should bring about the relief of most of the symptoms of depression.

A structurally diverse group of chemical compounds has been investigated *in vitro* for their ability to inhibit these transporters. A recent report noted more than twenty compounds were patented for their ability to simultaneously inhibit the transporter molecules, to varying degrees, in the period 2006-2012 [41]. Of the compounds investigated few have come to clinical trial and none have been registered with regulatory agencies for the treatment of depression. Some have been investigated and their development abandoned due to lack of efficacy. Thus

SEP-225289, RG-7166, BMS-820836 all failed early clinical trials [41]. The results of two double blind clinical trials of GSK372475 also reported that the drug failed to separate from placebo [42]. In these studies comparator agents, paroxetine and venlafaxine, were statistically superior to placebo indicating the validity of the trials. Amitifadine (EB-1010, formerly DOV 21,947) has shown promising efficacy in early clinical trials and is undergoing further evaluation of both efficacy and safety.

Amitifadine

Pharmacology

Amitifadine is the single isomeric form of the racemic drug DOV 216,303 [43]. Chemically it is described as (+) 1-(3, 4-dichlorophenyl)-3-azabicyclo-[3.1.0] hexane hydrochloride and the structure is reported in Fig. (**1**).

The ability of the drug to inhibit reuptake of serotonin, noradrenaline and dopamine was assessed in HEK-293 cells expressing recombinant forms of the human transporter molecules [43, 44]. The affinity of amitifadine for the transporter proteins was also assessed in membranes prepared from those cells. There was a dose dependent effect on $[^3H]$-amine reuptake as well as binding to the human transporter proteins (Table **1**). Amitifadine displayed a somewhat higher potency for inhibition of serotonin reuptake than for either noradrenaline or dopamine [44]. However, the drug was less potent than either fluoxetine or imipramine but more potent than nomifensine or desipramine (Table **1**). With respect to dopamine reuptake amitifadine was approximately equivalent in potency to nomifensine a selective dopamine reuptake agent. Binding to each of the amine transporter proteins showed a pattern of affinities similar to that of amine reuptake data (Table **1**).

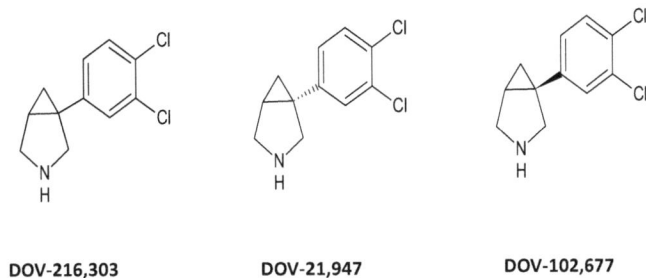

DOV-216,303   DOV-21,947   DOV-102,677

Fig. (1). Chemical structure of amitifadine (DOV-21,947) which is the (+) enantiomer of the racemic mixture DOV-216,303 and the (-)-enantiomer DOV-102,677.

Table 1: Inhibition of transporter binding and neurotransmitter reuptake.

Compound	h-SERT binding	h-NET binding	h-DAT binding	[^{3}H]-5HT	[^{3}H]-NA	[^{3}H]-DA
amitifadine	99	262	213	12.3	22.8	96
fluoxetine	1.1	1560	6670	7.3	1020	>10,000
desipramine	55	13.9	>10,000	64	4.2	>10,000
Imipramine	3.3	215	>10,000	8.0	70	>10,000
Nomifensine	2350	119	87	2780	42	7223

Data from Skolnick *et al.*, expressed as nM for K_i of radio-ligand binding and IC_{50} for inhibition of neurotransmitter reuptake

By contrast the rank order of effect of amitifadine on amine uptake kinetics in CHO cells was NET>SERT>DAT [45]. In rat native brain tissue an apparent separation in the binding affinities and occupancy ED_{50}s was observed across the three reuptake sites for amitifadine, *i.e.*, the affinity Ki/occupancy ED_{50} at SERT was 4- to 8-fold greater than at NET and 40- to 100-fold greater than at DAT [46]. Thus serotonergic effects are likely to be more prominent, particularly at lower therapeutic doses.

The effect of amitifadine (10mg/kg i.p.) on extracellular concentrations of monoamines and their metabolites in rat brain regions was investigated using an *in vivo* microdialysis technique [47]. Extracellular concentrations of 5-HT, NA and DA were increased to 412, 274, and 206% of their baseline at the peak effect, respectively. Increases were sustained for about 3h but returned to baseline concentrations by about 4h. In pre-frontal cortex extracellular concentrations of the metabolites of the neurotransmitters were decreased across a similar time frame. In the striatum and nucleus accumbens (NAc) DA concentrations were significantly increased about 20min after amitifadine administration. Maximum increases of 160-170% of baseline concentrations were observed and returned to near baseline concentrations in striatum by 3h, whereas in the NAc increases were sustained for about 4h. In a rat model of pain similar effects on DA concentrations in the NAc were noted [48]. Amitifadine blocked the effect of an intraperitoneal injection of dilute lactic acid which induces depression of NAc DA concentrations. In the absence of the noxious stimulus, amitifadine increased NAc levels of both DA and 5HT.

Amitifadine has been evaluated for antidepressant-like effects in some well characterised pharmacological tests. A dose dependent reduction in immobility time was demonstrated in the mouse tail suspension test [44]. Maximum reductions in immobility time were evident at 10-20mg/kg of amitifadine comparable with the reductions achieved by 20mg/kg of imipramine. Similarly the drug was effective in the rat forced swim test [44]. At a dose of 20mg/kg a maximum reduction in immobility time was achieved comparable to that of imipramine 15mg/kg. These results were replicated in an independent study using the mouse forced swim tests and tail suspension test [45]. Statistically significant effects on immobility time were evident from 2mg/kg. Both tests are predictive of antidepressant activity in the clinic.

Further confirmation of the potential antidepressant activity of amitifadine was obtained in a study of alcohol preferring rats [49]. A dose dependent reduction of binge alcohol drinking was observed over the dose range 6.3 to 50mg/kg amitifadine. Following 24h withdrawal from alcohol rats demonstrate an increased intracranial self-stimulation (ICSS) threshold. At doses of 12.5 to 50mg/kg oral amitifadine attenuated the effects of 24h alcohol withdrawal on ICSS parameters. Raised ICSS has been taken as a measure of anhedonia a core symptom of depression. In the FST amitifadine was active in reducing immobility time confirming the results of previous studies with this 'antidepressant' model.

Clinical Efficacy

A 6-week, multi-centre, randomized, double-blind, parallel group, placebo-controlled study evaluated the efficacy and tolerability of amitifadine in 63 patients with major depressive disorder [50]. Eligible patients were randomized to amitifadine 25 mg twice daily for 2 weeks, then 50 mg twice daily for 4 weeks or placebo. Mean baseline scores in the intent-to-treat population were 31.4 for the Montgomery-Åsberg Depression Rating Scale (MADRS), 29.6 for the HAMD-17. After the 6-week treatment, estimated least squares mean change from baseline (mixed-model repeated measures) in MADRS total score was statistically significantly superior for amitifadine compared to placebo (18.2 vs. 22.0; P< 0.05), with an effect size of 0.60 (Cohen's d). Differences in the HAM-D scores were not statistically different between drug and placebo. An anhedonia factor score, derived by grouping of different items from the MADRS score, demonstrated a statistically significantly superior effect for amitifadine compared to placebo (P < 0.05). Remission rates (defined as MADRS < 12) were not different between amitifadine and placebo.

The Triple Reuptake Inhibitor Anti-Depressant Effects or TRIADE study is a Phase II/III clinical trial comparing amitifadine to paroxetine and placebo in patients failing to respond to one course of first-line antidepressants. TRIADE is being conducted at 41 U.S. sites over a six week period. No formal peer reviewed publications are available but media release reports that amitifadine efficacy at 50 or 100 mg doses did not show a statistically significant difference from placebo in the primary endpoint of a change in MADRS scale [51]. However, paroxetine was significantly different from placebo suggesting that the study was valid. Thus the dose range (50 to 100mg) for amitifadine may have been too low. The lack of side effects with the 100 mg dose supports this notion. Several post-hoc analyses reported some efficacy for the drug suggesting that higher doses may show benefit in the target MDD patient group. The usefulness or otherwise of amitifadine is thus unresolved and awaits more detailed clinical investigations.

Multimodal Agents

Since their introduction in the 1980's the SSRI class of antidepressant agents have become the mainstay of treatment of MDD and some anxiety disorders [52]. Better tolerability is usually regarded as the reason for this predominance. Despite this there has been little improvement in response rates (at least in clinical trial data sets) and the speed of onset of action is not demonstrably different from other agents [53]. The realisation of the pivotal role played by raphe nucleus somato-dendritic 5-HT1A auto-receptors in the negative feedback regulation of 5-HT neurotransmission offered a potential mechanism with which to enhance onset of action of SSRIs [54, 55]. Pre-clinical studies suggested that desensitization of 5HT1A receptors was responsible for the observed increase in 5-HT concentrations in neural projection areas after chronic, but not acute, SSRI treatment. It was argued that a drug which inhibited the serotonin transporter molecule (SERT) while simultaneously having an antagonist action at 5-HT1A auto-receptors or accelerated their desensitization, through direct receptor stimulation, and stimulated postsynaptic 5-HT1A receptors would be expected to produce an enhanced 5-HT neurotransmission compared to an SSRI alone. Consequently an enhanced clinical response might be expected. Such an effect could not be achieved with a silent 5-HT1A receptor antagonist, since the benefit of enhancing presynaptic 5-HT function would be cancelled by the simultaneous blockade of postsynaptic 5-HT1A receptors [56]. The combination of the β-adrenoceptor partial agonist and 5-HT1A antagonist with SSRIs was reported to enhance the speed of onset of antidepressant action in clinical trials [57-60], although the reputed benefits of this approach have been questioned [61]. Nevertheless, a number of agents which utilise this strategy have been synthesised and two agents, vortioxetine and vilazodone, have been approved for use in clinical practice.

Vortioxetine

Pharmacology

Vortioxetine (Lu AA21004; 1-[2-(2, 4-dimethylphenyl-sulfanyl)-phenyl]-piperazine; Fig. (**2**)) was chosen for further development from among a series of structurally related arylpiperazine derivatives. The key desired pharmacological property of these derivatives was the ability to inhibit SERT, 5HT1A receptor agonist properties and antagonism at 5HT3 receptors [62]. Vortioxetine displayed high affinity for recombinant human 5-HT1A (Ki = 15nM), 5-HT1B (Ki = 33nM), 5-HT3A (Ki = 3.7nM), 5-HT7 (Ki = 19nM), and noradrenergic β1 (Ki = 46nM) receptors, and SERT (Ki = 1.6nM). The compound had antagonistic effects at 5-HT3A and 5-HT7 receptors, partial agonist actions at 5-HT1B receptors, agonistic properties at 5-HT1A receptors, and potent inhibition of SERT.

In vivo microdialysis studies showed that vortioxetine following both acute and sub-chronic (3 days) treatment caused a dose dependent, significant increase in extra-cellular 5-HT levels in the brain of awake, freely moving rats [63]. The increases observed in 5-HT were more than that from the administration of SSRI drugs. Most notable was that the increases in brain 5HT occurred in areas which have been associated with depression, *e.g.*, medial prefrontal cortex and the hippocampus [64]. Additionally extracellular concentrations of noradrenaline (NA), dopamine (DA), acetylcholine (Ach), and histamine (HA) in rat brain were also increased by administration of vortioxetine [63]. It is likely that the elevations of extracellular 5-HT evoked by Vortioxetine in excess of those by SSRIs arise from the multimodal receptor effects of the drug. Several 5-HT receptors for which vortioxetine has high affinity participate in negative feedback mechanisms controlling the release of serotonin. Further these same receptors have been implicated in the mechanism of action of antidepressant drugs [65]. In particular 5-HT3 antagonist effects of vortioxetine may be important in these effects as ondansetron, a selective 5HT3 antagonist augments SSRI mediated extracellular 5-HT release in mPFC and hippocampus [64]. Based on these pharmacological experiments it can be concluded that vortioxetine exhibits actions at multiple serotonin receptor targets both *in vitro* and *in vivo* which are believed to be relevant to the treatment of depression.

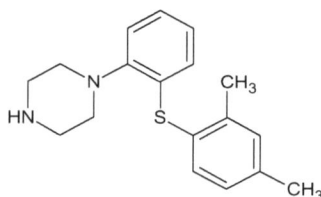

Fig. (2). Chemical structure of Vortioxetine.

The effects of vortioxetine in a variety of pre-clinical models commonly used to evaluate potential antidepressant and anxiolytic like activity has been reviewed in detail [66]. In general vortioxetine is active in the majority of antidepressant like and anxiolytic like models in which it has been tested. Thus in mice and Flinders Sensitive Line rats, the rat social interaction test, the rat conditioned fear-induced vocalization test, and the mouse novelty-suppressed feeding test and open-field test vortioxetine was active [66]. On the other hand vortioxetine was reported as inactive in the Chronic Mild Stress model. This model has been severely criticised by some authors as being unreliable and not readily translated between different laboratories [67]. Thus inactivity in the model is not regarded as a critical failure. Cell proliferation and survival in the dentate gyrus has also been demonstrated following repeated administration of vortioxetine in rats [68]. Increased hippocampal neurogenesis in the dentate gyrus of adult brain by drugs has been associated with antidepressant effects [69].

Clinical Efficacy

Vortioxetine was approved by the US Food and Drug Administration (FDA) for the treatment of major depressive disorder in September 2013. The basis for this decision has been reported in the literature [70]. Additionally a number of other jurisdictions have approved the drug for the treatment of major depression.

Short-term (6-12 weeks) efficacy of Vortioxetine has been evaluated in twelve trials [74-85]; (Table **2**), while longer term relapse prevention was evaluated in two studies. Studies varied in terms of the methodology used but the majority were double-blind, placebo controlled. Comparative agents, principally duloxetine and venlafaxine, were administered in some trials. While the majority of trials (8 of 12) were positive for vortioxetine compared to placebo [74, 76, 77, 80, 82, 84, 85], four studies failed to show statistically significant differences [75, 78, 79, 83]. Three studies were regarded as failed [75, 78, 79] while one was a negative trial as an active comparator also did not separate from placebo [83]. Detailed reviews of the data have been published recently, including a meta-analysis [66, 70-73]. It is not the intention here to review these studies again but rather to examine their results in the context of the multimodal hypothesis of antidepressant action using the criterion discussed above. Thus Table **2** reports the remission rates achieved by patient groups treated with vortioxetine in these trials. In moderately to severely depressed patients vortioxetine was more effective than placebo for most short term (6-12 weeks) trials. The active dose range appeared to be 5-20mg/day, with 1 mg/day being clearly inactive. In studies where an active comparator was included vortioxetine demonstrated equivalent efficacy to venlafaxine 225mg/day [74] and duloxetine 60mg/day [75, 77, 78, 80, 83] although arguably this dose is at the low end of the recommended dosage range. In a comparison with agomelatine, vortioxetine demonstrated somewhat superior efficacy [81].

In terms of the test of the multi-modal hypothesis, vortioxetine does not appear to offer any superior efficacy in terms of remission rates. Remission was assessed in all trials using accepted norms for such an evaluation: MADRS $\leq$ 10 or HAMD 17 item scale $\leq$ 7. In head to head comparisons vortioxetine demonstrated a numerical superiority for the number of patients in remission over agomelatine but was inferior or equivalent to duloxetine and venlafaxine (Table **2**).

Table 2: Clinical trials of Vortioxetine in major depressive disorder.

Remission Measure	Duration (weeks)	Active Comparator	Placebo	Number of Patients	Remission Rate	Reference
MADRS	6	Venlafaxine 225mg	Yes	429	49% (5mg) 49% (10mg) 55% (Ven) 27% (Pbo)	74
MADRS	8	Duloxetine 60mg	Yes	766	33% (2.5mg) 36% (5mg) 36% (10mg) 35% (Dul)	75
HAMD-17	8	No	Yes	560	21% (1mg) 24% (5mg) 24% (10mg) 11% (Pbo)	76
HAMD-17	6	Duloxetine 60mg	Yes	452	29% (5mg) 35% (Dul) 19% (Pbo)	77
MADRS	8	Duloxetine 60mg	Yes	611	33% (2.5mg) 32% (5mg) 51% (Dul) 33% (Pbo)	78

Table 2: cont....

HAMD-24	6	No	Yes	600	29% (5mg) 32% (Pbo)	79
MADRS	8	Duloxetine	Yes	608	35% (15mg) 38% (20mg) 54% (Dul) 19.0% (Pbo)	80
MADRS	12	Agomelatine	No	495	40% (10, 20mg) 29% (Ago)	81
MADRS	8	No	Yes	602	29% (10mg) 38% (20mg) 17% (Pbo)	82
MADRS	8	Duloxetine 60mg	Yes	614	27% (15mg) 29% (20mg) 26% (Dul) 27% (Pbo)	83
MADRS	8	No	Yes	462	21% (10mg) 22% (20mg) 14% (Pbo)	84
MADRS	8	No	Yes	469	27% (10mg) 24% (15mg) 22% (Pbo)	85

Abbreviations: Ago: Agomelatine; Dul: Duloxetine; HAMD-17: Hamilton depression rating scale, 17-item score; HAMD-24: Hamilton depression rating scale, 24-item score; MADRS: Montgomery-Asberg Depression rating Scale; Pbo: placebo; Ven: Venlafaxine.

An aspect of both remission and response not addressed in these studies was whether either was sustained *i.e.*, once a patient has achieved remission at one assessment point is this still the case at the next assessment? At face value, using a single assessment time as the outcome measure, the data with vortioxetine do not appear to support the multi-modal hypothesis. Generally remission rates are not different from those reported in the literature with other antidepressants. However there are some caveats to the interpretation of the data. The studies may not have been powered to demonstrate the statistical superiority of vortioxetine only to demonstrate non-inferiority. Secondly the doses of the drugs used may not have been equivalent. Potentially doses of vortioxetine exceeding 20mg/day may produce more depression remitters although to date there do not appear to be any published data on response and remission rates at such doses. Thirdly clinical depression is generally a chronic condition with relatively long episode durations. To judge a medication on the basis of short-term trials may be premature.

Nevertheless, data on remission rates with 'selective' or 'dual acting' antidepressant medications after equivalent short term trials suggests remission rates no different for that demonstrated with vortioxetine in the studies so far published.

Vilazodone

Pharmacology

Vilazodone (EMD 68843; 5-{4-[4-(5-cyano-3-indolyl)-butyl]-1-piperazinyl} - benzofuran-2-carboxamide hydrochloride; (Fig. **3**)) was developed by the modification of a series of indole-alkyl-phenyl-piperazine molecules [86, 87].

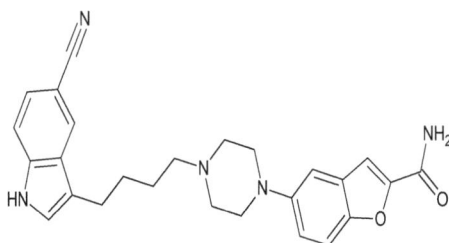

Fig. (3). Chemical structure of Vilazodone.

Key to the pharmacology of vilazodone is the ability to enhance serotonergic activity in the CNS through selective inhibition of serotonin reuptake coupled with the property of partial agonist action at serotonin 5-HT1A receptors [88]. Vilazodone exhibits potent inhibitory effects on the reuptake of 5-HT with an IC_{50} value of 0.5nM [86, 87]. The drug also exhibits high affinity for the 5-HT1A receptor (IC_{50}=0.2nM) and negligible affinity for other 5-HT receptors (5-HT1D, 5-HT2A, and 5-HT2C). The affinity for a range of other neurotransmitter receptors was tested *in vitro* and apart from a modest affinity for the dopamine D3 receptor (IC_{50}=71nM), Vilazodone had no appreciable affinity (IC_{50}>500nM) for adrenergic, dopaminergic, histaminergic or kappa opioid receptors [86, 87].

Two micro-dialysis studies reported that vilazodone increased serotonin concentrations in rat brain frontal cortex to an extent not achieved with selective serotonin reuptake inhibitors [88, 89]. Similar results were observed in the ventral hippocampus [89]. Challenge with the 5HT1A agonist 8-OH-DPAT reduced vilazodone induced extra-cellular serotonin concentrations more in the hippocampus than the frontal cortex. In the frontal cortex this might be explained by vilazodone acting at the 5-HT1A auto-receptor to limit the actions of 8-

OHDPAT as an auto-receptor agonist whereas in the hippocampus the 5-HT1A auto-receptor function may exert less regulation of extracellular serotonin [89].

Pharmacodynamic effects of vilazodone related to putative 5-HT1A receptor mediated changes have been examined in two studies. Administration of 5-HT1A agonists produces a characteristic behavioural syndrome (postural changes forepaw treading, head weaving, tremor) which was absent after vilazodone administration at doses which enhanced cortical serotonin output [89]. In contrast the classical 5HT-1A agonist 8-OH-DPAT produced all of the symptoms. Similarly basal body temperature in the rat has been shown to decrease following 5HT-1A agonist administration, an event reputedly mediated by post-synaptic receptors [90]. Vilazodone did not affect this response [91].

Vilazodone has been evaluated for activity in various rodent models thought to be predictive of anxiolytic activity such as the rat ultrasonic vocalizations test, the elevated plus maze, shock-probe burying tests and stress induced potentiated startle [91-93] Vilazodone demonstrated dose related efficacy in the shock probe test and attenuated startle, but was inactive in the elevated plus maze. In the Forced Swim Test (FST) [94] vilazodone was effective in both the rat and mouse versions of the model at a dose of 1 mg/kg but not at 3 and 10 mg/kg [89]. The drug reduced immobility and increased swimming behaviour a pattern of response repeatedly shown by SSRIs.

Clinical Efficacy

Vilazodone was approved by the US Food and Drug Administration (FDA) for the treatment of major depressive disorder in January 2011. The basis for this decision has been reported in the literature [95].

Short-term efficacy (8-weeks trials) of vilazodone was evaluated in five Phase II randomized, placebo-controlled studies in patients with MDD. Three studies included active comparator agents (fluoxetine or citalopram). All trials used change from baseline on the Hamilton Rating Scale for Depression-17 (HAM-D17) as the primary outcome measure. Doses of vilazodone ranged from 5 to 100 mg/day in the studies with the majority of patients dosed at ≤ 20 mg/day. No statistically significant differences were observed between vilazodone and placebo in intent-to-treat (ITT) analyses. Furthermore, where comparator agents were used these also failed to separate from placebo suggesting that these studies lacked assay sensitivity. Analysis of secondary end points (*e.g.*, MADRS, CGI scale, quality of life scales) provided some suggestion of efficacy in two of the trials when analysing patients assigned to higher dose groups [96].

Approval for marketing in the US was based on two pivotal phase III studies [95] which compared vilazodone to placebo, but did not include a comparator agent. A randomised, double-blind, parallel group, short-term (8 week) study was conducted in adult patients with DSM-IV diagnosed MDD [97]. Outcome was assessed using the HAM-D17, MADRS and Hamilton Anxiety (HAM-A) scales. The dose of vilazodone was titrated to 40mg/day over two weeks. Based on analysis of the ITT population at week 8, the mean change from baseline was significantly greater with vilazodone compared to placebo on the HAM-D (P<0.05; ANCOVA) and the MADRS (P<0.001; ANCOVA) scales. Vilazodone demonstrated separation from placebo beginning with the first measurements at week 1. Remission rates were not calculated for this trial but the response rates are reported in Table **3** [97]. While there was statistically significant difference from placebo for vilazodone, the rates (depending on the measure used) of ~45% do not appear to be much different than those reported in other antidepressant trials over the same evaluation time frame. Generally remission rates tend to be lower than response rates measured at the same time point (8 weeks in this study).

The second pivotal trial was of similar design to the first and was conducted in patients with DSM-IV-TR diagnosed MDD [98]. After 8 weeks vilazodone treated patients had significantly greater responses than those treated with placebo on the MADRS scale (P<0.01; ANCOVA). Similar results were noted on the HAM-D scale. In this study response rates were significantly higher for vilazodone than for placebo. However, the remission rates were not different between vilazodone and placebo (Table **3**). A pooled analysis of these two studies was undertaken [99] and confirmed the findings of efficacy *versus* placebo based on the MADRS scale. According to this analysis both response and remission based on either the MADRS or HAM-D scales were significantly higher for vilazodone than placebo at 8 weeks. Remission rates, based on a MADRS total score ≤10, were 28.6% for vilazodone and 20.1% for placebo. Using a more stringent criterion of remission, MADRS total score ≤9, a review of vilazodone reported remission rates of 25.4% for vilazodone and 18.1% for placebo [100].

Table 3: Clinical trials of Vilazodone in major depressive disorder.

Remission Measure	Duration (weeks)	Active Comparator	Placebo	Number of patients	Remission Rate	Reference
MADRS	8	No	Yes	410	40% 28% (Pbo)[§]	97
MADRS	8	No	Yes	481	27% 20% (Pbo)	98
MADRS	8	No	Yes	505	37%[§§] 17% (Pbo) [§§]	101
MADRS	8	No	Yes	505	34% 22% (Pbo)	102
MADRS	10	Citalopram 40mg	Yes	1162	30% (20mg) [§§] 33% (40mg) [§§] 31% (Cit) [§§] 26% (Pbo)[§§]	103

Abbreviations: Cit: citalopram; MADRS: Montgomery-Asberg Depression rating Scale; Pbo: placebo.
§ Response rates based on 40% change in MADRS score from baseline. Remission rates not reported.
§§ Sustained response rates reported defined as MADRS<12 over two consecutive double blind visits.

Since the pivotal trials two further evaluations of vilazodone in depression have been published and provide further evidence for the efficacy compared to placebo. An eight-week parallel group study compared vilazodone 40mg/day with placebo in DSM-IV-TR diagnosed patients [101]. Based on the change in the MADRS score from baseline vilazodone was significantly more effective than placebo (P<0.0001; mixed-effects model for repeated measures, MMRM). A post-hoc analysis of the trial [102] showed that remission rates (based on a MADRS score ≤10) after 8 weeks of treatment were 34% for vilazodone and 21.8% for placebo (P<0.003; logistic regression). Response rates (MADRS >50% change from baseline) were 50.6% for vilazodone and 33.3% for placebo (P<0.001; logistic regression).

The second trial compared two fixed doses of vilazodone (20 and 40mg) with citalopram (40mg) and placebo over 10weeks in patients with DSM-IV-TR MDD [103]. Change from baseline to week 10 in the MADRS score was analysed using a MMRM and showed vilazodone 20mg (P<0.01), vilazodone 40mg (P<0.005) and citalopram (P<0.005) to be more effective than placebo. Response rates were calculated for all four groups based on a ≥50% change in MADRS score from baseline at week 10. Under this criterion 50.5% of placebo treated patients were responders compared to 64.2% for 20mg vilazodone (P<0.01 v placebo); 64.6% for 40mg vilazodone (P<0.01 v placebo); 62.9% for 40mg citalopram (P<0.01 v placebo). Remission rates were not reported.

Taken as a whole the remission rates for vilazodone, where these have been reported in the literature, are not any greater than for other antidepressant medications judged over the same time frame.

CONCLUSION

Based on the criteria for the effectiveness of antidepressants outlined at the beginning of this chapter, the hypothesis of the multi-modal action increasing antidepressant drug response is not supported by the medications available to date. Some caveats apply to this conclusion. Importantly antidepressants are not generally used in the short-term. The clinical trial data-base from which the conclusions here have been drawn is therefore not representative of everyday clinical practice. Indeed relapse prevention studies frequently report an increase in response and remission rates with antidepressants with chronic use [4]. Depression is most often a chronic condition requiring long-term treatment, perhaps even life-long treatment in some cases. Remission after longer term administration (six or twelve-months or more of treatment) is clearly a better database on which to judge efficacy provided that comparable information is readily to hand. Short term trials, in all probability, are indicative of responses with continued treatment and they allow for strict comparability of responses within an equivalent time frame and using comparable methodology (with respect to trial design, rating scales and diagnostic criteria).

Remission (or response) rates are not the only criteria on which to judge the usefulness of the multi-modal hypothesis, although, arguably, they are the most important. Onset of action, or least the time to statistically significant change in depression scores, represents a potential method to evaluate the hypothesis. Rapid acting antidepressants, or at least drugs which act more rapidly than those which are currently available, are clearly of major clinical interest [104, 105]. Two issues which arise in this context are whether the extant methods of assessment are suitable for detecting subtle changes in mood and if statistically significant changes are meaningful clinically [106]. Leaving aside the second issue, many antidepressant drugs on *post-hoc* analysis of large, combined databases have been shown to produce significant changes within the first week of treatment [107-111]. Indeed it has been claimed that such responses are predictive of outcome on continued treatment [112]. An early onset of action (at week 1 the first time point evaluated) is also evident in some trials for vilazodone while for vortioxetine efficacy from 2-weeks onwards has been demonstrated. This result does not seem particularly surprising as improvements in vegetative symptoms of depression, such as improved sleep due to the sedative effects of the drugs, may often account for most or all of the improvement noted. Thus, the comparison of multimodal *versus* other antidepressants would require a nuanced approach with access to a large database and individual item scores of rating scales to tease apart effects on core depressive symptoms (*e.g.*, guilt, depressed mood) from symptoms likely to respond to the sedative effects of drugs.

Side effect profiles of medications encompassing treatment emergent effects and toxicity are also an important potential criterion for the usefulness of a medication in practice. Furthermore, side effects are intimately entwined with compliance, which, in turn, obviously influences response and remission rates. Compliance is a significant issue with psychotropic medications with partial and total non-compliance rates greater than 50% in the longer term [113, 114]. Generally it is accepted that the SSRIs are more acceptable to patients than tricyclic antidepressants suggesting that more selective modalities are less bothersome [115]. Other criteria may also be important standards by which to judge the success of the multi-modal hypothesis. It is claimed, for example, that vortioxetine has a significant effect on cognitive function which may arise as a result of its unique pharmacological profile (*vide supra*). It has long been recognised that cognitive dysfunction is a concomitant of MDD [116]. Clearly a medication which can restore cognitive function may prove useful in therapy. However few trials examine cognitive function as a specific outcome of treatment and therefore comparable studies are not available.

It would be premature to dismiss triple reuptake inhibitors as not providing better outcomes than conventional agents on the basis of one exemplar as described here. However, out of a plethora of triple reuptake agents so far devised, few have shown antidepressant efficacy and none have yet reached the market. Indeed it appears that the initial promise of amitifadine has not been borne out by more extensive studies (as reported in the media) although further studies are yet to be published. In the current climate of diminished investment by pharmaceutical companies in psychiatry research it would seem that further developments on this front are unlikely. More importantly, from a theoretical perspective at least, the triple re-uptake strategy may be doomed to pedestrian response and remission rates since it essentially remains within the confines of the monoamine hypothesis of depression. It does not represent a paradigm shift in the sense described by Kuhn [117]. If the monoamine hypothesis is to be abandoned then clearly an alternative explanatory paradigm needs to replace it. Some progress on this front has been made with respect to neurotrophic factors and inflammatory pathways as potential causal neurobiological factors in depression [118, 119]. Like the monoamine hypothesis however these approaches also suffer from the same logical fallacy of arguing from the effects of drugs back to a putative underlying abnormality.

Similar to triple reuptake inhibitors neither vortioxetine nor vilazodone have made a compelling case for increased efficacy despite their putative multimodal pharmacological profiles. One shortcoming of both agents is that, while they combine a number of actions within the same molecule, they still interact with monoamine receptors and transporter molecules. Thus the extent to which they can be regarded as genuinely multimodal might be questioned. For example, although vortioxetine has effects on the serotonin transporter and several 5HT receptors, it actions are still confined to the serotonin system and there are no direct potentially significant actions on receptors outside of this system. Vilazodone has similar defects in that it's most potent pharmacological effects are confined to the serotonin system with only weak effects on other receptors and transporter molecules. There may well be downstream effects on other systems as a result of these effects, but this is also true of agents not generally regarded as multimodal. Can it be said that these molecules have provided a stringent test of the multimodal hypothesis? It may well be this limitation which suggests that the multimodal hypothesis will survive, at least until there are genuinely medications that will provide a more stringent test.

ACKNOWLEDGEMENT

Declared none.

CONFLICT OF INTEREST

The author confirms that author has no conflict of interest to declare for this publication.

REFERENCES

[1]   Kirsch I, Deacon BJ, Huedo-Medina TB, *et al*. Initial severity and antidepressant benefits: a meta-analysis of data submitted to the Food and Drug Administration. PLoS Med. 2008; 5(2); e45. [doi: 10.1371/journal.pmed.0050045]
[2]   Moncrieff J, Kirsch I. Efficacy of antidepressants in adults. BMJ. 2005; 331 (7509); 155-157.
[3]   Ng CH, Schweitzer I, Norman TR, Eastel S. The emerging role of pharmacogenetics: implications for clinical psychiatry. Aust NZJ Psychiat. 2004; 38(7); 483-489.
[4]   Trivedi MH, Rush AJ, Wisniewski SR, *et al*. Evaluation of Outcomes With Citalopram for Depression Using Measurement-Based Care in STAR*D: Implications for Clinical Practice. Am J Psychiatry. 2006; 163 (1); 28-40.
[5]   Parker G, Fink M, Shorter E, *et al*. Issues for DSM-5: Whither melancholia? The case for its classification as a distinct mood disorder. Am J Psychiatry. (2010); 167(7); 745-747.

[6]   Parker G, Hadzi-Pavlovic D, Wilhelm K, *et al*. Defining melancholia: properties of a refined sign-based measure. Br J Psychiat. 1994; 164 (3); 316-326.

[7]   Parker G, Hadzi-Pavlovic D, Austin M-P, *et al*. Sub-typing depression, I. Is psychomotor disturbance necessary and sufficient to the definition of melancholia? Psychol Med. (1995); 25 (4); 815-823.

[8]   Parker G. Classifying Depression: Should Paradigms Lost Be Regained? Am J Psychiatry. (2000); 157 (8); 1195-1203.

[9]   Hirschfeld R. A History and evolution of the monoamine hypothesis of depression. J Clin Psychiatry. (2000); 61(Suppl6); 4-6.

[10]   Owens MJ. Selectivity of antidepressants: From the monoamine hypothesis of depression to the SSRI revolution and beyond. J Clin Psychiatry. (2004); 65(Suppl4); 5-10.

[11]   Palazidou E. The neurobiology of depression. Br Med Bull. (2012); 101 (1); 127-145.

[12]   Horden A. The antidepressant drugs. New Engl J Med. (1965); 272 (22); 1139-1169.

[13]   Burrows GD, Norman TR, Judd FK. Definition and differential diagnosis of treatment-resistant depression. Int Clin Psychopharmacol. (1994); 9 (Suppl 2); 5-10.

[14]   Thase ME, Rush AJ. Treatment-resistant depression. in: Bloom, FE, Kupfer DJ, Eds. psychopharmacology: The fourth generation of progress. New York: Raven Press Ltd 1995; pp 1081-1098.

[15]   Lam RW, Wan DD, Cohen NL, Kennedy SH. Combining antidepressants for treatment-resistant depression: a review. J Clin Psychiatry. (2002): 63(8); 685-693.

[16]   Shelton RC, Osuntokun O, Heinloth AN, Corya SA. Therapeutic options for treatment-resistant depression. CNS Drugs. (2010); 24 (2); 131-161.

[17]   Stahl SM. Enhancing outcomes from major depression: using antidepressant combination therapies with multifunctional pharmacologic mechanisms from the initiation of treatment. CNS Spectrums. (2010); 15; 79-94.

[18]   Richelson E. Multi-modality: a new approach for the treatment of major depressive disorder. Int J Neuropsychopharmacol. (2013): 16: 1433-1442.

[19]   Burrows GD, Norman TR. Treatment-resistant unipolar depression. In: Lader M, Naber D Eds. Difficult Clinical Problems in Psychiatry. London: Martin Dunitz 1999; pp. 57-73.

[20]   Dodd S, Horgan D, Malhi GS, Berk M. To combine or not to combine? A literature review of antidepressant.combination therapy. J Affect Disord. (2005); 89; 1-11.

[21]   Nelson JC, Mazure CM, Bowers MB, Jatlow PI. A Preliminary, Open Study of the Combination of Fluoxetine and Desipramine for Rapid Treatment of Major Depression Arch Gen Psychiatry. (1991); 48(4); 303-307.

[22]   Nelson JC, Mazure CM, Jatlow PI, Bowers Jr. MB, Price LH. Combining norepinephrine and serotonin reuptake inhibition mechanisms for treatment of depression: a double-blind, randomized study. Biol Psychiatry (2004); 55; 296-300.

[23]   Papakostas GI, Thase ME, Fava M, Nelson JC, Shelton RC. Are antidepressant drugs that combine serotonergic and noradrenergic mechanisms of action more effective than the selective serotonin reuptake inhibitors in treating major depressive disorder? A meta-analysis of studies of newer agents. Biol Psychiatry. (2007); 62; 1217-1227.

[24]   Blier P, Gobbi G, Turcotte JE, De Montigny C, Boucher N, He´bert C, Debonnel G. Mirtazapine and paroxetine in major depression: a comparison of monotherapy vs. their combination from treatment initiation. Eur Neuropsychopharmacol. (2009); 19; 457-465.

[25]   Blier P, Ward HE, Tremblay P, Laberge L, He´bert C, Bergeron R. Combination of antidepressant medications from treatment initiation for major depressive disorder: a double-blind randomized study. Am J Psychiatry. (2010); 167; 281-288.

[26]   Rush AJ, Trivedi MH, Stewart JW, *et al*. Combining Medications to Enhance Depression Outcomes (CO-MED): Acute and Long-Term Outcomes of a Single-Blind Randomized Study. Am J Psychiatry (2011); 168; 689-701.

[27]   Nutt DJ. Beyond psychoanaleptics - can we improve antidepressant drug nomenclature? J Psychopharmacol. (2009); 23; 343-345.

[28]   Chang T, Fava M. The future of psychopharmacology of depression. J Clin Psychiatry. (2010); 71; 971-975.

[29]   Pancrazio JJ, Kamatchi GL, Roscoe AK, Lynch C. Inhibition of neuronal Na_ channels by antidepressant drugs. J Pharmacol Exp Ther. (1998); 284; 208-214.

[30]   Ascher JA, Cole JO, Colin JN, *et al*. Bupropion: a review of its mechanism of antidepressant activity. J Clin Psychiatry. (1995); 56(9); 395-401.

[31]   Anttila SA, Leinonen EV. A review of the pharmacological and clinical profile of mirtazapine. CNS Drug Rev. (2001); 7(3): 249-264.

[32]   Stahl SM. Selective histamine H1 antagonism: novel hypnotic and pharmacologic actions challenge classical notions of antihistamines. CNS Spectr. 2008; 13; 1027-1038.

[33]   Stahl SM. Mechanism of action of trazodone: a multifunctional drug. CNS Spectr. 2009; 14; 536-546.

[34]   Millan MJ. Dual- and triple-acting agents for treating core and co-morbid symptoms of major depression: novel concepts, new drugs. Neurotherapeutics. (2009); 6; 53-77.

[35]   Drevets WC, Price JL, Furey ML. Brain structural and functional abnormalities in mood disorders: implications for neurocircuitry models of depression. Brain Struct Funct. (2008); 213; 93-118.

[36]   Maletic V, Robinson M, Oakes T, Iyengar S, Ball SG, Russell J. Neurobiology of depression: an integrated view of key findings. Int J Clin Pract. (2007); 61; 2030-2040.

[37]   Keller MB. Past, present, and future directions for defining optimal treatment outcome in depression remission and beyond JAMA. 2003; 289(23); 3152-3160.

[38]   Fava M, Rush AJ, Alpert JE, *et al*. Difference in treatment outcome in outpatients with anxious *versuS* non-anxious depression: a star*d report. Am J Psychiatry. (2008); 165; 342-351.

[39]   Kroenke K, Shen J, Oxman TE, Williams JW, Dietrich AJ. Impact of pain on the outcomes of depression treatment: Results from the RESPECT trial. Pain. (2008); 134 (1-2); 209-215.

[40]   Nutt DJ. Relationship of neurotransmitters to the symptoms of major depressive disorder. J Clin Psychiatry. (2008); 69 (Suppl E1); 4-7.

[41]   Shao L, Li W, Xie Q, Yin H. Triple reuptake inhibitors: a patent review (2006 - 2012) Expert Opin. Ther. Patents. (2014); 24(2); 131-154.

[42]   Learned S, Graff O, Roychowdhury S, *et al*. Efficacy, safety, and tolerability of a triple reuptake inhibitor GSK372475 in the treatment of patients with major depressive disorder: two randomized, placebo- and active-controlled clinical trials J Psychopharmacol. (2012); 26(5); 653-662.

[43]   Skolnick P, Krieter P, Tizzano J, *et al*. Preclinical and clinical pharmacology of DOV 216,303, a "Triple" Reuptake Inhibitor, CNS Drug Reviews. (2006); 12; 123-134.

[44]   Skolnick P, Popik P, Janowsky A, Beer B, Lippa AS. Antidepressant-like actions of DOV 21,947: A "triple" reuptake inhibitor. Eur J Pharmacol. (2003); 461; 99-104.

[45]   Fang X, Guo L, Jia J, *et al*. SKF-83959 is a novel triple reuptake inhibitor that elicits anti-depressant activity. Acta Pharmacol Sin. (2013); 34; 1149-1155.

[46]   Lengyel K, Pieschl R, Strong T, Molski T, Mattson G, Lodge NJ, Li Y-W, *Ex vivo* assessment of binding site occupancy of monoamine reuptake inhibitors: Methodology and biological significance Neuropharmacol. (2008); 55; 63-70.

[47]   Golembiowska K, Kowalska M, Bymaster F, Effects of the triple reuptake inhibitor amitifadine on extracellular levels of monoamines in rat brain regions and on locomotor activity, Synapse. (2012); 66; 435-444.

[48]   Miller LL, Leitl MD, Banks ML, Blough BE, Negus SS, Effects of the triple monoamine uptake inhibitor amitifadine on pain-related depression of behaviour and mesolimbic dopamine release in rats, Pain. (2015); 156; 175-184.

[49]   Warnock KT, Yang ARST, Yi HS, *et al*. Amitifadine, a triple monoamine uptake inhibitor, reduces binge drinking and negative affect in an animal model of co-occurring alcoholism and depression symptomatology, Pharmacol Biochem Behav. (2012); 103; 111-118.

[50]   Tran P, Skolnick P, Czoborb P, *et al*. Efficacy and tolerability of the novel triple reuptake inhibitor amitifadine in the treatment of patients with major depressive disorder: A randomized, double-blind, placebo-controlled trial,. J Psychiat Res. (2012); 46; 64-71.

[51]   http://www.businesswire.com/news/home/20130529006127/en/Euthymics-Reports-Top-Line-Results-TRIADE-Trial-Amitifadine#.VRjbKU0cTcs, accessed 30th March 2015.

[52]   Gartlehner G, Hansen RA, Morgan LC, *et al*. Comparative benefits and harms of second-generation antidepressants for treating major depressive disorder: an updated meta-analysis". *Ann Int Med.* (2011); 155 (11); 772-785.

[53]   Olver JS, Burrows GD, Norman TR. Third-generation antidepressants: do they offer advantages over the SSRIs? CNS Drugs. (2001); 15(12); 941-954.

[54]   Norman TR. The new antidepressants - mechanisms of action. Aust Prescr. (1999); 22; 106-108.

[55]   Stahl SM. Mechanism of action of serotonin selective reuptake inhibitors. Serotonin receptors and pathways mediate therapeutic effects and side effects. J Affect Disord. (1998); 51(3); 215-235.

[56]   Scorza MC, Lladó-Pelfort L, Oller S, *et al*. Preclinical and clinical characterization of the selective serotonin-1A receptor antagonist DU-125530 for antidepressant treatment. Br J Pharmacol (2012); 167; 1021-1034.

[57]   Bakish D, Hooper CL, Thornton ND, Vines A, Miller CA, Thebadoo CA. An open study of the treatment of major depressive disorder with nefazodone and pindolol combination therapy. Int Clin Psychopharm. (1997); 12; 91-97.

[58]   Berman RN, Darnal AM, Miller HL, Annand RN, Charney DS. Effect of pindolol in hastening response to fluoxetine in the treatment of major depression: A double-blind placebo controlled trial. Amer J Psychiat. (1997); 154; 37-43.

[59]   Perez V, Gilaberte I, Faries D, Alvarez E, Artigas F. (1997). Randomised double-blind placebo controlled trial of pindolol in combination with fluoxetine antidepressant treatment. Lancet 349: 1594-1597.

[60]   Tome MB, Isaac MT, Hart R, Holland C. Paroxetine and pindolol: A randomised trial of serotonergic autoreceptor blockade in the reduction of antidepressant latency. Int Clin Psychopharmacol. (1997); 12; 81-89.

[61]   Olver JS, Cryan JF, Norman TR, Burrows GD. Pindolol augmentation of antidepressants: a review and rationale. Aust N Z J Psychiat. (2000); 34; 71-79.

[62]   Bang-Andersen B, Ruhland T, Jørgensen M, *et al*. Discovery of 1-[2-(2,4-dimethylphenylsulfanyl)phenyl]piperazine (Lu AA21004): a novel multimodal compound for the treatment of major depressive disorder. J Med Chem. (2011); 54; 3206-3221.

[63]   Pehrson AL, Cremers T, Bétry C, *et al*. Lu AA21004, a novel multimodal antidepressant, produces regionally selective increases of multiple neurotransmitters - a rat microdialysis and electrophysiology study. Eur Neuropsychopharmacol. 2013; 23(2); 133-145.

[64]   Mørk A, Pehrson A, Brennum LT, *et al*. Pharmacological effects of Lu AA21004: A novel multimodal compound for the treatment of major depressive disorder. J Pharmacol Exp Ther. (2012); 340; 666-675.

[65]   Artigas F. Serotonin receptors involved in antidepressant effects. Pharmacol Ther (2013); 137; 119-131.

[66]   Sanchez C, Asin KE, Artigas F. Vortioxetine, a novel antidepressant with multimodal activity: Review of preclinical and clinical data. Pharmacol Ther. (2015); 145; 43-57.

[67]   Reid I, Forbes N, Stewart C, Matthews K. Chronic mild stress and depressive disorder: a useful new model? Psychopharmacol. (1997); 134; 365-367.

[68]   Li Y, Raaby KF, Sanchez C, Gulinello M. Serotonergic receptor mechanisms underlying antidepressant-like action in the progesterone withdrawal model of hormonally induced depression in rats. Behav Brain Res. (2013); 256; 520-528.

[69]   Banasr M, Duman RS. Regulation of neurogenesis and glio-genesis by stress and antidepressant treatment. CNS Neurol Disord Drug Targets. (2007); 6; 311-320.

[70]   Zhang J, Mathis MV, Sellers JW, *et al*. The US Food and Drug Administration's perspective on the new antidepressant Vortioxetine J Clin Psychiatry. 2015; 76(1); 8-14.

[71]   Berhan A, Barker A. Vortioxetine in the treatment of adult patients with major depressive disorder: a meta-analysis of randomized double-blind controlled trials. BMC Psychiatry (2014); 14; 276.

[72]   Katona CL, Katona CP. New generation multi-modal antidepressants: focus on vortioxetine for major depressive disorder. Neuropsych Dis Treatment. (2014); 10; 349-354.

[73]   Alvarez E, Perez V, Artigas F. Pharmacology and clinical potential of vortioxetine in the treatment of major depressive disorder. Neuropsych Dis Treatment (2014); 10; 1297-1307.

[74]   Alvarez E, Perez V, Dragheim M, Loft H, Artigas F. A double-blind, randomized, placebo-controlled, active reference study of Lu AA21004 in patients with major depressive disorder. Int J Neuropsychopharmacol. (2012); 15(5); 589-600.

[75]   Baldwin DS, Loft H, Dragheim M. A randomised, double-blind, placebo controlled, duloxetine-referenced, fixed-dose study of three dosages of Lu AA21004 in acute treatment of major depressive disorder (MDD). Eur Neuropsychopharmacol. (2012); 22(7); 482-491.

[76]   Henigsberg N, Mahableshwarkar AR, Jacobsen P, Chen Y, Thase ME. A randomized, double-blind, placebo-controlled 8-week trial of the efficacy and tolerability of multiple doses of Lu AA21004 in adults with major depressive disorder. J Clin Psychiatry. (2012); 73(7); 953-959.

[77]   Katona C, Hansen T, Olsen CK. A randomized, double-blind, placebo-controlled, duloxetine-referenced, fixed-dose study comparing the efficacy and safety of Lu AA21004 in elderly patients with major depressive disorder. Int Clin Psychopharmacol. (2012); 27(4); 215-223.

[78]   Mahableshwarkar AR, Jacobsen PL, Chen Y. A randomized, double-blind trial of 2.5 mg and 5 mg vortioxetine (Lu AA21004) *versus* placebo for 8 weeks in adults with major depressive disorder. Curr Med Res Opin. (2013); 29(3); 217-226.

[79]   Jain R, Mahableshwarkar AR, Jacobsen PL, Chen Y, Thase ME.A randomized, double-blind, placebo-controlled 6-wk trial of the efficacy and tolerability of 5 mg vortioxetine in adults with major depressive disorder. Int J Neuropsychopharmacol. (2013); 16(2); 313-321.

[80]   Boulenger JP, Loft H, Olsen CK. Efficacy and safety of vortioxetine (Lu AA21004), 15 and 20 mg/day: a randomized, double-blind, placebo-controlled, duloxetine-referenced study in the acute treatment of adult patients with major depressive disorder. Int Clin Psychopharmacol. 2014; 29(3); 138-149.

[81]   Montgomery SA, Nielsen RZ, Poulsen LH, Häggström L. A randomised, double-blind study in adults with major depressive disorder with an inadequate response to a single course of selective serotonin reuptake inhibitor or serotonin-noradrenaline reuptake inhibitor treatment switched to vortioxetine or agomelatine. Hum. Psychopharmacol. (2014); 29; 470-482.

[82]   McIntyre RS, Lophaven S, Olsen CK. Randomized, double-blind, placebo-controlled study of the efficacy of vortioxetine on cognitive function in adult patients with major depressive disorder (MDD). Int J Neuropsychopharm. (2014); 17; 1557-1567.

[83]   Mahableshwarkar AR, Jacobsen PL, Serenko M, Chen Y, Trivedi MH. A randomized, double-blind, duloxetine-referenced study comparing efficacy and tolerability of 2 fixed doses of vortioxetine in the acute treatment of adults with MDD. Psychopharmacol. (2014); DOI 10.1007/s00213-014-3839-0.

[84]   Jacobsen PL, Mahableshwarkar AR, Serenko M, Chan S, Trivedi MH. A randomized, double-blind, placebo-controlled study of the efficacy and safety of vortioxetine 10 mg and 20 mg in adults with Major Depressive Disorder. Poster presented at: American Psychiatry Association Annual Meeting; May 18-22, 2013; San Francisco, CA.

[85]   Mahableshwarkar AR, Jacobsen PL, Serenko M, Chen Y, Trivedi MH. A randomized, double blind, parallel, placebo-controlled, fixed-dose study comparing the efficacy and safety of 2 doses of vortioxetine (LU AA21004) in acute treatment of adults with major depressive disorder. 166th Annual Meeting of the American Psychiatric Association (APA), San Francisco, CA, USA, May 18-22, 2013. Poster NR9-02.

[86]   Heinrich T, Bottcher H, Gericke R, *et al.* Synthesis and structure-activity relationship in a class of indolebutylpiperazines as dual 5-HT1A receptor agonists and serotonin reuptake inhibitors. J Med Chem. (2004); 47; 4684-4692.

[87]   Heinrich T, Bottcher H, Schiemann K, *et al.* Dual 5-HT1A agonists and 5-HT re-uptake inhibitors by combination of indole-butyl-amine and chromenonyl-piperazine structural elements in a single molecular entity. Bioorg Med Chem. (2004); 12; 4843-4852.

[88]   Hughes ZA, Starr KR, Langmead CJ, *et al*. Neurochemical evaluation of the novel 5-HT1A receptor partial agonist/serotonin re-uptake inhibitor, vilazodone. Eur J Pharmacol. (2005); 510; 49-57.

[89]   Page ME, Cryan JF, Sullivan A, *et al*. Behavioral and neurochemical effects of 5-{4-[4-(5-cyano-3-indolyl)-butyl)-butyl]-1- piperazinyl}-benzofuran-2-carboxamide (EMD 68843): A combined selective inhibitor of serotonin re-uptake and 5-hydroxytryptamine(1A) receptor partial agonist. J Pharmacol Exp Therap. (2002); 302; 1220-1227.

[90]   O'Connell MT, Sarna GS, Curzon G. Evidence for postsynaptic mediation of the hypothermic effect of 5HT1A receptor activation. Br J Pharmacol (1992); 106; 603.

[91]   Bartoszyk GD, Hegenbart R, Ziegler H. EMD 68843, a serotonin re-uptake inhibitor with selective presynaptic 5-HT1A receptor agonistic properties. Eur J Pharmacol (1997); 322; 147-153.

[92]   Treit D, Degroot A, Kashluba S, Bartoszyk GD. Systemic EMD 68843 injections reduce anxiety in the shock-probe, but not the plus-maze test. Eur J Pharmacol. (2001); 414; 245-248.

[93]   Adamec R, Bartoszyk GD, Burton P. Effects of systemic injections of vilazodone, a selective serotonin re-uptake inhibitor and serotonin 1A receptor agonist, on anxiety induced by predator stress in rats. Eur J Pharmacol. (2004); 504; 65-77.

[94]   Borsini F, Meli A. Is the forced swimming test a suitable model for revealing antidepressant activity. Psychopharmacol. (1988); 94; 147-160.

[95]   Laughren TP, Gobburu J, Temple RJ, *et al*. Vilazodone: clinical basis for the US Food and Drug Administration's approval of a new antidepressant. J Clin Psychiatry. 2011; 72(9):1166-1173.

[96]   Khan A. Vilazodone, a novel dual-acting serotonergic antidepressant for managing major depression. Expert Opin Investig Drugs. (2009); 18; 1753-1764.

[97]   Rickels K, Athanasiou M, Robinson DS, Gibertini M, Whalen H, Reed CR. Evidence for efficacy and tolerability of vilazodone in the treatment of major depressive disorder: a randomized, double-blind, placebo-controlled trial. J Clin Psychiatry. (2009); 70; 326-33.

[98]   Khan A, Cutler AJ, Kajdasz DK, *et al*. A randomized, double-blind, placebo-controlled, 8-week study of vilazodone, a serotonergic agent for the treatment of major depressive disorder. J Clin Psychiatry. (2011); 72; 441-447.

[99]   Khan A, Sambunaris A, Edwards J, Ruth A, Robinson DS. Vilazodone in the treatment of major depressive disorder: efficacy across symptoms and severity of depression. Int Clin Psychopharmacol. (2014); 29; 86-92.

[100]   Citrome L. Vilazodone for major depressive disorder: a systematic review of the efficacy and safety profile for this newly approved antidepressant - what is the number needed to treat, number needed to harm and likelihood to be helped or harmed? Int J Clin Pract. (2012); 66; 356-368.

[101]   Croft HA, Pomara N, Gommoll C, Chen D, Nunez R, Mathews M. Efficacy and safety of vilazodone in major depressive disorder: a randomized, double-blind, placebo-controlled trial. J Clin Psychiatry. (2014); 75; 1291-1298.

[102]   Citrome L, Gommoll CP, Tang X, Nunez R, Mathews M. Evaluating the efficacy of vilazodone in achieving remission in patients with major depressive disorder: post-hoc analyses of a phase IV trial. Int Clin Psychopharmacol. (2015); 30; 75-81.

[103]   Mathews M, Gommoll CP, Chen D, Nunez R, Khan A. Efficacy and safety of vilazodone 20 and 40 mg in major depressive disorder: a randomized, double-blind, placebo-controlled trial. Int Clin Psychopharmacol. (2015); 30; 67-74.

[104]   Stassen HH, Angst J, Delini-Stula A. Delayed onset of action of antidepressant drugs? Survey of recent results. Eur Psychiatry. (1997); 12; 166-176.

[105]   Katz MM, Tekell J, Bowden CL, *et al*. Onset and early behavioral effects of pharmacologically different antidepressants and placebo in depression. Neuropsychopharmacol. (2004); 29; 566-579.

[106]   Leon AC, Blier P, Culpepper L, *et al*. An ideal trial to test differential onset of antidepressant effect. J Clin Psychiatry. (2001); 62; 34-36.

[107]   Taylor MJ, Freemantle N, Geddes JR, Bhagwagar Z Early onset of SSRI antidepressants: Systematic review and meta-analysis. Arch Gen Psychiatry. (2006); 63(11); 1217-1223.

[108]   Posternak MA, Zimmerman MD. Is there a delay in the antidepressant effect? A meta-analysis. J Clin Psychiatry. (2005); 66; 148-158.

[109]   Papakostas GI, Perlis RH, Scalia MJ. A meta-analysis of early sustained response rates between antidepressants and placebo for the treatment of major depressive disorder. J Clinical Psychopharmocol. (2006); 26; 56-60.

[110]   Machado-Vieira R, Luckinbaugh DA, Manji HK, Zarate CA. Rapid onset of antidepressant action: a new paradigm in the research and treatment of major depressive disorder. J Clin Psychiatry. (2008); 69; 946-958.

[111]   Parker G. On brightening up: Triggers and trajectories to recovery from depression. Brit J Psychiatry, (1996); 168(3); 263-264.

[112]   Szegedi A, Jansen WT, van Wugenburg AP. Early improvement in the first two weeks as predictors of treatment outcome in patients with major depressive disorder: a meta-analysis including 6,562 patients. J Clin Psychiatry 2009; 70: 344-53.

[113]   Demyttenaere K, Haddad P. Compliance with antidepressant therapy and antidepressant discontinuation symptoms. Acta Psychiatr Scand. (2000); 403 (Suppl); 50-56.

[114]   McManus P, Mant A, Mitchell P, Dudley J. Length of therapy with selective serotonin reuptake inhibitors and tricyclic antidepressants in Australia. Aust N Z J Psychiatry. (2004); 38(6); 450-454.

[115]   Hansen HV, Kessing LV. Adherence to antidepressant treatment. Expert Rev Neurother. (2007); 7(1); 57-62.

[116]   Gotlib IH, Joormann J. Cognition and depression: current status and future directions. Annu Rev Clin Psychol. (2010); 6; 285-312.

[117]   Kuhn TS. The structure of Scientific Revolutions, 3rd ed. Chicago; The University of Chicago press 1996.

[118]   Groves JO. Is it time to reassess the BDNF hypothesis of depression? Mol Psychiatry. (2007); 12; 1079-1088.

[119]   Miller AH, Maletic V, Raison CL. Inflammation and Its Discontents: The Role of Cytokines in the Pathophysiology of Major Depression. Biol Psychiatry. (2009); 65(9); 732-741.

Subject Index

A

AADC 62, 63, 64
  activity 63, 64
  additional source of 64
  encoding gene 62, 63
  gene 63, 64
Abnormal protein clearance 69
Accumulation of misfolded proteins 145
Acetylcysteine 105, 131
Acid(s) 8, 9, 20, 25, 27, 28, 29, 30, 31, 37, 57, 105,
  110, 113, 114, 115, 117, 123, 124, 129, 130,
  131, 137, 150, 194
  ascorbic 129, 130, 131
  formation, sulfinic 129
fatty 57, 113, 114, 194
  lipoic 150
  PNAs Peptide nucleic 8, 9, 37
  (QA) Quinolinic 20, 25, 27, 28, 29, 30, 31
  sulfenic 105, 110, 117, 123, 124, 129, 130
  sulfonic 105, 110, 115, 137
Aconitase 119
Activation, immune 188, 193
Activity 9, 14, 49, 52, 53, 59, 65, 66, 67, 71, 75,
  108, 109, 111, 117, 118, 119, 122, 125, 126,
  128, 130, 133, 135, 136, 137, 138, 140, 145,
  146, 150, 151, 174, 186, 187, 191, 192, 194,
  196, 210, 216, 220
  enzymatic 9, 117, 130, 137, 146
  enzyme 14, 49, 109
  mitochondrial complex IV 151
  network 66, 67
  neuronal 187, 192
  neuroprotective 52
  neurotrophic 53, 59
Adenine nucleotide translocator (ANT) 120
Adeno-associated virus (AAV) 4, 14, 15, 25, 29,
  48, 63, 64, 72, 73, 75
Advanced PD 48, 54, 55, 63, 64, 67, 174, 176, 180
Agents 3, 5, 6, 12, 20, 46, 135, 140, 176, 177, 203,
  204, 206, 208, 209, 210, 214, 215, 224, 225
  anti-parkinsonian 176
  multi-modal 204, 208, 210
  neuroprotective/neurorestorative 5, 6
  neurorestorative 3, 6
  neurotherapeutic 12
Aggregates 38, 44
  intracellular 44
  large 38
Aging and age-related diseases 107, 131

Aging process 131, 140
Agomelatine 217, 218
Alleles, mutant HTT 3, 36
Alpha 7 nicotinic 186, 187
Alzheimer's disease 4, 28, 107, 132, 138, 149, 186,
  187, 188, 191, 194, 195, 198
Antidepressant 203, 204, 206, 207, 208, 209, 216,
  223, 224
  effects 207, 208, 216
  medications 203, 204, 206, 223
  response 203, 204
  204, 206, 207, 208, 216, 218, 223, 224
  multi-modal 203, 208, 209
  tricyclic 204, 206, 208, 224
Antioxidants, thiolic 147, 148
Antisense 7, 8, 10
  oligonucleotides 7, 8
Apolipoprotein isoform 139
Apoptosis 42, 120, 122, 128, 130, 137, 138, 140,
  144, 147, 148, 152, 154, 192
  regulation of 120, 128
Apoptotic markers 150
Artificial miRNAs 8, 10, 11
ASO-based approaches 33, 34
Astrocyte end-feet 12
Astrocytes and microglia 187, 192
A-synuclein 43, 69, 70, 71, 72
Atomic force microscopy (AFM) 70
Atrophy, striatal neuron 23, 26
Autism 193, 194
Autophagy 39, 192, 193, 195, 196, 197
  regulation of 193

B

Bacterial artificial chromosome (BAC) 22
Barrier, blood brain 135, 144
Basal ganglia circuitry 65, 66
Basal lamina 12
Baseline concentrations 212
BDNF gene 25, 26, 57, 58, 59
BDNF levels 25, 57
BDNF over-expression 26, 27
BDNF protein 25
BDNF-secreting fibroblasts 27, 57, 58
Behavioral deficits 22, 27, 28, 29, 32, 58
Behavioral disorders, abnormal 174, 176
Bioenergetic capacity 119, 140, 152, 153
Biological underpinnings 194, 195
Biosynthetic enzymes 62

H

I

www.ingramcontent.com/pod-product-compliance
Lightning Source LLC
Chambersburg PA
CBHW050828220326
41598CB00006B/332